T0177167

Eruptions that Shook the World

In April 2010 Eyjafjallajökull volcano on Iceland belched out an ash cloud that shut down much of Europe's airspace for nearly a week. Although only a relatively small eruption, this precipitated the highest level of air travel disruption since the Second World War and it is estimated to have cost the airline industry worldwide over two billion US dollars.

But what does it take for a volcanic eruption to really shake the world? Did volcanic eruptions extinguish the dinosaurs? Did they help humans to evolve and conquer the world, only to decimate their populations with a super-eruption 73,000 years ago? Did they contribute to the ebb and flow of ancient empires, the French Revolution, and the rise of fascism in Europe in the nineteenth century? These are some of the claims made for volcanic cataclysm.

In this book, volcanologist Clive Oppenheimer explores rich geological, historical, archaeological and paleoenvironmental records (such as ice cores and tree rings) to tell the stories behind some of the greatest volcanic events of the past quarter of a billion years. He shows how a forensic approach to volcanology reveals the richness and complexity behind cause and effect, and argues that important lessons for future catastrophe risk management can be drawn from understanding events that took place even at the dawn of human origins.

CLIVE OPPENHEIMER is a Reader in Volcanology and Remote Sensing at the University of Cambridge, and a Research Associate of 'Le Studium' Institute for Advanced Studies at ISTO (University of Orléans/CNRS). His research focuses on understanding the chemistry and physics of volcanism, and the climatic and human impacts of eruptions in antiquity. He has carried out fieldwork worldwide in collaboration with archaeologists, atmospheric scientists and other geologists. Since 2003, he has studied the lava lake of Erebus volcano with the US Antarctic Program. In 2005, the Royal Geographical Society presented him with the Murchison Award 'for publications enhancing the understanding of volcanic processes and impacts'. Dr Oppenheimer is a co-author with Peter Francis of a leading volcanology textbook, and has contributed widely to television and film documentaries on volcanoes, including Werner Herzog's 'Encounters at the End of the World', and most recently, for Discovery, the History Channel, the BBC, Teachers' TV and National Geographic.

Eruptions that Shook the World

CLIVE OPPENHEIMER
University of Cambridge
Le Studium Institute for Advanced Studies
University of Orléans

Shaftsbury Road, Cambridge CB2 8EA, United Kingdom
One Liberty Plaza, 20th Floor, New York, NY 10006, USA
477 Williamstown Road, Port Melboarne, VIC 3207, Australia
314–321, 3rd Floor, Plot3, Splendor Forum, Jasola District Centre, New Delhi–110025, India
103 Penang Road, #05–06/07, Visioncrest Commercial, Singapore 23846

Cambridge University Press is part of Cambridge University Press & Assessment
a deparment of the University of Cambridge.

We share the University's mission to contribute to society through the pursuit of education, learning and research at the highest international levels of excellence.

www.cambridge.org
Information on this title: www.cambridge.org/9780521641128

First published 2011 (version 6, March 2023)

Printed in the United Kingdom by TJ Books Limited, Padstow Cornwall

A catalogue record for this publication is available from the British Library

Library of Congress Cataloguing in Publication data
Oppenheimer, Clive.
Eruptions that shook the world / Clive Oppenheimer.
p. cm.
ISBN 978-0-521-64112-8 (hardback)
1. Volcanism – Effect of environment on. 2. Volcanism – History. 3. Volcanology. I. Title.
QE522.O58 2011
551.21–dc22

2011004246

ISBN 978-0-521-64112-8 Hardback

Contents

Preface

The largest volcanic salvo of the last century took place in a remote part of the Alaska Peninsula in 1912. The eruption of Mount Katmai expelled around 28 cubic kilometres (nearly seven cubic miles) of ash and pumice, projecting roughly two-thirds of it into the air and the remaining third as ground-hugging hurricanes of dust and rock. The only event to have come close to it in more recent times is the 1991 eruption of Mt Pinatubo in the Philippines. Had an eruption the size of Katmai's 1912 outburst occurred in more densely populated regions of the 'lower 48' or, say, in Italy, Indonesia or the Caribbean, the event would be much better known outside of the volcanological coterie. In case you are wondering how to envisage 28 cubic kilometres of volcanic rock, it is sufficient to form a blanket seven centimetres thick (nearly three inches) over California, or 11 centimetres across the UK!

However, the Katmai eruption was a fairly trivial demonstration of volcanic fury viewed from either geological or human evolutionary perspectives. Around 7700 years ago, an eruption twice the size *did* strike the conterminous USA (in Oregon). Remarkably, the memory of the eruption, which formed the magnificent landform known as Crater Lake, lingers in the oral traditions of the Klamath native American tribe. Another eruption, more than twice as large again, struck the eastern Mediterranean only 3600 years ago. It may have had a devastating 'slow-fuse' impact on the Minoans, one of the great early civilisations. Stretching back 73,000 years ago, a volcanic cataclysm more than 200 times larger than Katmai's blast left a hole up to 80 kilometres across, in northern Sumatra. Some claims suggest that the event almost exterminated our ancestors! These comparisons demonstrate why we need to examine the records of much larger historic and prehistoric eruptions, if we wish to anticipate the full spectrum of possible future volcanic activity. What is more, the deep time perspective sheds light on the gamut of societal responses to volcanic disasters, again providing vital clues to assist preparation for future volcanic catastrophes. It also reveals the creative responses to both the resources and threats associated with volcanism, which have promoted positive developments in human society and culture.

Probing into Earth's past environmental changes has always been a primary objective of geology but geologists today work increasingly alongside climatologists, palaeo-oceanographers, ice-core specialists, dendrochronologists, anthropologists and archaeologists to understand how climate change and natural disasters have shaped human origins, migrations, replacements and the growth of society and culture. A recurring theme is the quest to understand how abrupt changes in the environment influenced human behaviour. Why, for instance, did ancient societies abandon their territory or start to decline?

Theories to explain such issues display cycles of popularity and disdain. Catastrophism, environmental determinism and the narratives of 'dark nature' have long pedigrees rooted in philosophy, geography, evolutionary biology, religion and popular fiction. In the Western tradition, the Creation story and Noah's battle with the Flood are especially significant. In the nineteenth century, however, catastrophism's pre-eminence diminished as the geologists of the day began to view 'the past as the key to the present', arguing that natural processes acting over very long periods of time constructed mountain ranges, ocean basins, deserts and ice caps.

However, catastrophism has never truly gone out of fashion – a cursory look at the television schedules of natural history channels proves the point. Among the 'documentaries' on excruciating toxins, dirtiest jobs, weirdest sharks and deadliest asteroid impacts, shows on volcanoes surface frequently. Often, they portray worst-case scenarios, encouraged surely by the recurrent publication of academic papers reporting volcanic catastrophes, both ancient and anticipated (see table). A primary aim of this book is to examine the claims that volcanism shaped prehistoric and historic social trajectories. To do this, we need to look at how volcanoes act on a very large scale, and how often they do it. Lifespans of volcanoes are variable but can exceed a million years, far in excess of the time that the species *Homo sapiens* has lived on Earth. Even an individual volcano might exert an intermittent influence on human ecology, demography and migration.

Such enquiry into the record of past volcanism and its impact is not only of interest to understanding archaeology and ancient environmental change. In considering the full range of risks posed by future volcanic activity it is vital to recognise that volcanoes can unleash disasters of a scale not seen for generations. In the field of flood defence, for instance, neglecting the effects of the one-in-a-hundred-year event has led to very substantial losses. What are the chances of a 'super-volcano' such as Yellowstone in the USA producing another 'super-eruption' in the next decades, and what would its impacts be? Might global climate change actually trigger volcanic eruptions? Could artificial volcanoes be used to control climate change? As well as considering these questions, this book also delves into the deeper geological record to explore the links between volcanism and mass extinctions identified in the fossil record.

Chapter 1 sets the scene by reviewing the most pertinent concepts of volcanology. It reviews the kinds of volcanoes and eruptions that are capable of 'shaking the world' and how often they do it. Then, the broad structure of the book is as follows: Chapters 2 and 3 provide the necessary background for understanding how volcanoes can abruptly change the environment and impact human societies across a spectrum of spatial and temporal scales. Some hazards are obvious – a glowing pyroclastic current entering through the back door for instance – but others are more insidious and potentially far more pervasive. These include the cold summers experienced after certain large eruptions due to the associated emissions of chemically reactive gases into the atmosphere. These two chapters thus distinguish between the immediate (but lasting), local-to-regional scale impacts of an eruption, and the hemispheric- to global-scale repercussions of eruption-induced climate change. One rather common (and useful) element – sulphur – turns out to be behind some of the most extravagant and far-reaching claims for volcano catastrophism. Chapters 4 and 5 provide further preparatory reading by explaining how we can reconstruct past volcanic events, environments and human responses.

Chapters 6 through 13 supply the main case studies. They are arranged to provide a time travelling experience, embarking in the deep geological past (why did the dinosaurs perish?) and ending in the second decade of the nineteenth century, when the largest and deadliest known historic eruption (of a volcano in eastern Indonesia) apparently contributed to social unrest and outbreaks of epidemic disease in Europe. In between, I review cases of eruptions that had major repercussions on human societies, reaching back to the first migrations of modern humans out of Africa, and the prehistory of Europe, Asia, Oceania and the Americas.

One reason for this progression through time is to aid reflection on lessons for the future. The final chapter builds from an understanding of the human ecology of natural disasters, and highlights key issues for managing volcanic catastrophe risks in the world to come. Human society might be more technologically advanced than it was a millennium ago but that does not in itself bring greater security in confronting potential environmental catastrophes. Indeed, the trivially sized Eyjafjallajökull eruption in Iceland in 2010 dramatically exposed some of the specific vulnerabilities of a globalised world.

I wrote this book because I became fascinated by the intersections of geology, climatology, ecology, archaeology and anthropology. In fact, it is this plexus of themes that makes volcanology such a great subject – just about anyone can get involved: mathematicians, physicists, architects, atmospheric scientists, civil protection managers, health professionals, risk analysts, engineers, archaeologists, oceanographers and planetary scientists, among others. This reflects the relevance of the subject to an equally wide range of academic, practical and vital issues and topics, including the origins of life, human evolution, climate change, food security, geothermal energy and worldwide aviation ... It has been a challenge to synthesise such a diverse and complex field. I hope that, notwithstanding the errors and omissions I have surely made, and the inevitable revisions of hypotheses that will emerge in the light of forthcoming data and models, that at least the book will convey the excitement of volcanology, and help to stimulate further research that overruns traditional disciplinary boundaries. My overall message is that, beyond the attention-grabbing claims of volcano catastrophism, what we actually know is far more nuanced (and speculative) but much more interesting.

For the sake of the forests (and the cover price), referencing has been minimised but a thorough listing of (hyperlinked) sources, plus a selection of colour images from the book, can be found at http://www.geog.cam.ac.uk/research/projects/eruptions.

Notable eruptions and some of the more extreme claims made for their effects.

Eruption(s) and date(s)	Magnitude (M_e)[1]	Impact scale	Extreme claims	Chapter
Siberian Traps, 250 million yeras ago	(11.9) 3 million km^3 lava	Global	Mass extinction	6
Deccan Traps, 65.5 million years ago	(11.6) 1.5 million km^3 lava	Global	Mass extinction (including dinosaurs)	6
East African Rift Valley, repeated eruptions over last millions of years	7–8	Regional	Migrations of archaic and modern humans	7
Toba, 73,000 years ago	8.8	Hemispheric–continental	Severe global climate change and near extinction of *Homo sapiens*	8
Campanian Ignimbrite, 39,300 years ago	7.4–7.7	Continental–regional	Acceleration of the European Palaeolithic Transition, demise of the Neanderthals	9
Mystery eruption, 17,000 years ago	?	Regional–local	Extinction of *Homo floresiensis* ('the Hobbit')	8
Laacher See, 10,970 BCE	6.2	Regional	Migration and cultural de-evolution of populations	9
Kikai, c. 5480 BCE	7	Regional	Abandonment of southern Kyushu and cultural replacement	4
Witori and Dakataua, repeated eruptions over last thousands of years	5.8–6.5	Regional	Migrations of Lapita people; spectrum of response from adaptation to continuity	5
Santorini, c. 1640 BCE	7.2	Regional	Decline of Minoan civilisation	9

Notable eruptions and some of the more extreme claims made for their effects. (cont.)

Eruption(s) and date(s)	Magnitude (M_e)[1]	Impact scale	Extreme claims	Chapter
Arenal, last thousands of years	~4	Local	Adaptation and continuity	5
Popocatépetl, c. 50 CE	5.3	Local	Rise of Teotihuacán	10
Mystery eruption (?Ilopango), 536 CE	6.9 (if Ilopango)	Hemispheric–regional	Justinian plague, fall of Teotihuacán	10,11
Mystery eruption, 1258 CE	?7–8	Hemispheric	Famine and pestilence in Europe, religious fervour	11
Laki, 1783–4	6.6	Continental	Famine, heat wave, severe cold, flooding, air pollution, crop damage	12
Tambora, 1815	6.9	Hemispheric	Famine, poor harvests, social unrest in Europe, rise of extremism and introduction of social reforms in Europe	13
?Yellowstone, 2100	?8.8	Global	Transfer of human civilisation to a safer place in the Solar System . . .	14

[1] Introduced in Chapter 1.

Acknowledgements

I originally planned to finish writing this book in 1999! I am very grateful, therefore, to Cambridge University Press and the editorial team – especially Laura Clark, Susan Francis, Chris Hudson and Matt Lloyd – for maintaining enthusiasm for such a prolonged project. The advantage of the slow pace was that the book's trajectory came to influence my research. I doubt otherwise that I would have ended up working with Quaternary scientists in Ethiopia and Eritrea, palaeopathologists excavating graves in Iceland, archaeologists in Yemen and India, or with atmospheric scientists in Italy and Antarctica.

Most of the text was reviewed in sections by friends and colleagues, who offered much sound advice that helped to improve the narrative. For this I thank Anna Barford, Peter Baxter, Amy Donovan, Hans Graf, Susanne Hakenbeck, Karen Holmberg, Phil Kyle, Christine Lane, Stephen Oppenheimer, Patricia Plunkett, Felix Riede, Alan Robock, Payson Sheets, Chris Stringer, Þorvaldur Þórðarson, Robin Torrence and Paul Wignall. Several people kindly provided illustrations or data including Mike Baillie, Stuart Bedford, Keith Briffa, Alain Burgisser, Richard Ernst, Marco Fulle (that's his spectacular photograph on the front cover), Emma Gatti, Evgenia Ilyinskaya, Kateřina Krylová, Steffen Kutterolf, Christine Lane, Patricia Plunkett, Felix Riede, Mike Salmon, Andrey Sinitsyn, Jørgen Peder Steffensen, Robin Torrence, Claire Witham and Sabine Wulf. David Watson skilfully prepared maps and diagrams. I thank, too, the following for additional comments and discussions: Frank Ackerman, Nick Barton, Clive Gamble, Emmanuel Garnier, Michael Herzog, Peter Jackson, Sveinbjörn Rafnsson, Janice Stargardt, Jørgen Peder Steffensen, Chris Stringer and Rachel Wood. I also thank the four (anonymous) reviewers of the original book proposal for their valuable critiques (even if they can't recall their contribution by now!). Chapters 8 and 13 are thoroughly overhauled versions of papers published in *Quaternary Science Reviews* and *Progress in Physical Geography*, respectively.

I owe a particular debt to the 'hall-of-fame' volcanologists, archaeologists, historians, Quaternary scientists and climatologists who have investigated the larger eruptions in history and prehistory. In particular, the works of Steve Carey, Craig Chesner, Peter Francis, Hans Graf, John Grattan, Peter Kokelaar, Patricia Plunkett, John Post, Mike Rampino, Alan Robock, Bill Rose, Steve Self, Payson Sheets, Haruldur Sigurdsson, Dick Stothers, Þorvaldur Þórðarson, Robin Torrence, Colin Wilson and Greg Zielinski have been a particular source of inspiration. I was also stimulated by a series of seminars staged in the mid-1990s by the King's College Research Centre in Cambridge on the topic of human evolution and diversity. I thank, too, John Lowe and Rupert Housley for inviting me to attend a 2010 meeting of their 'RESET' project, which is using volcanic ash layers found in sediment sequences and archaeological sites to understand the responses of ancient human societies to sudden environmental changes (http://c14.arch.ox.ac.uk/reset).

I spent 2010 at 'le Studium' Institute for Advance Studies in Orléans (http://lestudium.cnrs-orleans.fr/). It has been a pleasure living and working in France and my apartment just outside the old city has been a perfect bolthole to conclude work on the book. I am extremely grateful to le Studium and the University of Orléans for support and especially to Paul Vigny and Bruno Scaillet for enthusiasm and encouragement. I thank, too, the Leverhulme Trust, which supported some of the research presented here.

All projects of this endurance surely benefit from the support of side-kicks and soul mates. I particularly thank Peter Baxter, John and Sue Binns, Pierre Delmelle, Phil Kyle, Agnès Berthin and Bruno Scaillet in this regard, and above all, Anna Barford who has cheered me through the final mile of the writing marathon!

I hope you enjoy the book. I welcome feedback. Lastly, thank you, Iceland, for giving volcanology its 15 minutes (two weeks?) of fame!

1

Fire and brimstone: how volcanoes work

> 'Some volcanos are in a state of incessant eruption; some, on the contrary, remain for centuries in a condition of total outward inertness, and return again to the same state of apparent extinction after a single vivid eruption of short duration; while others exhibit an infinite variety of phases intermediate between the extreme of vivacity and sluggishness.'
>
> G. P. Scrope, *Volcanos* (1862) [1]

The Earth is cooling down! This has nothing to do with contemporary global warming of the atmosphere and surface. I refer instead to the Earth's interior – the source of the molten rocks erupted by volcanoes throughout the planet's 4.567 billion year history. Aeons before the continents drifted to anything like their familiar positions, and as early as 3.34 billion years ago, parts of the Earth were already colonised by primitive bacterial life forms. At this time, volcanoes erupted lavas with a much higher content of an abundant green mineral called olivine than found in most modern volcanic rocks. This testifies to much higher eruption temperatures for the ancient lavas compared with present-day eruptions from Mt Etna or the Hawaiian volcanoes. In turn, it reveals that the Earth's largest internal shell, the olivine-rich mantle, which comprises about 84% of the Earth's volume, used to be considerably hotter, too (anywhere between 100 and 500 °C depending on whom you believe). While the Earth changes irreversibly through time, it also exhibits behavioural cycles acted out on vastly different timescales, such as the amalgamation and break-up of supercontinents; the clockwork advance and retreat of ice ages steered by oscillations in the Earth's axis of rotation and orbit around the Sun; the seasons; and the tides.

A glance at a global map of active volcanoes, earthquake epicentres and plate boundaries (Figure 1.1) provides compelling evidence

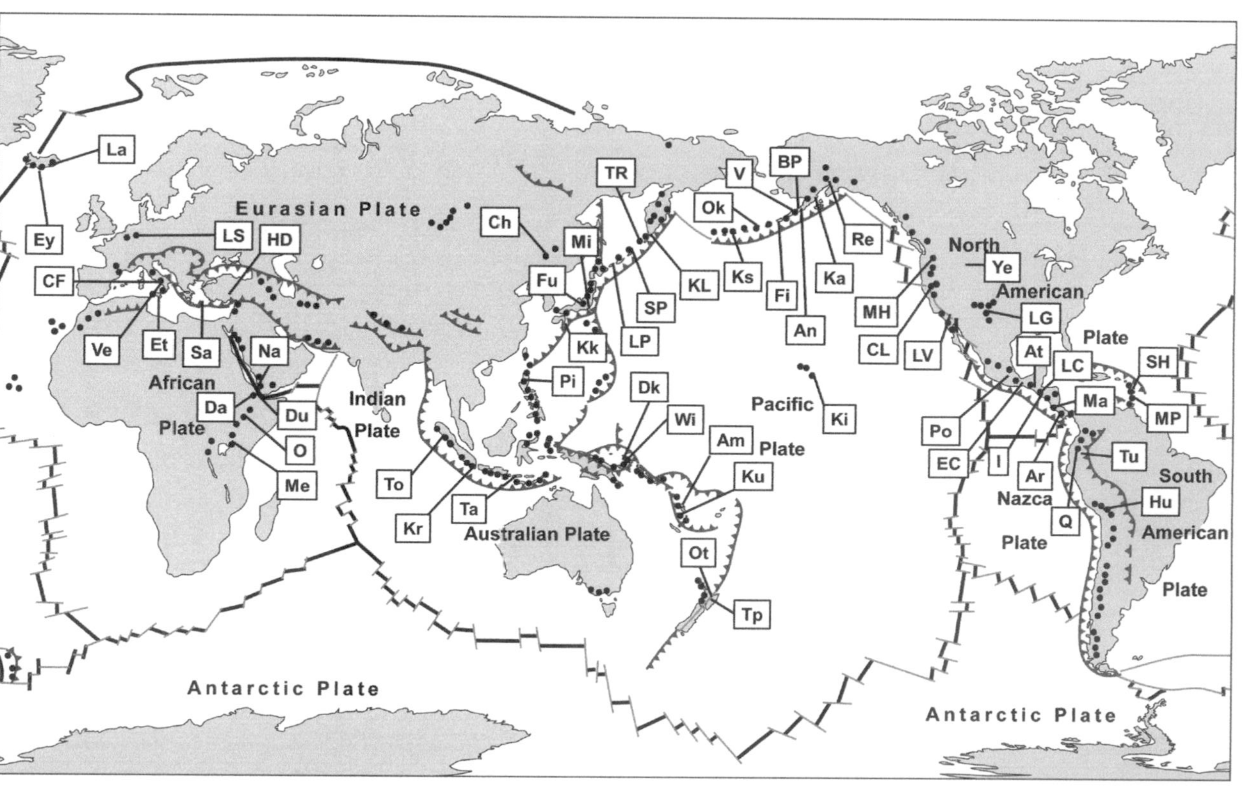
Eurasian Plate
African Plate
Indian Plate
Australian Plate
Pacific Plate
North American Plate
South American Plate
Nazca Plate
Antarctic Plate
Antarctic Plate
La
Ey
CF
Ve
Et
Sa
LS
HD
Na
Da
Du
O
Me
Ch
Mi
Fu
Kk
Pi
TR
SP
LP
KL
Ok
Ks
V
BP
Fi
An
Ka
Re
Dk
Wi
Am
Ku
Ot
Tp
To
Kr
Ta
Ki
MH
CL
LV
Ye
LG
At
LC
SH
Ma
MP
Po
EC
I
Ar
Tu
Hu
Q

Figure 1.1

for the coupling of tectonic and eruptive processes. Most volcanoes lie on the oceanic ridges formed as tectonic plates separate from each other. The volcanoes here exist in perpetual darkness except for their own magmatic glow. They erupt unobserved except by bizarre life forms that thrive on volcanic nutrients, and, just occasionally, by cameras on deep-diving research submarines. Nevertheless, collectively, they erupt far more lava than all the land volcanoes. This ocean-ridge volcanism also provides a particularly good example of how external pressure can influence the characteristics of eruptions. The overlying 2.5 kilometres of water exerts a crushing pressure 250 times the air pressure at sea level. This inhibits anything like the kind of explosive volcanism observed at the Earth's surface.

As newly formed oceanic plate trundles away from the volcanically active ridge, it cools and increases in density. Around much of the Pacific, the plate sinks back into the Earth's interior at a 'subduction zone', associated with some of the most dangerous volcanoes of the 'Ring of Fire'. Yet other volcanoes are located in the middle of nowhere, far from any plate boundaries. Hawai`i, right in the centre of the Pacific plate is, perhaps, the best known example but there are other 'hotspot' volcanoes both in the oceans and on the continents. Finally, volcanoes also congregate along the axis and flanks of great tears in the continents like the East African Rift Valley. To understand these various occurrences we need first to plumb the depths of the Earth to consider

---

Caption for Figure 1.1 Map summarising tectonic plates, bounded by spreading ridges (black segments), transform faults (light grey lines) and subduction zones (toothed lines), and distribution of volcanoes (dots). For clarity's sake, only a few the 1300 or so volcanoes known to have erupted in the last 11,500 years are shown but most of those discussed in the text are labelled as follows: Am (Ambrym), An (Aniakchak), Ar (Arenal), At (Atitlán), BP (Black Peak), CF (Campi Flegrei), Ch (Changbaishan), CL (Crater Lake / Mazama), Da (Dabbahu), Dk (Dakataua), Du (Dubbi), EC (El Chichón), Et (Etna), Ey (Eyjafjallajökull), Fi (Fisher Caldera), Fu (Fuji), HD (Hasan Dağı), Hu (Huaynaputina), I (Ilopango), Ka (Katmai), Ki (Kīlauea), Kk (Kikai), KL (Kurile Lake), Kr (Krakatau), Ks (Kasatochi), Ku (Kuwae), La (Laki), LC (Loma Caldera), LG (La Garita), LP (Lvinaya Past), LS (Laacher See), LV (Long Valley Caldera), Ma (Masaya), Me (Menengai), MH (Mt St Helens), Mi (Miyake-jima), MP (Mont Pelée), Na (Nabro), O (O'a), Ok (Okmok), Ot (Okataina), Pi (Pinatubo), Po (Popocatépetl), Q (Quilotoa), Re (Redoubt), Sa (Santorini), SH (Soufrière Hills volcano), SP (Sarychev Peak), Ta (Tambora), To (Toba), Tp (Taupo), TR (Tao-Rusyr Caldera), Tu (Tungurahua), V (Veniaminof), Ve (Vesuvius), Wi (Witori), Ye (Yellowstone).

where all this lava comes from in the first place. Before proceeding, let us agree on one element of the often intimidating nomenclature of igneous petrology: magma is molten rock below the surface; lava is what comes out of a volcano.

1.1 ORIGINS OF VOLCANOES: THE MANTLE

Virtually all volcanism on Earth begins in the 'mantle', the largest of the shells that constitute the planet (Figure 1.2). It lies between the

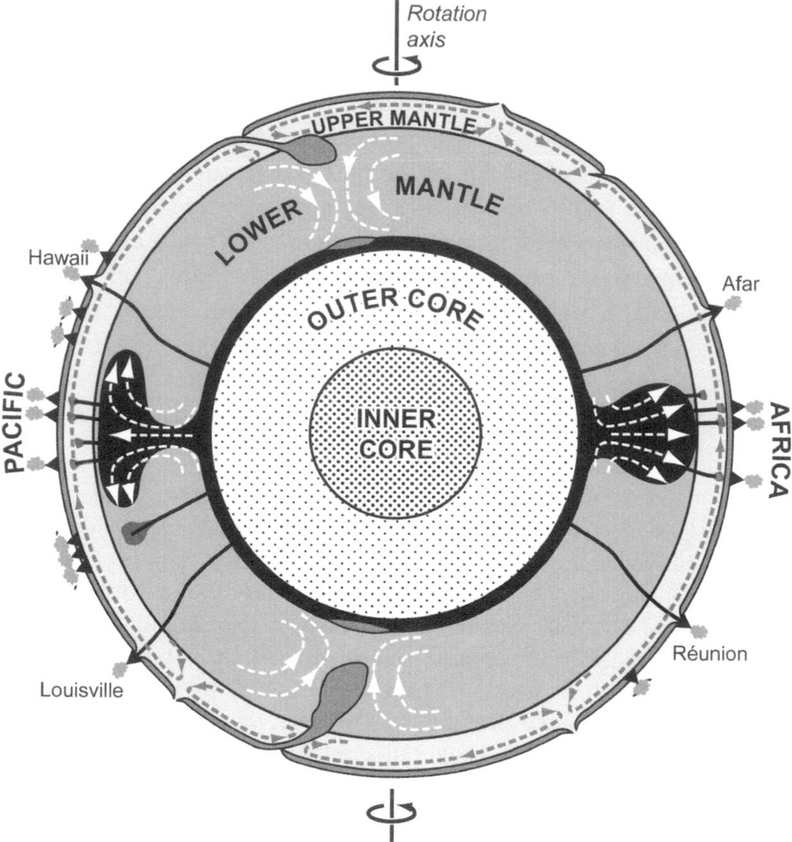

Figure 1.2 The Earth cut through its centre, illustrating 'primary' upwelling plumes thought to originate in the lowermost part of the mantle. Also shown are the plume tails beneath Hawai`i and Louisville (part of a seamount chain in the Pacific Ocean), Afar (northeast Africa) and Réunion (Indian Ocean), and subduction zones where the Earth's tectonic plates are recycled in the mantle. Modified from reference 2 with permission from Elsevier.

silica-rich crust (on which we live), and the dense, iron-rich core. The mantle is composed largely of a rock called peridotite which, in turn, is comprised of a number of crystalline minerals. Along with olivine are other silicate minerals including two kinds of pyroxene, garnet and plagioclase feldspar, and small quantities of metal oxides. A handful of elements – oxygen, silicon, magnesium, iron, aluminium and calcium – compose over 99% of the mass of peridotite. Although the mantle is solid – and we can be certain of this because it transmits certain kinds of earthquake waves that could not pass through a liquid – it is hot enough that it can flow by a slow process, called creep, in which crystals slip past each other, and atoms and ions diffuse from one place to another. (Ice is a more familiar example of a solid that can flow when it is thick enough, as attested to by glaciers and ice sheets.)

A combination of heat and gravity causes the mantle to flow. The Earth is hot inside – this is obviously the case since the lavas pouring out of volcanoes can reach temperatures well over 1100 °C; anyone who has approached within a few metres of a lava flow will be familiar with their searing radiance. Less prosaic is the question of where the heat comes from. Some of it, amazingly, dates back to the formation and infancy of the Earth. This primordial heat came from several sources including the kinetic energy of meteorite hails, chemical reactions, and the decay of some very ephemeral but fiercely radioactive elements. In addition, crystallisation of the Earth's core and radioactive decay of lingering isotopes of uranium, potassium and thorium continue to release heat into the Earth's interior.

Meanwhile, deep space is exceptionally cold. In fact, the electromagnetic radiation filling the cosmos indicates a background temperature of −270.43 °C (close to the absolute limit of −273.15 °C). The Earth is thus way out of thermal equilibrium with space, and consequently loses heat to it. Although the large size of the Earth renders this a slow process, hence the longevity of the primordial heat, the heat is transferred by convection out of the Earth to its surface. Like a pot of soup on the stove, the mantle is heated from the core beneath it while being cooled from above by heat radiation into space. Like most substances, the hotter the mantle, the lower its density; thus, under the action of gravity, hotter regions of mantle rise, while colder regions sink. This circulation of the solid mantle is essential to the melting that gives rise to magmas, and without it there would be no volcanoes on Earth.

If it still seems odd to think of the solid mantle flowing, there is a wonderful illustration of its fluid nature to be observed today in regions

Table 1.1 *Key properties of magmas and some comparative materials.*

Material	Silica content (% by mass)	Typical temperature (°C)	Viscosity (pascal seconds)
Water	-	20	$\sim 10^{-3}$
Ice	-	<0	$\sim 10^{12}$
Honey	-	20	~10
Peanut butter*	-	20	~200
The mantle**	~45	>1300	7×10^{18}
Basaltic magma	45–52	1100	10^{2}–10^{3}
Intermediate magma	52–63	1000	10^{3}–10^{5}
Silicic magma	>63	800	10^{5}–10^{10}

* Smooth not crunchy.

** The solid but convecting upper mantle known as the asthenosphere.

of Scandinavia, Siberia and North America that were covered in thick ice during the last ice age, which peaked 18,000 years ago. The weight of up to three kilometres thickness of ice was enough to push the Earth's crust down into the mantle, which then flowed away from the zones of greatest ice accumulation. It was the slow process of solid mantle creep that allowed this fluid behaviour. When the ice disappeared, the mantle crept back and the land surface started rising, and this continues today. By dating past shorelines using radiocarbon techniques (Section 4.1.3) it is possible to determine the pace of uplift, which continues at peak rates of around one centimetre per year). This rate of 'glacial rebound' yields an estimate of the mantle's viscosity (a measure of how well a material will flow when a force is applied to it; Table 1.1). It is 35 quadrillion times stickier than peanut butter!

Volcanoes exist because the mantle melts. But what causes melting? Two key processes are involved: one occurring at oceanic ridges and hotspots, the other at subduction zones. Interestingly, neither process is associated with heating. The first is the depressurisation that occurs as mantle convection currents rise to within 300 kilometres or so of the surface. Before exploring 'decompression melting' further, we need to recall that peridotite, like many rocks, is composed of several minerals. The different minerals have different melting temperatures; in fact, individual minerals themselves display a range of melting point according to their chemistry – olivine, for example, comes in a compositional spectrum between iron-rich and magnesium-rich varieties, which melt at different temperatures. Melting points are not only sensitive to chemical composition, they are also strongly dependent on pressure. With falling pressure, melting point drops.

As hot, solid mantle rises up in a convection current, and decompresses due to the reduced weight of rock above it, it can begin to melt. Crucially, the ascending mantle current is not so hot that all of it melts by this process. Instead, it is just those mineral constituents with the lowest melting points that melt; the high-melting-point minerals remain solid. Typically, somewhere between 1 and 20% of the peridotite melts, and hence the process is known as 'partial melting'. It is extremely important in the Earth since, over the course of geological time, it has changed the mantle's composition (by preferentially extracting certain magma-loving elements), and led to the growth of the crust and continents. The minerals pyroxene and plagioclase feldspar have lower melting temperatures than olivine, so a typical decompression event yields a liquid whose content best approximates a mixture of pyroxene, plagioclase feldspar and a little olivine. This melt is typically referred to as 'basaltic' and contains around 45% silica (SiO_2) by mass. The great pressure squeezes the basalt 'melt' from the crystals remaining in the mantle, a process a bit like depressing the plunger in some coffee makers. The melt percolates upwards forming pools of magma, which continue to rise owing to their lower density than their surrounds. Basalt contains all the ingredients needed to generate new oceanic crust at mid-ocean ridges.

Decompression melting is also responsible for the hotspot volcanoes, which are distinguished from oceanic ridges by their association with localised and especially hot upwelling zones known as mantle plumes [2] (Figure 1.2; Chapter 6). Their higher temperature sometimes results in a larger degree of partial melting. Volcanoes appear where mantle plumes blowtorch through the plates – this is how the Hawaiian Islands and the trail of seamounts to their north formed over the last 70 million years. Mantle plumes today account for something like 5–10% of the heat and magma extracted from the Earth's mantle. When mantle plumes impinge on continents they can initiate the kind of rifting for which East Africa is justly famous (Section 7.1). If sustained, the stretching of the continent can end up with the formation of a new ocean basin. One spectacular location where this occurs today is in the Danakil Depression of Ethiopia (Figure 1.3). Iceland is also generally considered to be the result of

Figure 1.3 Aerial photograph of the Da'Ure eruption site in Ethiopia close to Dabbahu volcano. This must rank as one of the world's smallest explosive eruptions! It was triggered by the passage of a 60-kilometre-long dike of basaltic magma, which destabilised a silicic magma body relatively close to the surface. This view shows part of the fissure, a small lava dome, and the blanket of fine ash produced by the explosive activity.

hotspot volcanism, and mantle plumes have been responsible for the greatest outpourings of lava known in the geological record, sometimes called 'large igneous provinces' (Section 6.2).

The creation of new oceanic crust at ridges and its consumption at subduction zones represents the Earth's main means of cooling its infernal depths (hotspot volcanoes also contribute). The total length of ridges worldwide is around 50,000 kilometres. Taking an average spreading rate of five centimetres per year (comparable to the growth rate of human hair and fingernails) indicates that around 2.5 square kilometres of new ocean floor are born every year.

While the association between volcanism and rising currents of hot mantle seems logical, the reason why volcanoes develop at subduction zones, where old, cold oceanic plate plummets back into the mantle, is less intuitive. The answer is the second key process that causes the mantle to melt: hydration. To understand this, we need to begin at the oceanic rift. One of the most remarkable features of active oceanic ridges are the chimneys, known as black smokers, which belch hot fluids charged with minerals rich in sulphur. This brew of chemical nutrients feeds bacteria that, in turn, nourish an entire ecosystem of bizarre creatures thriving in the stygian waters. The discharges result from the percolation and circulation of seawater deep into the brand-new oceanic crust. The seawater reacts with the hot volcanic rocks, extracting sulphur but at the same time hydrating minerals such as olivine. The crystals end up accommodating a quantity of water molecules. The result is to transform basalt into a slippery green rock called serpentinite. As the oceanic plate trundles sideways from the volcanic ridge on its journey to a subduction zone, it carries this incarcerated seawater with it. Meanwhile, the seabed also accumulates water-rich clays and other waterlogged sediments. Much of this water is ultimately drawn down into the subduction zone.

The sinking oceanic plate carrying its complement of seawater experiences ever greater pressures the deeper it penetrates the Earth's interior. Once it reaches a depth of around 100 kilometres, the clay minerals, along with the olivine and pyroxene crystals that had trapped seawater when the crust was created at the ridge, now find themselves under phenomenal pressure, and their regular frameworks can no longer contain the water. It is expelled, along with seawater trapped in pores between minerals, and the resulting fluid percolates into the overlying mantle.

The addition of water to the mantle dramatically depresses its melting point, causing partial melting. If this seems unusual consider an analogous process. In parts of the world that experience cold winters, the authorities grit icy roads with salt. This addition lowers the freezing point of water by a few degrees, which is enough to turn ice into water, so long as it is not too cold (it is even possible to use this principle to make ice cream). In the case of a subduction zone, the melts and water-rich fluids that are produced migrate upwards. Unlike oceanic ridges, subduction zones may source magmas that rise into thick overlying continental crust (as in the Andes). This typically provides much greater opportunity for chemical and physical evolution of the initial magma composition than is the case for oceanic volcanoes, and results in an amazing variety of magma types and volcanic activity.

1.2 MAGMA

Magma is a fascinating and remarkably complex substance. It represents the building material of volcanoes. The challenges of

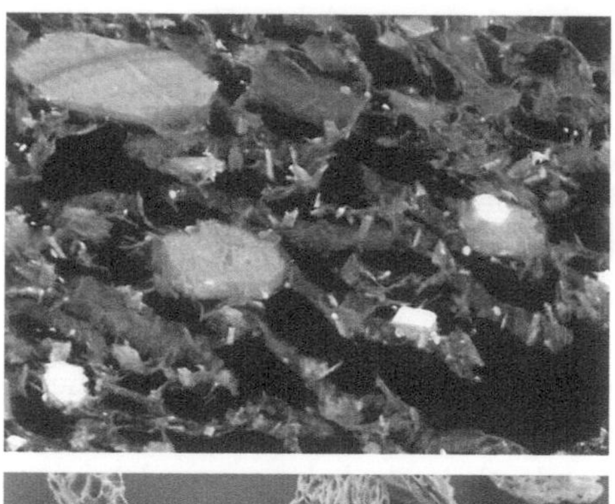

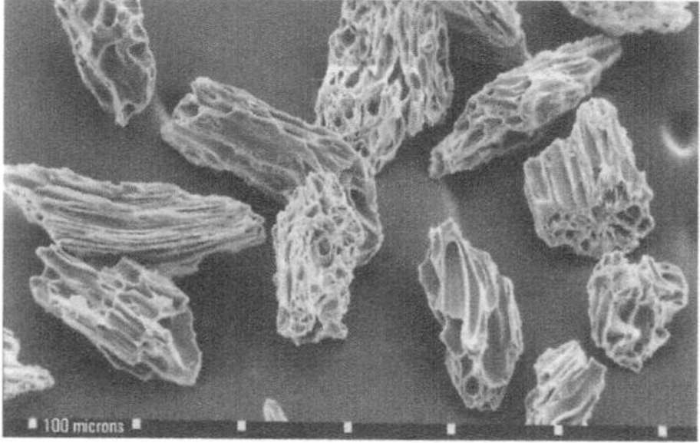

Figure 1.4 Images of ash and pumice: (top) an X-ray image of a sample of pumice (just half a millimetre across) that was erupted by Soufrière Hills volcano on Montserrat in 1997. The larger crystals within the lozenge-shaped sample are of the mineral amphibole and the minute, needle-like crystals are plagioclase feldspar. The black regions are bubbles; the remainder is glass (melt cooled too rapidly to crystallise). Image courtesy of Alain Burgisser. (Bottom) Scanning electron microscope image (0.6 millimetres across) of ash from a very large eruption 600,000 years ago of Brokeoff volcano, California. Note the shapes of gas-bubble holes (vesicles) – some have been stretched out into tubes by the explosivity of the eruption. Credit: A. M. Sarna-Wojcicki, US Geological Survey.

understanding its properties stem partly from the extraordinarily complex physical behaviour of molten rock with changing temperature, and the additional complications that arise from its constitution by all three phases of matter: solid, liquid and gas (Figure 1.4). The solid component is in the form of crystals of one or more minerals (such as olivine, feldspar, pyroxene and quartz). These are generally suspended in a silicate melt, which is dominated by loose arrangements of silicon and oxygen atoms and a brew of other elements including aluminium, sodium, potassium, calcium, magnesium and iron. In addition, the melt contains 'volatile' components, such as water, carbon dioxide, sulphur, and lesser amounts of halogens (chlorine, fluorine, bromine and iodine), and trace metals including lead and mercury. (They are called volatiles because they readily form vapours; the word is derived from the Latin for 'flying'.)

If it were possible to view a silicate melt at the atomic level we would discern that the silicon and oxygen atoms bind together to form pyramid-like tetrahedra, with a silicon atom at the centre and oxygen atoms at each of the four corners. These tetrahedra can share electrons with each other, establishing chemical bonds that make magma far more viscous than it would otherwise be. In fact, the extent to which this 'polymerisation' of silica tetrahedra develops and evolves exerts a major influence on how magmas move around in the Earth's crust, how they erupt, and their associated volcanic hazards (Chapter 2). The third phase of matter in magmas is gas. In fact, bubbles are one of the most interesting and complex aspects of magmas, and they, too, play a crucial role in triggering eruptions, eruptive behaviour and the environmental and climatic impacts of volcanism.

The chemical species in bubbles derive from the aforementioned volatile constituents. When the mantle melts, volatiles are preferentially extracted out of the mantle rock into the newly formed liquid. Under the very high pressures experienced in the mantle these volatiles are typically dissolved in the melt, and they can constitute several per cent of the mass of the magma. But as the nascent magma ascends into the crust it feels less of the weight of the overlying rocks, and the reduced pressure allows the volatiles to form bubbles, a process known as 'exsolution'. Deep in the crust these bubbles are generally dominated by carbon dioxide, and indeed, on geological timescales, the output of carbon dioxide from volcanism is the primary source of this greenhouse gas to the atmosphere. However, as magmas continue to rise and decompress, other volatiles exsolve, including water, sulphur dioxide and hydrogen fluoride. The result, when emitted at the surface, is usually a pungent and choking cocktail of gases mixed with miniscule particles of sulphuric acid and metal chloride condensates.

Although the degree to which the mantle melts when subjected to compression or hydration dictates the starting composition of a magma, the overriding influences on the chemical composition of erupted rocks are processes occurring during transport and storage of magmas in the crust. Magmas rise in the first place because of their buoyancy – they are less dense than the surrounding rocks. However, the density of the crust reduces upwards, so an ascending magma will generally find a level, anywhere from 3 to 30 kilometres below the surface, where it has the same density as the host rock. It has reached what is known as the level of neutral buoyancy, and gravity no longer acts to propel it further upwards. With sustained melting of the mantle source, magmas drip feed into this zone, accumulating to form magma chambers. The more magma that collects, the greater is the potential for a copious eruption. The most excessive eruptions - super-eruptions - expel thousands of cubic kilometres of pumice [3]. This is only possible if comparable volumes of magma have amassed in the chamber.

Of course, squeezing ever more magma into a chamber will increasingly pressurise it if the chamber walls cannot deform enough to make room. This is one of the mechanisms that can lead to eruption. If the pressure in a chamber is high enough to break the enveloping rocks, then a crack filled with magma can propagate outwards and upwards. These magma-filled cracks are referred to as dikes, and if they reach the surface, then eruption ensues. Fissure eruptions, such as the first phase of the 2010 Eyjafjallajökull eruption in Iceland, represent particularly good examples of the surface expression of dikes.

It may take a long time before a magma body erupts. Magma can accumulate and brew in a chamber for hundreds, thousands, even tens of thousands of years before erupting, especially in regions of thick continental crust and under tectonic stress regimes that reinforce the crust. During such long periods, it is far from inert. The surrounding rocks are much colder and extract heat from the magma causing it to crystallise. This leads to one of the most important processes that take place in the Earth's crust, known as fractional crystallisation. In essence, it is a kind of subterranean distillation and acts in the opposite way to partial melting: as the magma cools down, the first crystals to grow are composed of the minerals with the highest melting (i.e. freezing) points. These include olivine, which on precipitation may sink to the bottom of the magma chamber, or be plastered on to its walls. The silica content of olivine is less than 40% by mass, so its extraction must leave behind a melt progressively enriched in silica (recalling that basalt has at least 45% silica by mass), and also depleted in iron and magnesium.

As cooling proceeds, other minerals precipitate out but always enriching the remaining melt in silica, and, crucially, volatiles. This is why, counter-intuitively, cooling of a magma chamber can actually trigger an eruption. As it crystallises, leaving behind more and more volatiles, the proportion of dissolved water, carbon dioxide and other species increases. Eventually, the magma may become saturated in these volatiles, and further concentration due to cooling and fractional crystallisation will result in formation of bubbles. Once formed, bubble expansion acts to pressurise the magma chamber, which affords

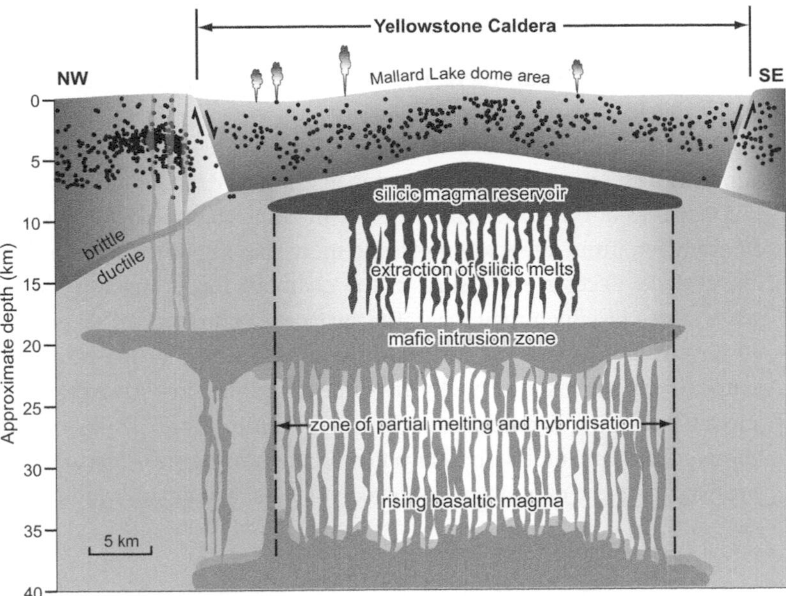

Figure 1.5 Schematic cross-section through Yellowstone 'super-volcano' in northwest Wyoming. The dots in the shallow part of the crust represent earthquake locations. The silicic magma reservoir 5–10 kilometres below the surface fed 'super-eruptions' 2.1 million years ago and 640,000 years ago. The caldera is more than 50 kilometres across. Modified from reference 3 and used with permission of the Mineralogical Society of America.

another route towards fracturing of the containing rock walls and initiation of an eruption.

Fractional crystallisation can account for much of the wide spectrum of volcanic rocks found on Earth. Sometimes, very pristine basalts are found, indicating rapid passage of partial melts from the mantle to the surface. At the other end of the spectrum, long-lived magma chambers, like Yellowstone's, can erupt rhyolites, whose silica content exceeds around 73% by mass (Figure 1.5). Any text on igneous petrology will elaborate on the geochemical character and evolution of magmas, and will entertain the reader with descriptions and definitions of the extraordinary array of volcanic rocks from picrites to phonolites to pantellerites and everything in between [4]. For our purposes it is sufficient to consider three classes of magma according to their silica content (and degree of fractional crystallisation): basaltic (45–52% SiO_2 by mass), intermediate (52–63% SiO_2 by mass) and silicic (> 63% SiO_2 by mass).

1.3 ERUPTION PARAMETERS

Beneath every volcano there is a source region where mantle rocks melt, 100 kilometres down at a subduction zone, typically less than 50 kilometres beneath ocean ridges. The melt rises by percolating upwards and eventually accumulates at the level in the crust where the densities of the magma and the surrounding rocks match. This can be anywhere from a few kilometres to more than ten kilometres below the surface. Excessive pressure in the chamber forces magma out in dikes, some of which may reach the surface via conduits a few metres or tens of metres in diameter causing eruptions; others freeze in the crust to form intrusions. So why are some eruptions violently explosive, propelling ash and gases into the upper atmosphere, while others involve the torpid emission of lava flows, lakes and domes? The answer, once again, is the content of dissolved volatiles, principally water.

1.3.1 Explosive and effusive volcanism

Magmas at high pressure in the Earth's crust can contain considerable proportions of dissolved volatile components – more than 10% by mass in some cases. Subduction-zone magmas tend to have the highest quantities of dissolved volatiles since they derive plenty of water, sulphur and chlorine from the old subducted oceanic rocks. In addition, because they tend to reside in thicker crust for longer periods, they typically melt the rocks surrounding the magma chamber and, in so doing, acquire more volatiles (for instance carbon dioxide from limestone, water from granite). Hotspot volcanoes on both oceanic and continental crust can still have plenty of fizz in them, though.

As magma ascends towards the surface, it decompresses and the melt increasingly struggles to contain the volatiles in solution. They exsolve, forming bubbles, which lower the density and increase the volume of the magma. This acts to accelerate the magma towards the surface, potentially in a runaway process. But there is a competing force, especially important for magmas of intermediate to silicic composition; when water is dissolved in the melt, it inhibits the bonding between silica tetrahedra (Section 1.2). So, as water moves from the melt into bubbles, the tetrahedra increasingly string together into chains. This can dramatically increase magma viscosity such that it moves more and more sluggishly towards the surface. The ever accumulating and expanding bubbles coupled with the increasing resistance to magma flow can culminate in highly explosive conditions if the accumulated pressure is released suddenly, for instance when the chamber walls fail and dikes zip to the surface.

The opening of a bottle of soda pop or, better still, the uncorking of a bottle of champagne provides a well-rehearsed analogy. The fizz comes from dissolved carbon dioxide (which, along with water, is a dominant component of magmatic volatiles). Easing the cork off reduces the pressure inside the bottle, reducing the solubility of carbon dioxide in the drink, releasing carbon dioxide gas. The runaway process, especially beloved of award ceremonies at elite sporting events, leads to rapid exsolution and expansion of carbon dioxide to such an extent that a foam of champagne jets out of the bottle. The process can be so efficient that little champagne remains in the Nebuchadnezzar to be enjoyed. This is pretty much what happens in some eruptions. In the pipe feeding an explosive eruption there will be a region, perhaps a few hundred metres down, where magma with a small proportion of bubbles rising at around one metre per second (equivalent to a strolling pace) fragments into a gas-dominated mixture containing shattered glassy ash, crystals and pumice that, by the time it reaches the vent, has accelerated to speeds of several hundred kilometres per hour.

An alternative trigger of explosive volcanism arises when magma meets water. This could occur when magma intrudes rocks near the surface that are saturated with ground water. Depending on how much water and magma end up coming into contact with each other, the sudden production of steam and accompanying expansion can yield explosions of tremendous violence. Such activity is broadly termed 'hydrovolcanic', and it is commonly associated with the reawakening of long dormant volcanoes. It can also develop in the course of an eruption of an island volcano, should seawater suddenly gain access to the vent area. In the case of the 2010 Eyjafjallajökull eruption in Iceland, the hydrovolcanic character was greatly enhanced by the interaction between erupting magma and glacial melt-water (Figure 1.6). This ensured high explosivity, and the very fine fragmentation of the ash contributed to the threat to aviation that resulted in closures of airspace across much of the north Atlantic and Europe.

Alternatively, if a magma only has a low content of volatiles or if it is hot and runny (such as a typical basalt), gas bubbles may be able to escape freely, leaving a languid flow of mostly melt and crystals that erupts peacefully in the crater or down the flanks of the volcano. These more passive events are known as effusive, or lava, eruptions. In reality, most eruptions go through phases of explosive and effusive activity,

Figure 1.6 Close-up aerial view of the summit crater and eruption of Eyjafjallajökull in 2010. Note the white streaks of condensed steam arising from magma–ice interaction. Photographed by Evgenia Ilyinskaya.

even simultaneously, but the distinction in eruptive style remains useful.

Further, we can broadly distinguish between the products of these two kinds of activity: explosive eruptions produce fragmented rocks (pumice, ash, bombs) collectively known as tephra (from the Greek for 'ash') or pyroclasts (from the Greek for 'broken by fire'); effusive products are simply referred to as lavas. As one would expect, explosive volcanoes are more predisposed to cause trouble but, as we explore in Chapters 6 and 12, large and intense effusive eruptions can also have widespread and substantial impacts.

Volcanology is no stranger to nomenclature. Indeed, the discipline is strewn with abstruse petrological and technological jargon, and dry classification schemes that confuse volcanologists let alone anyone else. Among the latter is the division of the spectrum of volcanic eruption styles that describe the physical nature of an eruption as it might be observed by an eyewitness. Volcanologists often describe eruptions as 'Hawaiian', 'Strombolian', 'Vulcanian', 'Plinian' or 'Surtseyan' assuming that their colleagues all share the same view of what the nomenclature refers to (which is not necessarily the case). These terms derive from particular historic eruptions (for example, Vulcanian refers to the 1888–90 eruption of Vulcano; Plinian to the 79 CE eruption of Vesuvius recorded by Pliny the Younger; Surtseyan to the hydrovolcanic explosions of Surtsey off the coast of Iceland in the 1960s), or to the characteristic behaviour of individual volcanoes (Strombolian refers to Stromboli volcano's indefatigable propensity for modest pyrotechnics; Hawaiian to the 'fire fountains' exemplified by Kīlauea volcano).

However, a volcano can erupt for weeks, months or years displaying changes in behaviour that cannot be accurately conveyed by a single term. Another issue with the terminology is that it has come to be applied to two different though related phenomena – to the eruption itself, and to volcanic deposits (for example, a 'Plinian eruption' and 'Plinian pumice fallout'). Even present day volcanic eruptions go unobserved, requiring application of the same techniques as are used in reconstructing ancient eruptions from their associated deposits. In the case of a lava flow, so long as it has not become too eroded, it is fairly straightforward to relate the visible landform to the process that created it. However, reconstructing eruption characteristics from widely dispersed, possibly chemically altered, weathered and only partially exposed pyroclastic rocks can be far more challenging.

We therefore start by considering two fundamental eruption parameters – 'magnitude' and 'intensity' – before worrying too much about how to describe and interpret eruptive styles. This approach has the advantage of highlighting measurable physical properties of eruptions.

1.3.2 Magnitude

Magnitude is an expression widely used in science. One of the most familiar magnitude scales is Charles Richter's earthquake spectrum. Just as the Richter scale expresses the energy released in a seismic event (giving a preliminary idea of the area likely to have been affected and, where knowledge of the building stock is available, the likely levels of damage), so would a useful eruption scale signify energy release. In practice, a closely related but more readily measured quantity is used – the mass or volume of erupted products. (The use of mass is preferable since the volume of different eruptions is only directly comparable given knowledge of the densities of the erupted products.

These can vary by a factor of up to four depending on how full of bubbles ('vesicles') the lavas or tephra are. Pumice can be so full of holes that it will float on water, a property that is responsible for the dispersal of coral larvae across thousands of kilometres of open ocean enabling them to colonise new habitats.) Sometimes, eruption magnitudes are reported as a 'dense rock equivalent', taking into account the actual density of the deposit and the density of bubble-free magma (approximately 2600 kilogrammes per cubic metre).

In this book, we shall use an eruption magnitude scale akin to the Richter scale except that it is based on the total mass of erupted materials, m (expressed in kilogrammes):

$$M_e = \log_{10}(m) - 7 \tag{1.1}$$

It is impossible, of course, to weigh an entire volcanic deposit but some cunning methods for estimating m are explained in Section 4.1.2. Because of the logarithmic scale, a unit increment in M_e corresponds to a tenfold change in actual size. An M_e 7 eruption, for example, is ten times larger than an M_e 6 event. One fortunate property of the logarithmic scale is that it hides the considerable uncertainty inherent in most estimates of eruption size. For instance, five cubic kilometres of magma with a density of 2600 kilogrammes per cubic metre corresponds to an M_e of 6.1. An eruption twice the size (ten cubic kilometres) of the same magma is equivalent to an M_e of 6.4. Although we shall use this scale throughout the book, there is a prior classification that should be mentioned since it has been influential and remains in use: the Volcanic Explosivity Index (VEI). The constant value '7' in Equation 1.1 was deliberately selected to bring the two scales broadly into line.

The first application of a magnitude scale is in comparing the overall size of different eruptions. For instance, the Mt Pinatubo (Philippines) eruption of 1991 was an M_e 6.1 event, while Krakatau's famous 1883 outburst in the Sunda Strait had an M_e of 6.5. The 1980 eruption of Mt St Helens (USA), though extraordinarily destructive, registered an M_e of just 4.8. However, just as usefully, we can begin to look at the recurrence rates of eruptions of different sizes – for an individual volcano, for a region, or for the whole Earth. 'Frequency–magnitude curves' exist for many kinds of phenomena, including earthquakes and floods, and they are important for revealing physical processes, and for underpinning long-term hazard assessment and emergency planning (Chapter 14). Figure 1.7 shows the frequency–magnitude statistics for all known eruptions on land. As

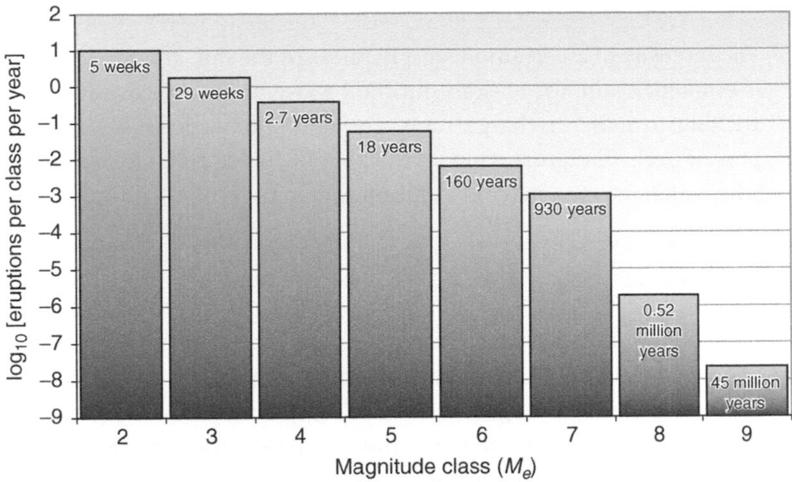

Figure 1.7 Frequency versus magnitude plot for volcanic eruptions based on records for the last 300 years for eruption magnitudes M_e of between two and six; for the last 2000 years for M_e between six and eight; and for all known M_e 'super-eruptions' of the past 45 million years [5]. The average return period for each magnitude class is labelled on the plot. The only M_e 9 eruption in this compilation is the 28-million-year-old Fish Canyon Tuff (Colorado, USA). The available data are imperfect because it is probable that not all very great eruptions in the last few million years or so have yet been identified. Nevertheless, the apparent fall off in the frequency of very great eruptions (M_e classes of eight and nine) is probably real. It suggests that rather different physics apply to very large eruptions (all of which are associated with caldera formation), which could have something to do with a threshold magma chamber size above which dike formation is suppressed and magma chambers grow rather than erupt.

is the case for many kinds of natural (and unnatural) phenomena, large events are more infrequent than small events. One interesting feature of the frequency–magnitude curve is its apparent tailing off towards the high end of the magnitude scale.

Another property of the curve is that it is possible to convert the magnitude into an energy scale, and then to compare the spectrum of volcanic activity with the energy involved with other kinds of phenomena. One finding of this approach is that very large volcanic eruptions are more common than huge meteorite strikes of equivalent energy. There has been some serious debate concerning what we should do about the threat from impacts of very large 'Near Earth Objects' but it could be that the threat from within is a more imminent one [5].

1.3.3 Intensity

The intensity of an eruption simply refers to the rate at which magma is erupted. Again, for reasons outlined above, there are advantages in sticking to mass discharge rates, though volumetric fluxes tend to prevail in the literature. One intensity scale in use is the following, in which *MER* refers to magma eruption rate in units of kilogrammes per second:

$$Intensity = \log_{10}(MER) + 3 \tag{1.2}$$

Intensity could refer to the average value for a given event or, if enough information is available, it might be possible to chart variations in intensity through time revealing waxing and waning magma discharge. In this case, the magnitude of the eruption is given simply by integrating the intensity with respect to time.

Eruption intensity is a vital parameter for an explosive eruption, since it strongly influences the height in the atmosphere to which the column of ash and gases will rise. The importance of plume altitude was demonstrated by the Eyjafjallajökull eruption in 2010. Reliable forecasting of the ash cloud trajectory, which was essential to manage the aviation hazard, required accurate measurements of the plume altitude. These proved very difficult to obtain. The most intense eruptions develop ash columns that penetrate the stratosphere, reaching heights in excess of 20–30 kilometres above sea level. The reason why intensity scales with column height is because both are related to the heat flux of the eruption. From theoretical considerations, the height H in metres reached by a sustained atmospheric plume is a function of the energy flux, Q (in units of watts):

$$H = 8.2 \times Q^{0.25} \tag{1.3}$$

Q is related to the thermal energy of the erupting magma and the mass eruption rate. Thus, as eruption intensity increases, so does the rate at which heat is pumped into the atmosphere. As hot tephra mixes with air a vigorous atmospheric plume develops. The physics are rather similar to those of a hot-air balloon: increase the heating and the balloon climbs; increase the intensity of an explosive eruption, and the volcanic cloud ascends further.

For an effusive eruption, intensity is just as important, since it will strongly influence the speed and distance over which a lava flow will advance. It also has a bearing on the eventual morphology and surface texture of lava-flow landforms.

1.4 SUMMARY

There is remarkable variety to volcanic eruptions and their associated phenomena. Eruptions owe their origin to heat convection currents that stir the Earth's solid – but fluid – mantle. Even on relatively short timescales, such as the period of 12,500 years or so since the end of the last glaciation, there is evidence of the mantle's ability to flow (seen in raised shorelines associated with glacial rebound). Where hot mantle rises within a few hundred kilometres of the surface it experiences partial melting, which ultimately feeds magma chambers sitting some kilometres or a few tens of kilometres below the Earth's surface. These may erupt frequently (usually as relatively benign effusions) or, instead, stew and brew for millennia, resulting in sporadic but more violent explosive eruptions that propel ash columns high in the sky.

Both the behaviour of volatiles dissolved in magmas and the magma composition (silica content) play a crucial role in dictating whether eruptions are explosive or effusive. This basic picture explains why subduction-zone volcanoes tend to be more explosive – their magmas have inherited much higher water contents during their formation. Different styles of eruption eject different kinds of volcanic materials. It is primarily (though not exclusively) what comes out of a volcano that represents the volcanic hazard (Chapter 2).

Eruptions show typical characteristics of many kinds of phenomena in the relationship between magnitude and frequency of events: smaller eruptions are more common than larger ones. The very largest explosive eruptions expel thousands of cubic kilometres of magma, and occur more frequently (over geological time) than meteorite impacts of equivalent energy. Even the much more modest Plinian eruptions witnessed in the modern period have propelled eruption columns to altitudes of 30 kilometres or more, forming high altitude veils of fine sulphurous dust that can encircle the planet. It is this haze that can result in some of the most widespread and devastating impacts of volcanic disasters (Chapter 3).

2

Eruption styles, hazards and ecosystem impacts

> And the streams thereof shall be turned into pitch, and the dust thereof into brimstone, and the land thereof shall become burning pitch. It shall not be quenched night nor day; the smoke thereof shall go up for ever: from generation to generation it shall lie waste; none shall pass through it for ever and ever.
>
> Isaiah 34:8–10

One of the major complications of managing the risks posed by volcanoes arises from the variety of weapons in a volcano's armoury. Volcanoes can unleash ash and toxic gas clouds, lava flows and the exceptionally destructive, searing avalanches known as pyroclastic currents. Even after decades or more of lying dormant, volcanoes may emit harmful gases and particles, and, on account of their construction from inter-layered rocks and propensity towards steep slopes, can continue to pose a threat in the guise of mudflows, gigantic landslides, and tsunamis. The intensity and magnitude of eruptions only correlate in a loose sense with human impacts, since the exposure and vulnerability of societies vary greatly from one place to another.

According to a review of available records, nearly five hundred volcanic events in the twentieth century impacted people, with up to six million people evacuated or left homeless [6]. Fatalities occurred in around half of the events, with an estimated total death toll of up to 100,000. The risk of catastrophic human and economic losses from future eruptions is significant, especially given the barely restrained urban growth that has taken place in many volcanic regions. A further notable feature of the statistical record is that the number of injured (about 12,000 in the twentieth century) is much lower than the number of deaths – volcanic phenomena are often associated with low survival rates.

A crucial point in understanding 'eruptions that shook the world' is the distinction between impacts that arise when a community finds itself at ground zero at the wrong time, and those that are mediated via climate change wrought by the sulphurous dust lofted into the middle atmosphere by certain eruptions (Chapter 3). In other words, an eruption might devastate a society because of its proximity to the active volcano (though this can still be hundreds, potentially thousands of kilometres away in the case of ash fallout). Even if the direct impacts of an eruption are not overwhelming in themselves, they might yet upset regional power balances leading to political or economic upheaval and internal or external conflict. One society's misfortune might be another's opportunity. Such volcano-induced regime changes have been argued for in connection with eruptions of Santorini (Greece) in the Bronze Age (Section 9.3), and Ilopango (El Salvador) around the sixth century CE (Section 10.2).

This chapter elaborates on the distinctions drawn between effusive and explosive volcanism by examining the range of primary and secondary volcanic phenomena, and provides an overview of the immediate and localised effects and hazards with which they are associated.

2.1 ERUPTION CLOUDS

The classic manifestation of an explosive volcanic eruption is a cloud of ash, rock and gases (Figure 2.1). These clouds actually engage in rather different behaviours but we begin with the more straightforward processes. One of the most reliable, though not risk-free, ways to observe explosive volcanism is to head for the island of Stromboli situated in the Tyrrhenian Sea between the coasts of Sicily and mainland Italy. Even a short time spent on the island will leave little doubt that it is (a) a volcano and (b) active. It is a near perfect, declivitous cone of dark basalt lavas topped with a collection of overlapping, perpetually steaming craters. Every ten minutes or so – and Stromboli has been doing this at least since the days of Aristotle – one of the craters fires a salvo of lava bombs accompanied by a reverberating detonation. These fulminations result when large gas bubbles violently rupture the lava ponded in vents on the crater floor. An ephemeral ash cloud rises a few hundred metres above the crater before being swept away by the wind.

Impressive though it is, this kind of activity – named Strombolian after its famous exhibitor – along with the more sustained fire-fountaining characteristic of the Hawaiian volcanoes, occupies the

Figure 2.1 Close-up of the classical development of a Plinian eruption column at Mt St Helens (USA) on 18 May 1980. Photograph by Robert Krimmel, courtesy of USGS/CVO.

lower end of the explosive eruption scale. At much higher intensities, and with magma discharge sustained for hours or tens of hours, eruption clouds can soar 20 or more kilometres above the surface, projecting vast quantities of fine ash and sulphurous gases into the middle atmosphere. In 2008, after more than 9000 years of remission, Chaitén volcano in southern Chile burst into the headlines with such a Plinian eruption. Until then, Chaitén had gone essentially unrecognised by the volcanological community, underlining a crucial lesson: the biggest eruptions tend to occur at volcanoes we know nothing about. They are latent for centuries or millennia, biding their time, and accumulating magma. The paroxysms of Mt St Helens (1980), El Chichón (Mexico, 1982) and Mt Pinatubo (1991) provide further compelling evidence for our scientific ignorance. Eruptions of this scale have the potential to change global climate for several years by forming stratospheric veils of fine sulphuric particles that intercept sunlight. The aftermath of the Pinatubo eruption demonstrates this very clearly and Section 3.1 reviews the case in detail.

The behaviour of these soaring Plinian eruption clouds is quite well understood in terms of the physics of thermal currents in the atmosphere. They rise to such great heights (up to four times the cruising altitude of commercial jet aircraft), not so much due to the momentum obtained from being blasted at great velocity out of the vent by the thrust of expanding magmatic gases, but primarily because they are

Figure 2.2 An exceptional view down the length of an eruption column fortuitously observed from the International Space Station. The volcano in question is Sarychev Peak in the Kurile Islands (Russia) and the date is 12 June 2009. Pyroclastic currents can be seen shooting down the sides of the volcano while a column of ash rises into the atmosphere, mushrooming out at its top. The ash cloud caused significant disruption to air traffic. Photograph ISS020-E-9048 from NASA's Earth Observatory.

hot. Indeed, kinetic energy amounts to much less than 10% of the thermal energy of an eruption plume. Immediately after the eruption mixture leaves the vent, it has typical densities of around ten kilogrammes per cubic metre. This is obviously very considerably less than the density of silicate melt (about 2600 kilogrammes per cubic metre) reflecting the tremendous expansion of bubbles in the magma (Figure 1.4). What erupts at the vent is essentially a rock foam whose density is about ten times greater than that of air. However, the nascent plume, travelling at speeds in excess of 350 kilometres per hour, very rapidly ingests the surrounding air and heats it up (Figure 2.2). The reduced density of hot air compensates for the dense ash particles and pumice suspended in the plume. Once sufficient air is sucked in and heated up, the plume becomes less dense than the ambient air and convects upwards. Its buoyancy will loft it to the height where it has the same density as the surrounding air. It may even overshoot this point thanks to its momentum but it will then sink back under gravity to

Figure 2.3 Magnificent mushroom cloud formed above pyroclastic currents during an eruption of Redoubt volcano, Alaska on 21 April 1990. Cloud height is about 12 kilometres above sea level. Photograph by J. Warren, USGS.

its 'neutral density level', flowing downwards and outwards to form a mushroom or umbrella-like cloud (Figure 2.3), before being dispersed by the wind.

Plinian eruptions typically involve intermediate or silicic magmas but basaltic events of this scale are recognised in the geological record. Mt Etna (Italy), eminently basaltic, erupted in Plinian fashion in 122 BCE. At very much the same time, so did another basaltic volcano, Masaya (Nicaragua). Significantly, both are famed in volcanological circles for their astonishing and sustained present-day emissions of sulphur and halogen gases into the atmosphere (Etna even emits 700 kilogrammes of gold into the atmosphere every year).

Without air entrainment and expansion, fledgling eruption columns would quickly run out of puff and cascade under gravity like fountains. Stable, convecting columns are favoured by high eruption velocities of gas-charged magma through narrow vents, since these factors promote air entrainment. Conversely, high mass eruption rates, low exit velocities, low gas contents and wide vents all favour collapse of the eruption column, since they inhibit the consumption and mixing of air.

During the course of an eruption, intensity may increase such that the vent widens due to erosion by the discharging magma. The outer regions of the plume may continue to mix turbulently with air and become buoyant, but the interior can remain dense and then collapse when it runs out of kinetic energy, gushing on to the flanks of the volcano as a 'pyroclastic current'. If fragmentation of the magma in the eruption conduit is less developed – as is often the case for basaltic volcanoes – then larger pyroclasts will be ejected, many of which follow ballistic trajectories to the ground as if fired from a cannon. These will be less efficient in transferring their heat energy to the atmosphere (because magma is a rather poor conductor of heat) and thus not scale the same atmospheric heights as Plinian eruption columns (typically, basaltic plumes ascend much less than ten kilometres). In the case of fire fountains, the magma fragments are typically baseball sized, and the height they reach has a lot to do with the momentum of the erupting mixture at the vent. The clots of lava fall back to the ground, still molten inside having failed to impart their thermal energy to the atmosphere to generate anything like a Plinian eruption plume. Such behaviour lies at the end of the spectrum of eruption column collapse. At higher eruption intensities, collapsing eruption clouds routinely generate the most infamous weapon in the volcanic arsenal: pyroclastic currents (Section 2.3).

2.1.1 Hazards

In the past, airborne ash would have posed limited risk to people – only in the heaviest ash falls is suffocation a real threat. Certainly, the Cimmerian shade beneath thick ash clouds is terrifying, and will have dramatic immediate impacts on temperatures on the ground, but such effects are generally ephemeral. On the other hand, the 2010 eruptions of Eyjafjallajökull in Iceland have highlighted the frailties of global aviation, and today it is recognised that one of the more serious direct threats of airborne ash clouds is to aircraft in flight (see also Section 14.1.2). Turbofan and turbojet engines are especially prone to failure when they are operating in dilute ash clouds. Although it could be considered a technological risk, the threat of volcanic ash clouds to aviation is significant in thinking about the multidimensional hazards of large eruptions in future, and merits a brief examination. A related consideration is the potential challenge that airborne ash can pose to search-and-rescue operations following a volcanic disaster – ash clouds not only impede aviation, they also disrupt radio communications.

There have been many reported encounters between aircraft and volcanic clouds and two near disasters in which Boeing 747s, with full complements of passengers and crew, temporarily lost all engine power. Even remote volcanoes far from any large populations on the ground, such as those in the Kamchatka Peninsula

(Russia) and the Aleutian Islands (Alaska), can pose a serious risk to the busy air corridors within their range. One well-known instance struck an almost-new Boeing 747–400 aircraft en route between Amsterdam and Tōkyō in December 1989. It was descending to make a scheduled stop at Anchorage in Alaska when it encountered an ash cloud at an altitude of 25,000 feet (7.6 kilometres), some 280 kilometres from the erupting Redoubt volcano (Figure 2.3). The conversation between the pilot and air traffic control in Anchorage is straightforward to track down on the Internet. The concern on the flight deck, as the situation quickly, and more or less literally, spiralled out of control, is viscerally palpable in the gathering pitch of the pilot's communications:

AIR TRAFFIC CONTROL:	*'Do you have good sight of the ash plume?'*
PILOT:	*'It's just cloudy it could be ashes. It's just a little browner than a normal cloud.'*
PILOT:	*'We have to go left now. It's smoky in the cockpit at the moment, Sir.'*
AIR TRAFFIC CONTROL:	*'KLM 867 heavy, roger, left at your discretion.'*

The flight crew immediately powered up in a full-thrust climb to escape the cloud but after a minute-and-a-half all four engines stalled, and half the instruments in the cockpit flickered on and off.

PILOT:	*'We're climbing to level 390. We're in the black cloud, heading 130.'*
PILOT:	*'. . . we have flame out all engines and we are descending now.'*
AIR TRAFFIC CONTROL:	*'KLM 867 heavy . . . Anchorage.'*
PILOT:	*'KLM 867 heavy we are descending now. We are in a fall! . . . we need all the assistance you have, Sir. Give us radar vectors, please.'*

The plane glided for five minutes, dropping 10,000 feet before the flight crew managed to restart two engines and regain control of the aircraft. The descent was so rapid that objects in the cabin appeared to float in a state of weightlessness. The damage to the plane was extensive and included abrasion of the compressor blades, and accumulation of ash in the combustor and inlet to the turbines, which had caused the flame-out. The cockpit windshields were completely crazed, and the avionics compromised. In its official report, the National Transportation Safety Board reported that 'the lack of available information about the ash cloud to all personnel involved' also contributed to the severity of the incident.

2.2 TEPHRA FALLS

For large explosive eruptions, much of the fallout of ash and pumice comes from the base of the umbrella cloud. Probably the first sensation for anyone caught in such fallout will be the disorientation and terror of being plunged into complete darkness. Even the darkest nights do not compare: in a dense fall of ash, there is no glimmer of light from Sun, sky, Moon or stars, and the chances are that power lines will have been cut, so no response from light switches ... Section 4.1 will examine in more detail the characteristics of tephra deposits but the key observations are that they thicken towards the volcano responsible, and that larger-magnitude events produce thicker deposits. In volcanological usage, 'ash' refers to particles less than two millimetres across, and thus represents the finest sized material in tephra; next come 'lapilli' (from the Greek for stones; up to 6.4 centimetres across) and 'bombs' or 'blocks' (> 6.4 centimetres; bombs are distinguished by having more intriguing shapes than blocks). Other terms in use include 'scoria' (from the Greek for dung), which describes the frothy, cindery product typical of Strombolian activity, and, of course, 'pumice' (from the Latin for foam), the typically silicic equivalent to scoria.

2.2.1 Hazards

Volcanic ash might sometimes look light and fluffy as it falls through the air but it is composed of volcanic glass and crystals and, thus, as it accumulates it exerts a considerable load on whatever lies beneath. In larger events, several metres' thickness of ash might accumulate near the volcano (as at Pompeii in 79 CE), and even a few centimetres can pile up hundreds-to-thousands of kilometres downwind. Ash fallout can adversely affect both the built and natural environments in many ways. By crushing crops and contaminating pasture, they can even result in major loss of life through starvation and pestilence [7] (Section 13.4). Tephra loads on the roofs of buildings can lead to structural failure. For instance, during the 1991 eruption of Pinatubo (Philippines), the combination of heavy ash fall and rainfall from a typhoon resulted in a dense, concrete-like mixture that caused heavy loss of life as buildings collapsed on their occupants. Even modest ash falls can severely hamper rescue and relief efforts during volcanic eruptions by putting roads, airports, and electrical power lines and telecommunications systems out of action. They can also precipitate public health crises if the functioning of water treatment plants is compromised.

Table 2.1 *Impacts of tephra falls on plants and soils.*

Dusting < 0.1 cm ash	Potentially beneficial supply of nutrients such as selenium to the soil.
Shallow burial < 0.5 cm ash	No plant burial or breakage but burning and loss of foliage by acids leached from ash by rainfall. Ash is mechanically incorporated into the soil within one year. Vegetation canopies recover within weeks.
Moderate burial 0.5–2.5 cm ash	Buried algae may survive and recover. Larger grasses are damaged but not killed. Soil remains viable and is not so deprived of oxygen or water that it ceases to act as topsoil. Vegetation canopies recover in the next growing season.
Deep burial 2.5–15 cm ash	Complete burial and elimination of algae. Small mosses and annual plants will only reappear after re-colonisation. Widespread breakage and burial of grasses and other non-woody plants; some plants do not recover. Large proportion of plant cover eliminated for more than one year. Plants may extend roots from the surface of the ash layer down to the buried soil, helping to turn over the ash on a timescale of about five years. Vegetation canopy recovery takes decades. Mixing of new ash into the old soil by people or animals greatly speeds recovery of plants.
Very deep burial >15 cm ash	All non-woody plants are buried. Burial will sterilise soil profile by isolation from oxygen. Soil burial is complete and there is no communication from the buried soil to the new ash surface. Several hundred (to a few thousand) years may pass before new equilibrium soil is established but plants can grow within years to decades.

Source: New Zealand Ministry of Agriculture and Forestry.

Anything more than about one centimetre of ash depth is liable to cause serious damage to plants [8] (Table 2.1). Depending on climate factors, even a few centimetres of ash can render a terrain agriculturally sterile for generations. Tephra can also carry significant quantities of aggressive chemical species adsorbed on to the surface of ash particles. Fluorine is of particular relevance in this regard. Although it is initially released from magmas in the form of hydrogen fluoride gas, it can be readily scavenged by tephra during eruption, and is thereby delivered to the Earth's surface as the ash sediments out of the plume. On account of their feeding habits (they don't wash their food before eating it), grazing animals ingest large amounts of soil. For instance, the average cow, on an average pasture, takes on board something like a kilogramme of soil a day. (No wonder they have four stomachs!) When contaminated ash and foliage lies on the ground, grazing livestock are liable to consume great quantities of it, and can be quickly poisoned when the fluorine is released inside their alimentary tracts. It is an excruciating process leading to abnormal tooth and bone growth, haemorrhage and organ failure. During recent eruptions in Chile and New Zealand, thousands of sheep and cattle died from such fluorosis. Tephra falls may also corrupt drinking water and it is feasible this could even lead to fluorine poisoning in human populations (Section 12.4).

Ash-fall deposits may pose a long-term health risk in areas where soil development is slow (in arid regions, for instance, or during prolonged volcanic eruptions) since they represent a source of potentially toxic dust. This is particularly relevant in deposits that contain a variety of quartz known as cristobalite that has carcinogenic properties and is associated with lung disease [9].

2.2.2 Ash fertilisation

It is not all bad. Volcanic soils tend to have a good reputation for fertility, and there is evidence that tephra fallout, as occasional dustings, provides nutrients such as sulphur and selenium to soils thereby having a beneficial impact on agriculture [10]. One factor suggested to have contributed to the fragility of Easter Island's environment following human occupation was its remoteness from sources of volcanic ash that would have helped to replenish nutrients lost from the soil by erosion [11]. Ash fallout in the oceans can also supply macronutrients and bioactive trace metals such as iron that are vital to phytoplankton growth near the surface. Good evidence for this effect was obtained following the 2008 eruption of Kasatochi in Alaska. Satellite imagery collected over the northeast Pacific Ocean shortly after the eruption revealed a striking phytoplankton bloom across a wide zone that coincided in time and space with the calculated ash fallout and iron complement [12]. The potential larger-scale effects of this phenomenon – substantial removal of carbon dioxide from the atmosphere – are explored in Section 6.6.

2.3 PYROCLASTIC CURRENTS & CALDERA FORMATION

Pyroclastic currents are searing mixtures of ash, rock and gases that flow under gravity down the flanks of a volcano at speeds of up to 200 kilometres per hour. Their behaviour and impacts simultaneously embody the qualities of an atomic bomb blast, an immense avalanche, and a Category Five hurricane, with the added complication that their temperatures can reach hundreds of degrees Celsius. Even relatively small examples can readily travel 10 or 20 kilometres across the ground, while the most intense eruptions identified in the geological record disgorged pyroclastic currents that travelled over 100 kilometres from the vent. The pyroclastic-current deposits from large-magnitude eruptions are sometimes called 'ignimbrites' or 'ash-flow tuffs'. Although the nomenclature is contentious, we shall refer to pyroclastic-current deposits from eruptions of M_e 6 and upwards as ignimbrites.

Cinematography of pyroclastic currents (examples of which are naturally scarce), studies of their deposits and some ingenious simulations have all helped to understand the deadly efficiency of these phenomena: they move rapidly, silently, can carry chunks of lava the size of trucks, and they readily ignite anything flammable. The deposits often contain carbonised tree trunks testifying to the high temperatures. Pyroclastic currents can form from the collapse of an explosive eruption column (this is what happened at Vesuvius in 79 CE). However, they are also commonplace on volcanoes with active lava domes, as witnessed on countless occasions since 1996 at Soufrière Hills volcano on Montserrat. Here, the currents can be initiated by gravitational collapse of portions of the hot lava dome or by detonations of pressurised gas close to the surface.

Another important feature of pyroclastic currents readily appreciated from photographs and cinematography is their vertical extent (Figure 2.4). It is impossible to see inside an active pyroclastic current but we can be fairly sure that most (excepting the most dilute and turbulent) consist of a denser basal flow of ash, blocks of lava and other debris picked up en route, which travels beneath a soaring cloud of finer ash. The upper parts actually behave like Plinian eruption columns owing to their heat energy and turbulence. They entrain air and, enhanced by the sedimentation of the denser components of the flow, develop buoyant thermal plumes that punch up into the sky. These phenomena are known as 'co-ignimbrite plumes', or more

Figure 2.4 One of the earliest photographs of a pyroclastic current, descending the flank of Mont Pelée on 16 December 1902, and taken by pioneering volcanologist Albert Lacroix. Following the valley of the Rivière Blanche, the wall of cloud reaches 4000 metres high, while the toe of the current has just reached the sea. A similar current had destroyed the town of St Pierre just over six months earlier. From Lacroix, A. (1904) *La Montagne Pelée et ses éruptions*, Paris: Masson.

lyrically as 'phoenix clouds'. Unlike eruption columns developed above erupting vents, phoenix clouds do not have a lower gas-thrust region, and, for a given intensity, they do not reach so high into the atmosphere as a Plinian plume (though they can still readily soar to altitudes of 20 kilometres or more). As eruptions get larger (beyond M_e 6), their plumes are likely to be more intense and prone to column collapse. The very largest eruptions are, in essence, pyroclastic-current eruptions and their deposits consist of roughly equal measures of ignimbrite and fallout from phoenix clouds. They are invariably associated with the formation of circular or elliptical craters, usually termed calderas, which can be tens of kilometres across (Figure 2.5). These result from subsidence of the crust into the void space left by the evacuated magma chamber.

2.3.1 Hazards

On 11 May 1902, Theodore Roosevelt received the following cable transmitted to the White House:

> DISASTER COMPLETE. CITY WIPED OUT. CONSUL PRENTIS AND FAMILY DEAD. GOVERNOR SAYS 30,000 DEAD, 50,000 HOMELESS, HUNGRY.

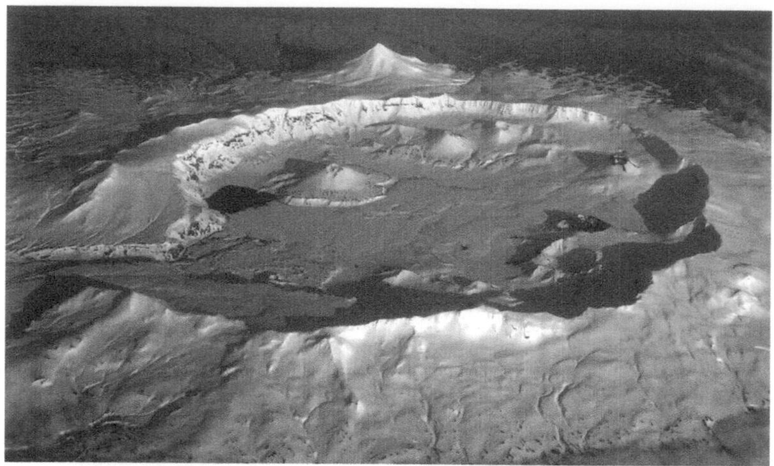

Figure 2.5 This almost looks like a three-dimensional model but it is a lucky photograph taken during a commercial flight above Okmok volcano on Umnak Island in the eastern Aleutian Islands of Alaska. The caldera is 500 metres deep and around nine kilometres in diameter, and formed as a result of a M_e 6.9 eruption around 2000 years ago. Subsequent lesser activity has constructed numerous cones and craters and lava flow fields – the last eruption was in 2008. Another large explosive eruption, around M_e 6.7, took place about 12,000 years ago, making Okmok an especially productive and destructive volcano. Photograph taken in June 2007 by Cyrus Read, AVO/USGS.

ASK RED CROSS CODFISH FLOUR BEANS RICE SALT MEATS BISCUITS QUICK AS POSSIBLE. VISIT OF WAR VESSELS VALUABLE.

Three days earlier, following a modest eruption of Mont Pelée on the Caribbean island of Martinique, pyroclastic currents had swept into the town of St Pierre (Figure 2.6) at an estimated speed of 160 kilometres per hour. In minutes, 99.997% of the population (29,000 people) perished. The statistics for other eruptions also bear out the general rule that being engulfed by pyroclastic currents, indoors or outside, confers very meagre chances of survival: for each person injured by a pyroclastic current, ten are killed. These are substantially higher odds of mortality than for just about any other type of natural disaster. In the case of St Pierre, there were only two survivors – one of whom, Louis-Auguste Cyparis, had been kept in solitary confinement in the town's jail. He suffered terrible burns as the pyroclastic currents entered through a small grille at the top of his cell and was only found after

Figure 2.6 St Pierre on Martinique (top) before the 1902 eruption, and (bottom) in ruins the month after the eruption. From Lacroix, A. (1904) *La Montagne Pelée et ses éruptions*, Paris: Masson.

four days. (Although subsequently released and pardoned, his fate was to be exhibited around the USA in the Barnum and Bailey 'Greatest Show on Earth', billed as 'the only living object that survived in the "Silent City of Death"'.)

Pyroclastic currents thus represent one of the most lethal manifestations of volcanism. The main causes of death in victims of pyroclastic currents are heat-induced shock, asphyxiation, thermal injury of the lungs and burns. Survivors tend to have been exposed to only the more dilute parts of the current or sheltered in some way, but can be critically injured due to respiratory and skin burns. A display of the impacts of pyroclastic currents, which remains deeply poignant and disturbing even two millennia after the event, can be viewed at the excavations of Herculaneum and Pompeii (Figure 2.7). For a long time it had been thought that all the residents of Herculaneum had fled but, in the 1980s, a row of arched chambers, open towards what would have been the shoreline in Roman times, were exhumed. Hundreds of skeletons were found inside these boat sheds. Presumably the victims had sought refuge as the eruption became increasingly threatening and had hoped for rescue by sea. The positions and postures of the bodies, the articulation of the skeletons and fracturing of bones, and the signs of incineration indicate that the victims died instantaneously due to the intense heat (up to 500 °C) of the pyroclastic

Figure 2.7 Victims of pyroclastic currents from the 79 CE eruption of Vesuvio uncovered at Herculaneum.

currents that swept into the chambers. A similar fate befell more than 300 people on Merapi volcano (Java) in November 2010.

Numerous factors dictate the mobility and run-out distance of pyroclastic currents, including the generating process (for example, dome failure or eruption-column collapse), eruption intensity, volcano height and slope, and the nature of terrain over which the currents travel (for instance, forest, tundra or open water). However, on average, the pyroclastic currents associated with eruptions of magnitude (M_e) 4, 5, 6, 7, 8 and 9 travel 5, 10, 20, 40, 75 and 140 kilometres, respectively. At the higher end of this scale, pyroclastic currents are typically disgorged in all directions. Thus, an eruption of M_e 8 would likely cover an area of about 20,000 square kilometres in ignimbrite - more than enough to bury an area the size of Switzerland or New Jersey. An M_e 7 event would spread pyroclastic currents across an area of roughly 5000 square kilometres.

2.4 LAVA FLOWS AND DOMES

The term 'lava flow' usually refers to erupting lava that has the opportunity, and sufficiently low viscosity, to travel down the flanks of a volcano or cross open ground. The expression is used both for active flows during emplacement, and for the resulting landform. In the course of a long-lived eruption, a 'lava flow field' may develop by the superposition of many individual flow units. The current eruption of Kīlauea (Hawai`i) has yielded well over two cubic kilometres of lava and built up a flow field of around 100 square kilometres in area since it began in 1983 (an average eruption rate of about three cubic metres per

Figure 2.8 Pele at play: lava flow on Kīlauea volcano, Hawai`i, in 2007. Note the magnificent (solidified) pāhoehoe lava field either side of the active flow.

second). An astonishing characteristic of moving lava flows is the accompanying noise – reminiscent of a crackling bonfire but sometimes with sharper reports that fling out fragments of chilled lava a metre or so into the air. The easiest way to track down active lava flows in daylight is to listen out for their distant clatter. But lava flows are surely most spectacular seen at dusk and dawn, their surface glowing dull red in the recesses of folds, bright orange at fractured margins or cracks (Figure 2.8).

Active lava flows radiate prodigious quantities of heat near the vent such that they rapidly form a surface crust. This can thicken sufficiently to insulate the core of the flow from heat losses, keeping it fluid, and thereby extending the distance it may travel. Basaltic flows quite often crust over completely, with lava continuing to flow in tunnels, which can grow in cross-section by melting back the walls. When the supply of lava at the vent ceases, the last slug of lava may drain down-slope leaving an empty lava tube or 'pyroduct'. On Kīlauea, much of the flow between the Pu`u` O`o vent and the coast, where lava pours into the sea (a distance of over ten kilometres), takes place via a tunnel network, with only sporadic 'breakouts' at the surface.

If the erupting lava is more viscous – as is typically the case for intermediate or silicic compositions – it tends to pile up around the vent to form a lava dome. These domes are inherently unstable because of their sheer faces.

2.4.1 Hazards

In rare cases of high-intensity eruptions of very hot, fluid magma, lava flows have resulted in loss of life. One of the most infamous volcanoes in this respect is Nyiragongo (Democratic Republic of Congo) whose lava flows probably claimed hundreds of lives during eruptions in 1977 and 2002. More commonly, lava flows represent more of a threat to property, roads, and power and telecommunication lines but they can also result in loss of agricultural land and forest by burial and conflagration. In some cases, cities have been engulfed by lava causing significant damage, again, recently in the case of the 2002 Nyiragongo eruption (which bisected the city of Goma, home to half-a-million people). One of the earliest cities of ancient México, Cuicuilco in the central highlands (now part of México City), was inundated by lava erupted from Xitle volcano around the third century CE.

Lava flows follow topography. Flows can thereby disrupt or divert water supply if they run down a river valley. Where they do interact with wet ground, secondary explosions can result, though these tend to be fairly localised events.

As mentioned in the preceding section, growing lava domes are prone to collapse and provide another source of pyroclastic currents. In 1997, catastrophic failure of part of a lava dome growing in the summit crater of Soufrière Hills volcano (Montserrat) initiated pyroclastic currents that flared down the cultivated flanks of the volcano killing 19 people.

2.5 ROCK AVALANCHES AND MUDFLOWS

Because of their steep slopes and typical construction from juxtaposed layers of lava rubble and loose ash and cinders, and through the weakening action of acidic gases and ground waters, volcano flanks are prone to gravitational failure. These events may be triggered in several ways: by the destabilising effects of magma intrusions into the cone; the ground-shaking detonations of explosive eruptions; local or large earthquakes; and even heavy rainfall. The largest but rarest events are called debris avalanches. These involve the collapse of an entire sector of a volcano, generating enormous gravity-driven rock avalanches that run out for tens of kilometres. Sometimes, as at Mt St Helens in 1980, such collapses trigger eruptions. At Socompa volcano in the central Andes of Chile, a huge debris avalanche travelled 40 kilometres before

Figure 2.9 Tungurahua in Ecuador epitomises the importance of gravitational potential energy on volcanoes. The summit crater reaches over 5000 metres above sea level, 3000 metres higher in elevation than the town of Baños, which lies in a valley to the north of the volcano and just eight kilometres away. Many pyroclastic currents and mudflows have descended the flanks of the volcano since activity resumed in 1999.

coming to rest. Because these phenomena are infrequent it is difficult to calculate the probability of their recurrence.

Larger debris avalanches are not confined by established channels. However, if the moving debris is water-saturated and does enter drainage channels, it is then termed a debris flow, and if it consists of a significant fraction of clay-sized particles, it is called a mudflow or (from the Indonesian word) a lahar. Such flows can pick up further water and debris along the way, while at the same time dropping their coarser, denser materials. Gradually, they transform into syrupy clay- and water-rich flows. Mudflows can also result when lava or hot tephra is erupted on to ice or snow; when explosive eruptions take place beneath volcanic lakes; and when there is intense rainfall on loose volcanic deposits. Their course is strongly controlled by topography (Figure 2.9).

2.5.1 Hazards

Even small volcanic landslides can be devastating in populated areas, and they can occur long after volcanic activity has ceased. Two similar disasters occurred in 1998 as the result of torrential rainfall on volcanic slopes and deposits, one in Nicaragua on the flanks of Casita volcano in which more than 2500 people were killed, the other in the Sarno mountains to the east of Vesuvius, which claimed over 150 lives. In both cases, the initiating landslides transformed rapidly into fast-moving slurries of rock and mud. The major hazards posed by the

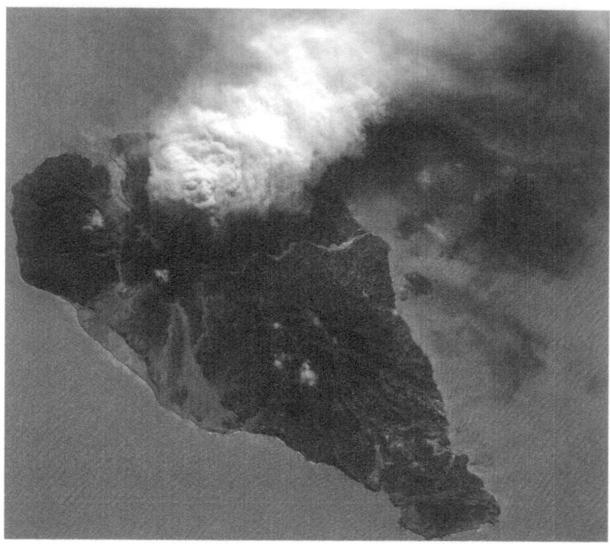

Figure 2.10 There's not much place to hide when your island is little bigger than the erupting volcano at one end of it . . . Soufrière Hills volcano on Montserrat seen here from the International Space Station. Note the column of ash and gases rising above the lava dome and then drifting downwind, and the lighter-toned deposits of mudflows and pyroclastic currents stretching down the flanks of the volcano (and extending the shoreline with tephra deltas). The former capital of the island (Plymouth) is completely buried in volcanic debris and located beneath the volcanic cloud. The remaining population live concentrated towards the thin end of the island. Dimensions of the island are approximately 16 kilometres × 9 kilometres. Credit: NASA/JSC.

various kinds of volcanic avalanche and flow phenomena are physical injuries related to burial and property damage, and drowning (Figure 2.10).

The most appalling volcanic tragedy in recent times took place in 1985 after a modest eruption (M_e 3.5) of Nevado del Ruíz in Colombia caused sudden snow melt around the crater, unleashing a devastating mudflow. It took an hour-and-a-half for the mud to reach the town of Armero, 60 kilometres away, but no one was prepared for it and an estimated 22,800 people died. Larger eruptions can dump so much pyroclastic sediment on the landscape – overloading river systems and changing the infiltration rates of rainfall at the surface – that it can take decades or even millennia for the post-eruption landscape to readjust and stabilise [13] (Section 2.9).

Following large explosive eruptions, debris flow and mudflow hazards may persist for years or decades simply because there is so much loose pumice and ash on the flanks of the volcano available for redistribution during heavy rains.

2.6 TSUNAMI

On volcanic islands and coastal volcanoes, landslides and avalanches as described above may initiate tsunami if the displaced material drops into the sea. In 2003, the usually more benign Stromboli volcano experienced a small landslide that caused a locally destructive tsunami, with peak wave heights of a metre or two. Other ways that volcanic eruptions can induce tsunami include the physical impact of pyroclastic currents hitting the water; the catastrophic collapse of the crust above a magma chamber during caldera formation on an undersea volcano; and the hydrovolcanic explosions that result when seawater gains access to the eruption vent. The precise mechanisms of the latter process are poorly known since there has not been a large eruption of a shallow marine volcano in modern times. Much of our knowledge of such events has thus been inferred from deposits of prehistoric eruptions and from outputs of computer models.

2.6.1 Hazards

The Indian Ocean earthquake that struck off the coast of Sumatra on 24 December 2004 demonstrated traumatically the immense geographic range and human cost of energetic tsunami. Many volcanoes are in coastal areas or form islands in the sea (and lakes) and, as we have seen, there are several mechanisms by which they can generate tsunami – both during and between eruptions. Such volcanoes have the potential to threaten distant shorelines with the run-up of giant waves of seawater. Since much human settlement today, and in the past, has been close to the sea, great eruptions of shallow-water volcanoes have potentially devastating consequences (deep-water volcanism tends to be effusive rather than explosive).

Ancient volcanogenic tsunami, such as those which accompanied the caldera-forming Minoan eruption of Santorini, are thought to have been highly destructive (Section 9.3.4). A less well-known case accompanied formation of the Kikai caldera in Japan around 7300 years ago, which generated tsunami that devastated the southern part of Kyushu. The pyroclastic currents travelled at least 100 kilometres while the expulsion of 70–80 cubic kilometres of magma (M_e 7.2) from the magma chamber resulted in a submarine caldera 17 kilometres × 20 kilometres across [14]. But perhaps the most famous volcanically induced tsunami are those that were associated with the 1883 eruption of Krakatau (Indonesia). These wreaked havoc on the facing coastlines of Sumatra and Java, and accounted for the majority of the estimated 36,600 deaths associated with the eruption.

2.7 EARTHQUAKES

Dormant volcanoes are often associated with high levels of background seismicity due to movements of magma and associated fluids below the surface, as well as processes such as bubble growth in magma (which leads to pressurisation) and heating of ground water. Larger eruptions are typically preceded by a crescendo in earthquake activity as magma ascends towards the surface or accumulates at shallow depth in the crust. Much of this ground shaking originates from the physical fracturing of the host rocks as they adjust to accommodate the magma. Such signs, though, are not certain predictors of eruption, since bodies of magma can often stall and freeze in the crust.

2.7.1 Hazards

Seismicity associated with magmatic and volcanic activity can result in substantial damage to the built environment. One of the most extraordinary recent instances afflicted the town of Pozzuoli in the Phlegrean Fields (Campi Flegrei), some 25 kilometres west of Vesuvius. Established as a Roman military colony, Pozzuoli (known then as Puteoli) went on to become one of the major trading ports of the Mediterranean in the Roman period. Its fortunes since have literally yo-yoed. For a millennium, the ground sank slowly at a rate of about a centimetre per year, totalling more than ten metres of subsidence. As the town became gradually inundated, the inhabitants resettled on higher ground. There followed five centuries of steady uplift, which accelerated dramatically in the days prior to an eruption a few kilometres away that constructed the cone of Monte Nuovo in 1538. Then, the land plunged again, until 1969, when the motion reversed dramatically. The main phases of recent uplift from 1969–1972 and 1982–1984 were accompanied by tens of thousands

Figure 2.11 Staircase at the archaeological site of Akrotiri on Santorini (Greece) cracked by earthquakes prior to the gigantic Minoan eruption (Section 9.3). This provides an important lesson concerning the precursors to large eruptions, which can include damaging earthquakes.

of small earthquakes. Progressively, the old town disintegrated from this relentless seismic onslaught, and it remains under reconstruction. The extent to which all this unrest reflects magmatic versus geothermal processes remains a topic of vigorous scientific debate. Seismic activity prior to large eruptions has even left its imprint in the archaeological record, for instance at Pompeii and Akrotiri (subsequently buried during the Minoan eruption of Santorini, Figure 2.11; Section 9.3).

Just as progressive earthquake activity increasingly weakens buildings, so it can render them more vulnerable in the event of an eruption. A building whose structural integrity is already compromised by seismic shaking is more prone to collapse under the weight of ash fallout or due to the dynamic pressures of a pyroclastic current or mudflow. Thus, the combined impacts of seismicity, tephra falls, pyroclastic currents and mudflows amplify volcanic risks in complex ways that can have different outcomes in terms of numbers of victims. In one respect, volcano seismicity might be considered a beneficial phenomenon: it can make life sufficiently unnerving, if not downright threatening, that people move away before volcanic action really picks up ...

Figure 2.12 Double trouble. Massive emissions of gases and particles from the two main craters of Ambrym volcano, Vanuatu (in Melanesia): Benbow (left) and Marum (right). At the time this photograph was taken in 2005, Ambrym was one of the largest point sources of sulphur, chlorine and fluorine on the planet. Sulphur dioxide emissions alone reached up to 20,000 tonnes per day. The acid gases damaged crops, and rainout of large quantities of fluoride from the plume contaminated water supplies resulting in pitting and staining of the teeth of many children living downwind of the crater.

2.8 VOLCANIC GAS EMISSIONS

The pre-eminent role of magmatic gases in triggering eruptions and controlling their deportment was outlined in Section 1.3.1. We also noted the consequences of volcanic fluoride emissions when carried to the ground on tephra (Section 2.2.1). In fact, there is a whole cocktail of gases and particles that is released to the atmosphere from volcanoes both during and between eruptions (Figure 2.12; Chapter 3). One of the most notorious instances of volcanic degassing accompanied the 1783–4 eruption of Laki in Iceland (Section 12.2), which discharged more than half a gigatonne of carbon dioxide, water vapour, sulphur dioxide, hydrogen fluoride and hydrogen chloride. The gas and aerosol clouds also deposited substantial amounts of lead, cadmium, zinc, bismuth and thallium on to the Greenland ice sheet.

2.8.1 Hazards

Among the species that can have particularly profound impacts on terrestrial and aquatic ecosystems, agriculture, infrastructure and human health are sulphur, fluorine, chlorine, bromine and metals (including mercury, arsenic, cadmium, copper and lead). Various components of volcanic emissions (including acid species, lead and mercury) can damage vegetation. The detrimental effects usually result from acidification of soils by volcanically polluted rainfall, or by fumigation of foliage and respiration of sulphur dioxide, hydrogen chloride and hydrogen fluoride through plant stomata. Chronic exposures even of rather low amounts of sulphur dioxide may injure leaves, impair plant ecosystems and decrease agricultural productivity.

One persistently degassing volcano renowned for its impact on agriculture is Masaya in Nicaragua. In the region downwind of the volcano, chemical burning of leaves and flowers is widespread, and substantial economic losses arise from acid scorching of coffee crops. The worst affected area resembles a badlands with sparse cover of shrubs. It is suitable agriculturally only for *pitaya* cactus (dragon fruit). Further long-term effects can result from chemical modification of soils by the acids leached out of tephra that falls on the ground [15]. The impacts of sulphate deposition on soils have been investigated widely in the context of anthropogenic pollution. A key factor in sulphate retention is soil mineralogy which is strongly influenced by the particular rock formation that is weathering to make the soil. Thus, soil type will strongly control the extent to which ecosystems are disturbed by sulphur deposition from volcanic clouds.

Section 2.2.1 introduced the specific threat of fluorine when it is delivered to the ground adsorbed on tephra. Rainfall through a fluorine-rich volcanic plume can also contaminate water supplies. One volcano where this occurs is Ambrym in Vanuatu, which is one of the world's largest point sources of a variety of gas species (Figure 2.12). Part of its fluorine emission ends up polluting drinking water causing dental fluorosis amongst communities living downwind of the crater [16]. Though not life threatening, it demonstrates the low levels of fluorine that can cause harm.

Several volcanic gas species are damaging on contact with the skin, if taken into the lungs or ingested. Only a handful of primary studies have been conducted into health effects of volcanic gases, and those that exist are limited in terms of exposure assessment, so the true extent of health effects from volcanic gases remains unclear. Most research to date relates to carbon dioxide, hydrogen sulphide and sulphur dioxide exposures. Sulphur gases and sulphuric acid particles can affect respiratory and cardiovascular health in humans. Air quality in Hawai'i is reportedly affected by 'vog' (volcanic fog) associated with sulphur dioxide [17] and sulphuric acid aerosol from Kīlauea's plume, and 'laze' (lava haze), composed of hydrogen chloride-rich droplets formed when active lavas enter the sea.

Accumulations of hydrogen sulphide in volcanic and geothermal areas, sometimes associated with faulty geothermal heating systems, have resulted in fatalities from poisoning. Even low concentrations of the gas, if sustained, may adversely affect the nervous system and respiratory and cardiovascular health. Emissions of carbon dioxide can also accumulate dangerously in low-lying areas and have resulted in deaths due to asphyxiation. Dissolved carbon dioxide may also accumulate in lake water in volcanic areas. Sudden displacement of such potentially effervescent water can release a cloud of carbon dioxide. Because the gas is denser than air, it can flow under gravity, suffocating people and animals in its path. Such a disaster occurred at lakes Nyos and Monoun in Cameroon in the 1980s. A further toxic volatile species encountered in volcanic and geothermal areas is radon, a known carcinogen.

2.9 RECOVERY OF ECOSYSTEMS

A large explosive eruption can turn a lush ecosystem into a sterile desert in a matter of hours. But when this happens, how long does it take the landscape and its ecology to recover? Much work has been carried out on the subject, not least since large volcanic events effectively reset the biological clock to zero in the worst-impacted areas, providing ecologists with fascinating natural experiments on colonisation and succession of flora and fauna. Pre-eminent amongst these have been investigations of Krakatau, Katmai (Alaska), Surtsey (Iceland) and Mt St Helens (Figure 2.13). What emerges from these studies is that recovery is a very complex process that cannot readily be formulated or predicted because of the importance of chance events and the intricate and interwoven site-specific ecological, climatic and landscape characteristics that can promote or inhibit regeneration. What is also abundantly clear is that the effects of major volcanic disturbance can be lasting – often centuries but, in some cases, a thousand years or more – as long-term landscape and microclimatic readjustments repeatedly disturb habitats and biodiversity.

Three decades after the devastating 18 May 1980 blast at Mt St Helens, soils have been renewed and nourish fungi and herbs; streams once choked with tephra are again home to trout and salmon; and many birds, mammals and reptiles have returned. However, the succession is far from mature. In the worst-hit zone, flattened by a sideways explosion and plastered in deep drifts of pumice from

Figure 2.13 Recovery of vegetation in the Toutle River Valley at Mt St Helens.

pyroclastic currents, vegetation cover is still a fraction of its original density, and is constituted of pioneer species such as lupins, grasses and occasional shrubs and saplings rather than the former forest species.

One lesson of the 1980 eruption was that being big was disadvantageous. For deer, elk, bears and goats caught out in the open by the blast, it was the end, whether by incineration, impact, abrasion or burial (likely all four). While any surviving carnivores probably managed quite well for a while, it was smaller fauna, especially burrowers, that fared best through the calamity. Gophers were probably amongst the first animals to glimpse the new-born landscape once the dust settled, after they bored up through the fresh tephra to peer out disbelievingly at the uniquely lifeless terrain. They also deserve as much credit as any animal in assisting with the recovery process. Thanks to their immense industry in constructing gopher towns consisting of mazes of burrows and tunnels, they gradually mixed the former soil back up to the ground level. This not only re-introduced nutrients into the surface layer but also excavated buried, but viable, seeds and spores that then had a chance to germinate.

Although recovery is complex, there are nevertheless some general rules. It occurs thanks to the combined efforts of survivors and opportunistic colonisers. Naturally, therefore, the size and intensity of the eruption and physical extent and nature of its deposits count enormously. Even the most determined gopher is not going to tunnel its way through several metres of pumice. Thicker tephra also take longer to erode and may deny any prospect of re-exposing former soils with their bio-banks of seeds and germs. At the same time, thick tephra sequences can upset the hydrological system for decades by choking lakes, rivers, estuaries and coastal waters as the landscape continues to re-adjust to the burden. Such prolonged redistribution of sediment, perhaps concentrated during the rainy season, can monotonously terminate each embryonic recovery by reburial.

Furthermore, the more widespread the disturbance, the more isolated the worst-affected terrain becomes from potential colonisers. In the case of Krakatau, following its 1883 eruption, recolonisation proceeded apace thanks to seabirds and sea currents of the Sunda Strait, and the volcano's proximity to the biologically rich islands of Java and Sumatra. The first colonist was apparently a lone spider found nine months after the eruption but, by 1896, 53 species of plant had established themselves, and a survey in 1908 discovered 200 animal species (mostly insects) amongst the pioneering grasses. Eventually, various trees took root, resurrecting forest cover. A more remote island would most likely undergo a much slower-paced recovery.

Climate is a further crucial factor determining restoration. While very fine ash in light sprinklings can supply nutrients and bring a beneficial mulching effect to soils, lava flows and thick deposits of coarse pumice and scoria are generally very nutrient-poor, drought-prone and unstable. Furthermore, they tend to have very low capacity for retaining water, which of course is vital for biology. Recovery of such landscapes benefits enormously from chemical and physical weathering of the tephra and influx of nutrients via windblown mineral dust (perhaps from ongoing but less dramatic volcanism in the region) and organic matter. Recovery, therefore, correlates with precipitation, temperatures, duration of the growing season, and so on, and thus, in the broadest terms, with latitude, altitude and continentality. Lava flows can put up a particularly strong fight in the face of erosive processes, but where they have cracks and crevices that trap airborne dust and moisture, ecological niches will be seized on by opportunistic lichens, mosses, ferns and flowering plants, which, in their turn, will accelerate the formation of soils so that ultimately shrubs and trees can get a foot-hold.

Finally, chance factors can influence regeneration. For instance, the season of an eruption, and even the time of day it occurs, can determine whether animals and plants are more, or less, vulnerable. The time of year of an eruption – especially at higher latitudes – dictates the growth status of plants and animals in a way that can strongly influence survival and recovery. For example, it would determine whether migratory species were present, whether hibernating species were in caves and dens where they could be more protected and, likewise, whether plants and trees were hunkered down for the winter or had just come into bud. The timing during the day and week has a major effect on human casualties during sudden disasters such as earthquakes, eruptions and floods. Similarly, whether an eruption happens during the day or night will have a strong influence on the relative chances of survival of nocturnal and diurnal animals.

2.10 VOLCANIC DISASTERS

We can gain some useful insights into the impacts of volcanic eruptions by reviewing the available statistical data covering the last few centuries [18]. Table 2.2 catalogues the ten most lethal volcanoes in recorded history, with a breakdown of fatalities by phenomenon. While the records are certainly incomplete and increasingly inaccurate further back in time, a number of striking observations emerge. Firstly, that an eruption does not need to be big to be a mass killer. The minor explosion of Nevado del Ruíz in 1985 (M_e 3.5) resulted in a comparable death toll to the 600-times larger eruption of Krakatau a century earlier. Both lahars (mudflows) and pyroclastic currents recur in the list. Such flow phenomena can strike very suddenly and, because of their mobility, affect areas distant from the volcano (especially in the case of lahars). They are especially destructive in the built environment, and the chances of survival for anyone caught in their path can be extremely low.

Also prominent in Table 2.2 are the indirect consequences of eruptions, including the devastation of food crops and contamination of lakes and rivers. These have in the past resulted in greater

Table 2.2 *The ten deadliest volcanoes on record**

Volcano	Total fatalities	Event year (death tolls and cause where known)
Tambora (Indonesia)	> 60,000	**1815** (> 11,000 P, T; > 49,000 F, E)
Krakatau (Indonesia)	36,600	**1883** (4600 P; 32,000 W)
Mont Pelée (Martinique)	29,000	**1902** (28,600 P; 400 M)
Nevado del Ruíz (Colombia)	24,436	**1595** (636 M); **1845** (1000 M); **1985** (22,800 M)
Kelut (Indonesia)	15,444	**1586** (10,000); **1848** (21 M); **1864** (54 M); **1919** (5110 M); **1951** (7 M); 1966 (215 M); **1986** (1 M); **1990** (32 P, T); **1990** (4 M)
Unzen (Japan)	14,598	**1644** (30 M); **1792** (13,800 W; 724D); **1991** (43 P); **1993** (1 P)
Laki Craters (Iceland)	10,521	**1783–4** (F, E)
Awu (Indonesia)	8330	**1711** (3000 P); **1812** (953 P, M); **1856** (2806 P); **1892** (1532 P, M); **1966** (39 P, M)
Vesuvio (Italy)	8296	**79** CE (3500 P, T); **1631** (4000 P); **1682** (4); **1737** (2 T); **1794** (400); **1805** (4); **1872** (9 L, G); **1906** (350 T); **1944** (27 T, L)
Santa Maria (Guatemala)	>7000	**1902** (2000–3000 T; 5000–10,000 E); **1929** (200–5000 P, M)

* Excludes long-range impacts mediated through climate change. P = pyroclastic density currents; T = tephra falls and ballistics; F = famine; E = epidemic disease; M = mudflows (lahars); W = tsunami; D = debris avalanches; L = lava flows; G = gas. Data from references 6 and 18.

human loss than the direct physical impacts of pyroclastic currents and other primary volcanic phenomena. Malnourishment and pestilence in the aftermath of the 1783 Laki eruption (Chapter 12) and Tambora 1815 (Chapter 13) (respectively, the largest effusive and largest explosive eruptions since Medieval times) claimed many tens of thousands of lives. Nor should we imagine this is just a reflection of a pre-modern age. Today, around one billion people −15% of the total population – remain malnourished according to World Health Organization (WHO) statistics. Chronically hungry people are more at risk when additional stresses are placed on them, whether they are increased food prices, climate extremes, emerging infectious diseases or volcanic eruptions.

Lastly, notwithstanding the concentration of volcanic risk in the world's fourth most populous country, Indonesia, the simple statistics in Table 2.2 reveal the worldwide threat of volcanic disasters. The top ten most notorious volcanoes span the Caribbean and Central America, the North Atlantic, Southern Europe, and Northeast and Southeast Asia. Note also that disasters have affected cities (St Pierre on Martinique; Armero in Colombia; Naples in Italy), coastal areas (partly reflecting the tsunami hazard) and rural regions (famines in Iceland and Indonesia). All human communities, large and small, complex or egalitarian, have their specific vulnerabilities. In the case of cities, their concentrated populations and reliance on the state can result in heavy losses especially when essential services such as water supply and sanitation fail. This is an increasingly important lesson for the future given that half the world's population now lives in cities (Chapter 14).

2.11 SUMMARY

The character of an eruption is influenced by many physical and chemical processes, together with complex feedback mechanisms that render volcanic systems highly nonlinear. Volcanic eruptions also involve a wide spectrum of primary and secondary manifestations; can last for many decades; and can impact regions thousands of kilometres distant from the crater in the case of ash fallout or tsunami, potentially decimating natural resources over vast areas. These factors severely compound the hazards faced by communities around the world in places both near and far from active volcanoes and, combined with our uncertainties in how volcanoes work and interpreting their signs of unrest, they make forecasting volcanic activity inherently challenging.

Post-eruption famine and epidemics, pyroclastic currents, mudflows and volcanogenic tsunamis account for the majority of recorded deaths arising from volcanism. Few deaths have been associated with lava flows, however. While they may result in considerable damage to infrastructure and property they usually move sufficiently slowly to permit evacuation of threatened residents. The different manifestations of volcanism are not only a direct threat to life but can result in such violent disturbance to an ecosystem that full recovery can take a millennium. Following some of the larger eruptions in prehistory, humans abandoned the impacted zones for centuries (Chapter 5).

The effects we have been discussing broadly manifest themselves at local-to-regional scales. In the next chapter, we focus on the global consequences of volcanism.

3

Volcanoes and global climate change

Had the fierce ashes of some fiery peak
Been hurl'd so high they ranged about the globe?
For day by day, thro' many a blood-red eye . . .
The wrathful sunset glared.

Tennyson, St. Telemachus (1892)

'For those trying to understand the natural and anthropogenic processes for global change, the eruption [of Mt Pinatubo in 1991] presented perhaps a once in a lifetime opportunity.'

M. P. McCormick *et al.*, *Nature* (1995) [19]

Volcanic eruptions can expel many cubic kilometres of rock. Typically, a few per cent of the total mass of erupted materials is made up of gases. The rocks are just rocks – essentially inert bodies of lava or accumulations of tephra. The gases, on the other hand, become part of the atmosphere. Even non-erupting volcanoes can emit significant quantities of gas from fumarole vents. Volcanic gas emissions are by no means inert, either chemically or in their direct and indirect effects on the Earth's heat budget. Indeed, the global-scale climatic and environmental impacts of large eruptions arise principally through the action of this minor component of magmas – gas – and one chemical species in particular, sulphur. This chapter reviews the chemical and physical processes by which volcanic clouds affect the Earth's atmosphere and climate, emphasising the potential widespread and diverse knock-on impacts of major eruptions mediated through global climatic change. In many cases in the past, afflicted communities will almost certainly have had no idea that a volcanic eruption was the root cause of environmental stress.

One eruption, more than any other, has taught us how volcanism affects the Earth's climate system, that of Mt Pinatubo in the Philippines in 1991. It was responsible for the greatest loading of particles into the layer of the atmosphere known as the stratosphere (which begins 11–17 kilometres above the Earth's surface) for more than a century. A wide range of ground-based and satellite remote-sensing techniques measured the development, composition and effects of the volcanic cloud. Samples of the plume were even collected from converted US military spy planes capable of stratospheric flight. Especially fortuitously, within a month of the eruption, NASA's Upper Atmospheric Research Satellite (UARS) became operational, providing unprecedented information on the chemistry and dynamics of a volcanic cloud. These studies characterised the nature of the volcanic materials entrained into the stratosphere and demonstrated their profound impacts on atmospheric chemistry and on the Earth's heat budget and climate.

Attempts to model these radiative and chemical effects have provided acid tests for climate models and for our understanding of how the atmosphere and climate work. When the volcanic signal in climate change is removed accurately, the human impact on climate becomes even clearer. More importantly for our purposes, the eruption and its quantified effects serve as benchmarks for understanding the global reach of past (and future) volcanic eruptions.

3.1 PINATUBO'S GLOBAL CLOUD

On 2 April 1991, steam explosions were observed on a little known volcano called Mt Pinatubo on the island of Luzon, in the Philippines. Until these rumbles, Pinatubo had not been identified by scientists as a potentially active volcano – it was completely off the volcanological radar screen. This serves as an important reminder of just how ignorant we likely remain about the volcanoes that, in the future, will produce the largest and most consequential eruptions. Pinatubo provides a further salutary lesson in that the only forewarning of the eruption appears in storytelling of the indigenous population that once lived on the volcano (Section 5.2.1).

Within 11 weeks, on the afternoon of 15 June, and following a crescendo in activity (Figure 3.1), the volcano erupted ten billion tonnes of pumice (equivalent to around ten cubic kilometres in bulk volume). Its magnitude, M_e, of 6.1 makes it the largest eruption in a century (since that of Katmai in Alaska in 1912). An area of 400 square

Figure 3.1 Pinatubo clears its throat on 12 June 1991, three days before the climactic eruption. Here, an 18 kilometre-high plume rises above the volcano, photographed from Clark Air Base (20 kilometres east of the volcano). Photo credit: Rick Hobblitt, CVO/USGS.

kilometres was utterly devastated and much of southeast Asia was veneered with ash. The atmospheric layer known as the stratosphere was punctured both by the Plinian eruption column fed directly from the crater, and by phoenix clouds that lofted above immense pyroclastic currents gushing on to the western and southern flanks of the volcano. The eruption intensity peaked around mid afternoon, based on recordings of shock waves made as far away as Japan. At this time, it is estimated that magma was discharging at the rate of 1.6 megatonnes per second, with speeds of up to 1000 kilometres per hour (an intensity of 12 on the scale described in Equation 1.2).

Due to the vast scale of Pinatubo's eruption plume, satellite observations provided the only effective way to track its growth and dispersal around the globe. By 16:40 local time, the umbrella cloud forming in the stratosphere was 500 kilometres across. It covered an area of 300,000 square kilometres at 19:40 hours, and reached a maximum diameter of over 1100 kilometres (Figure 3.2). The umbrella cloud was 10–15 kilometres thick and its top reached 35 kilometres above sea level. Thirty-six hours after the start of the eruption, the cloud covered a staggering area of 2.7 million square kilometres. By this stage the plume was travelling west–southwest with the ash concentrated at an altitude of 16–18 kilometres, and producing a layer of fallout up to ten centimetres thick across 400,000 square kilometres of the floor of the South China Sea. Thus, after injection into the atmosphere, the ash and gases soon followed different trajectories – most of the coarser ash fell out in a matter of hours, whereas the gases and very fine ash remained aloft.

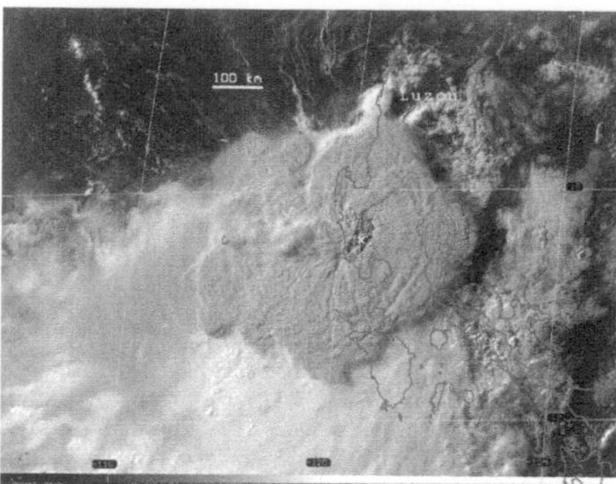

Figure 3.2 Umbrella cloud of the 15–16 June 1991 Mt Pinatubo eruption as seen by the GMS weather satellite. The cloud is over 1000 kilometres across. The cross marks the volcano's location. Image processed by Rick Holasek.

Several satellites mapped the initial spread of the cloud around the tropics and its gradual expansion to high latitudes in both hemispheres. Weather satellites tracked the cloud for two days, beyond which it became too dilute (from spreading and sedimentation of ash) to discern. Another spaceborne instrument, the Total Ozone Mapping Spectrometer (TOMS), picked up on the continuing dispersal of the plume (Figure 3.3). Although TOMS was designed to measure ozone abundance in the stratosphere, it proved to be very effective at detecting large volcanic sulphur dioxide (SO_2) clouds thanks to its combination of ultraviolet channels. Combined with data from additional ultraviolet and microwave satellite sensing instruments [20], the best estimate for the total mass of sulphur dioxide released into the stratosphere by the 15 June 1991 eruption of Pinatubo is an astonishing 18 megatonnes. That is equivalent to the weight of more than ten million typical automobiles! Although subsequent eruptions did manage to entrain material into the lower region of the stratosphere they were of negligible magnitude by comparison.

It took 22 days for the volcanic cloud to complete its first circumnavigation of the globe. A striking observation emerging from analysis of the TOMS data was that the amount of sulphur dioxide present decreased day by day. In the first week, three megatonnes of the sulphur dioxide just disappeared! Of course, it had to be transforming into something else, namely a mixture of 75% sulphuric acid (H_2SO_4) and 25% water in the form of minute particles. The chemical reaction involved requires a powerful oxidant, a role filled admirably by the highly reactive hydroxyl radical. (So reactive in fact that it is sometimes referred to as the atmosphere's detergent on account of its capacity to

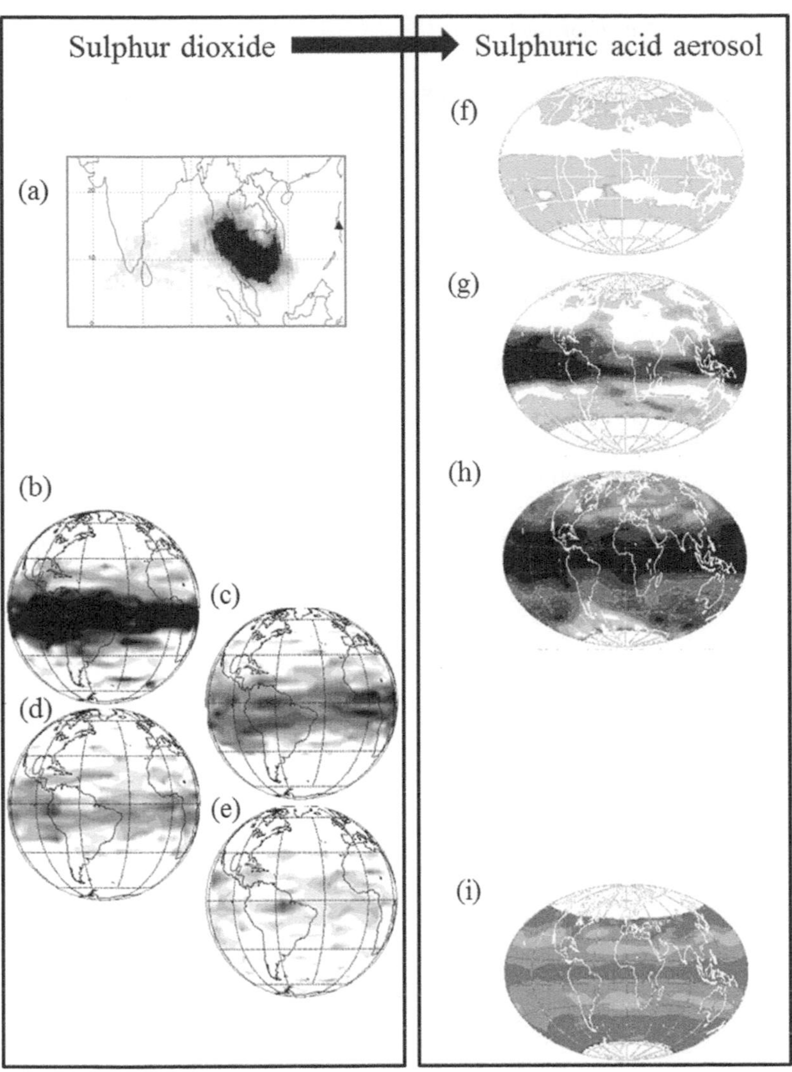

Figure 3.3 Satellite observations of sulphur dioxide (on the left) and sulphuric acid aerosol (on the right) in the stratospheric cloud generated from the 1991 Mt Pinatubo eruption. Time runs down the page. Sulphur dioxide was measured by (a) the TOMS sensor on 17 June (http://toms.umbc.edu), and then the MLS sensor [20] on (b) 21 September, (c) 2 October, (d) 16 October and (e) 17 November. Aerosol measurements were made by the SAGE sensor [19] (f) before the eruption for the period 10 April–13 May, (g) 15 June–25 July, (h) 23 August–30 September and (i) 5 December 1993–16 January 1994. Note the equatorial and then latitudinal spread of the cloud and the dissipation of sulphur dioxide through time, and growth then global spread of sulphuric acid aerosol. Note in (h) that most of the Earth is now girdled by volcanic aerosol, and in (i) the preferential removal of aerosol from the stratosphere at mid latitudes due to enhanced mixing of air across the tropopause. Scales are omitted but darker tones indicate higher abundances of sulphur dioxide or aerosol.

Figure 3.4 Double layer of sulphuric acid aerosol generated following the Pinatubo eruption seen on 8 August 1991 above Central Africa from the Space Shuttle (on mission STS 43). Layers are approximately 20 and 25 kilometres above sea level. Large thunderstorm cloud tops also in view below the volcanic haze. Photograph from the Image Science and Analysis Laboratory, NASA-JSC.

destroy assorted pollutants.) The key step in the oxidation process is thus:

$$SO_2 + OH \rightarrow HOSO_2 \tag{3.1}$$

In the stratosphere, the abundance of hydroxyl radicals results in a half-life for sulphur dioxide of approximately three weeks (Figure 3.4). Various chemical pathways then led to the formation of H_2SO_4:

$$HOSO_2 + O_2 \rightarrow SO_3 + HO_2 \tag{3.2}$$

$$SO_3 + H_2O \rightarrow H_2SO_4 \tag{3.3}$$

Sulphuric acid initially forms in the gaseous state but it condenses spontaneously under stratospheric conditions to form particles. These particles then grow from a few millionths of a millimetre across to less than one thousandth of a millimetre across but their enduring small size enables them to stay aloft in the stratosphere for months.

A ground-based remote-sensing technique called lidar (standing for 'light detection and ranging' and the optical equivalent of radar) provided valuable information on the cloud's vertical distribution. Whereas radars emit microwave energy, lidars pulse laser light up into the sky, some of which is scattered by air molecules and particles back to an attached telescope. The travel time of the light pulses provides an accurate measure of the height of a given reflecting layer.

Lidar stations have been established at many observatories and institutes worldwide and their observations showed that the bulk of Pinatubo's sulphuric acid aerosol was initially travelling at altitudes between 20 and 27 kilometres above sea level.

Over the next months, the cloud climbed higher into the stratosphere, its top reaching around 35 kilometres above sea level by October 1991. This ascent has been attributed to warming of the aerosol layer as it absorbed infrared radiation emitted from the Earth's surface and lower atmosphere. The warming may also have assisted the cloud's penetration into the southern hemisphere by modifying stratospheric wind patterns. Despite this early jump across the equator, the cloud only made gradual progress in expanding further into both hemispheres, remaining trapped within a band between latitudes 30° N and 20° S through mid 1991. Fortunately for twilight watchers across much of the globe, some aerosol at lower altitudes did mix further polewards providing spectacular sunsets and afterglows (Section 3.1.1). Then, as the jet streams in the middle atmosphere strengthened and meandered in the run up to the northern hemisphere winter, further horizontal mixing of the volcanic aerosol occurred. Strengthening of another large-scale poleward motion in the stratosphere known as the Brewer–Dobson circulation also helped to mix the volcanic particles around the planet.

The aerosol cloud was tracked and measured by several satellite sensors including NASA's Stratospheric Aerosol and Gas Experiment II (SAGE II; Figure 3.3). The estimated total mass of particles generated was around 30 megatonnes. Gradually, the specks of sulphuric acid settled to the surface and, by the end of 1993, only around five megatonnes of aerosol remained airborne. The half-life for the aerosol layer works out to about nine months, comparable to that measured after the previous major stratospheric aerosol disturbance due to the eruption of El Chichón in Mexico in 1982. Lidar observations from Japan and New Zealand detected the presence of Pinatubo aerosol for up to five years.

3.1.1 Optical illusions

The widespread atmospheric diffusion of Pinatubo aerosol led to many spectacular optical effects, including vividly coloured sunsets and sunrises, crepuscular rays and a hazy, whitish appearance to the Sun. These phenomena occurred as the aerosol veil absorbed and scattered sunlight.

In the months following the eruption, the opacity of the stratosphere reached the highest values recorded by modern techniques, peaking in August and September 1991. In fact, the sulphurous dust layer was so dense that the SAGE II satellite could not see through it! Astronauts on board Space Shuttle mission STS-43 in early August also noted the murkiest atmosphere since NASA's space photography began in 1962. The haze blurred the Earth's surface and distorted its colour. Oblique views through the Earth's limb revealed the layered structure of the aerosol (Figure 3.4), and sunlight reflecting off the volcanic aerosol resulted in illusory 'early' sunrises.

The haze was observed from the ground, too. The pioneering meteorologist Hubert Lamb, who did much to advance understanding of the climate change wrought by volcanic eruptions (Section 5.3), noted Pinatubo's sunsets from Holt in Norfolk (UK) from 24 August 1991. The Sun was tinged a fiery and livid orange-red, and a strong rose-red afterglow followed sunset, with red crepuscular rays radiating up to 40° above the horizon and lasting about half an hour after sunset on 28 August. Such twilight effects were observed from many parts of the world throughout the remainder of 1992. The first lunar eclipse after the eruption occurred on 9 December 1992; its appearance was 10% dimmer than usual because of the haze.

By 1993, the optical effects were diminishing, though as late as May and June 1993, reddish twilights were still being reported. At the time, I was part of a field team working at Socompa volcano in northern Chile. In the last week of May, we observed impressive afterglows that started as apricot before turning to a violet glow strong enough, up to 20 minutes after sunset, to cast shadows and illuminate the sinuous drifts of snow adhering to steep gullies on nearby Pajonales volcano.

3.2 ATMOSPHERIC AND CLIMATIC CHANGE

The aforementioned optical effects of Pinatubo's aerosol veil provide palpable evidence for its effect on the transmission of visible solar radiation to the Earth's surface. Since the peak energy from the Sun spans visible wavelengths of light, the abundant stratospheric aerosol can be expected to have influenced the Earth's heat budget and, hence, climate. Until Pinatubo's eruption, theoretical modelling of volcanic forcing of climate was hampered by uncertainties in key parameters such as the size and geographic distribution of volcanic sulphurous aerosol, and its optical properties (especially its efficiency at scattering or absorbing light at different wavelengths). Since the model outputs were only as reliable as the input parameters and assumptions, the theoretical results remained just that: theoretical. For Pinatubo, a multitude of observations from satellites, high-altitude balloons and aircraft, ground-based lidars and weather stations, provided the information necessary both to initialise computational models for climate, and the global-scale meteorological observations needed to validate the results.

3.2.1 Effects on light and heat radiation

The effects of stratospheric aerosol veils on electromagnetic radiation are highly complex because the haze can consist of variable proportions of minute glassy ash fragments and sulphuric acid droplets of different compositions, sizes and shapes (and hence optical properties). The different components also accumulate and sediment out at different rates according to their masses and shapes, so that any effects on the Earth's heat budget can be expected to change through time. Immediately after a major explosive eruption, there are likely to be significant quantities of silicate ash lofted into the stratosphere with dimensions exceeding a tenth of a millimetre. However, this ash sediments in a matter of days and weeks, while the sulphuric acid aerosol load increases as sulphur dioxide oxidises (Equations 3.1–3.3). Once most of the gaseous sulphur has been converted, little more aerosol is formed and the total stratospheric aerosol load decreases as the particles subside into the lower layer of the atmosphere called the troposphere, from which they are rapidly deposited to the surface by rainfall and other processes. As the total aerosol burden decreases, so the particle size distribution changes because bigger particles will be falling faster than smaller ones. Settling rates for two-thousandths-of-a-millimetre-diameter particles (with a density of two grammes per cubic centimetre) are about six centimetres per minute in the stratosphere. At this rate, it would take aerosol of this size around four months to fall ten kilometres. Measurements following the El Chichón and Pinatubo eruptions confirm the expected decrease in mean radius of stratospheric particles in the months after an eruption.

The stratospheric volcanic particles scatter incoming solar ultraviolet and visible radiation, directing some back into space and some sideways and forwards. The particles can also absorb radiation – visible and near-infrared wavelengths from the Sun, or long-wavelength infrared emission from the Earth – and warm up. The net effect on the Earth's radiation budget is thus complex but, broadly, for warming to outbalance cooling, the aerosol's effective radius (a size measure weighted by the particles' surface area) should exceed two thousandths of a millimetre. Prior to Pinatubo, the effective radius of stratospheric aerosol was about a tenth of this value. Pinatubo's contribution pushed effective radius up to about 0.5 thousandths of a millimetre but this was still below the threshold at which a greenhouse effect wins out over surface cooling.

The consequences of the stratospheric aerosol veil from Pinatubo were observed from space by the Earth Radiation Budget Experiment (ERBE) [21]. The ERBE sensor measured the radiation reaching it from 1000-kilometre-wide portions of the Earth in two wavebands - one dominated by reflected sunlight at short wavelengths spanning from the ultraviolet to the middle infrared region (Φ_{SW}), the other recording the total outgoing radiation deep into the infrared band (Φ_{TOTAL}, i.e. reflected sunlight plus thermally emitted radiation from the Earth's surface and atmosphere). This combination permitted monthly surveillance of the reflected shortwave and emitted longwave ($\Phi_{LW} = \Phi_{TOTAL} - \Phi_{SW}$) radiation. The monthly albedo, α, is then given by:

$$\alpha = \Phi_{SW}/\Phi_{SUN} \tag{3.4}$$

where Φ_{SUN} is the average incoming solar radiation for a given month. The monthly net radiation, Φ_{NET} (the difference between the absorbed shortwave radiation and the emitted longwave radiation) is given by:

$$\Phi_{NET} = (\Phi_{SUN} - \Phi_{SW}) - \Phi_{LW} \tag{3.5}$$

Other things being equal, if Φ_{SW} increases (i.e. if the Earth's albedo increases), then Φ_{NET} falls and the Earth's surface cools. On the other hand, a decrease in Φ_{LW} (for example, due to enhanced quantities of greenhouse gases in the atmosphere) increases Φ_{NET}, warming the Earth. The effects of clouds and aerosols are complex because they can both increase albedo (increasing Φ_{SW}) and trap outgoing longwave radiation (reducing Φ_{LW}). The net effect thus depends on several factors including the exact physical nature of the particles and their concentration and size distribution.

By August 1991, ERBE revealed that the backscattering of solar radiation by the aerosol had increased the global albedo by around 5%. This seemingly small difference actually corresponds to a cut in direct sunlight of 25–30%, which was partly compensated for by an increase in the diffuse light from the sky. The ERBE measurements were also able to show that the albedo increase was not uniform across the Earth but was most pronounced in normally 'dark', cloud-free regions, including the Australian deserts and the Sahara, and in typically 'bright' regions associated with deep convective cloud systems in the tropics such as the Congo Basin. The latter observation is initially puzzling since, over regions that have a naturally high albedo, the percentage increase due to volcanic aerosol in the stratosphere is small. It appears that fallout of Pinatubo aerosol across the tropopause (the boundary between the stratosphere and the troposphere) seeded clouds, or modified the optical properties of existing upper tropospheric cirrus clouds. Satellite observations support this interpretation since they also indicate a correlation between Pinatubo aerosol and increased cirrus clouds that persisted for more than three years after the eruption. The enhancement was especially noticeable at mid latitudes by late 1991, consistent with the time it would have taken for the fine sulphuric particles to sediment into the troposphere. The mid latitudes are dominant sites for transfer between stratosphere and troposphere across so-called tropospheric folds, which form in the 'surf zone' where planetary atmospheric waves (meanders in the jet stream) break. It is also likely that moisture transported to the upper troposphere by thick cumulonimbus (convective) clouds led to growth of the volcanic particles such that they eventually turned into cirrus cloud.

The ERBE measurements show that by July 1991 the outgoing shortwave heat flux had increased dramatically over the tropics. This corresponded to a change in the net flux of up to −8 watts per square metre in August 1991, twice the magnitude of any other monthly anomaly (the minus sign indicates more energy exiting the Earth than going in). The net forcing for August 1991 amounted to −4.3 watts per square metre for the region between 40° S and 40° N. Unfortunately, ERBE did not operate poleward of 40° latitude but even if there were no aerosol forcing at higher latitudes, then the globally averaged volcanic forcing still amounts to −2.7 watts per square metre, a significant sum (it compares with the roughly +2 watts per square metre forcing associated with anthropogenic greenhouse gases). This should represent the minimum global forcing because enhanced stratospheric aerosols were observed at higher latitudes by mid August 1991. The net radiation-flux anomalies seen by ERBE are the largest that have been observed by satellites. (The next largest net flux anomaly in the ERBE record, −3.5 watts per square metre, was recorded in June 1982, two months after the El Chichón eruption.) Following Pinatubo, the net flux returned to normal levels by March 1993.

3.2.2 Summer cooling, winter warming

For two years, Pinatubo's forcing effect on Earth's heat budget exceeded the positive forcing due to anthropogenic greenhouse gases (carbon dioxide, methane, chlorofluorocarbons and nitrous oxide). Climatologists at the Hadley Centre in the UK Meteorological Office were among the first to demonstrate the actual effects of Pinatubo on global climate via an examination of worldwide meteorological records of air temperature [22]. The observed global cooling was initially rapid but punctuated by a warming trend, predominantly over northern landmasses, between January and March 1992 (Figure 3.5). Cooling resumed and, by June 1992, amounted to about 0.5 °C. This may appear modest but, as a globally averaged figure, it hides much wider regional variations that included pockets of abnormally strong heating as well as cooling. For example, the Siberian winter was 5 °C warmer than normal, while the north Atlantic was 5 °C cooler than average. The overall cooling is also a fraction of a degree stronger if the warming effect of the prevailing El Niño is accounted for.

Globally, there was a significant drop in rainfall over land during the year following the eruption, making it the driest period in the half-century for which good records are available [23]. However, there were strong regional patterns. For instance, 1992 witnessed one of the coldest and wettest summers in the USA in the twentieth century – the Mississippi flooded its banks spectacularly – whereas drought affected much of sub-Saharan Africa, south and southeast Asia and central and southwestern Europe (enhanced by a prevailing El Niño). There was further relative warming in early 1993 and early 1995, and the mid-year cooling reduced each year. These trends are mirrored in average temperatures for the lower troposphere determined from space by the Microwave Sounding Unit (MSU). The cooling was concentrated over the continents away from the oceans. The response was stronger in the northern hemisphere where there is a greater land surface area. The globally averaged cooling effect on lower-troposphere and surface temperatures peaked during the first two years following the eruption but the effect persisted for a few more years until the signal disappeared into natural climate noise.

The winter warmth after the Pinatubo eruption was concentrated over Scandinavia, Siberia and central North America [24]. In fact, the effect shows up more clearly in climate data for the second winter compared with the first when El Niño conditions were

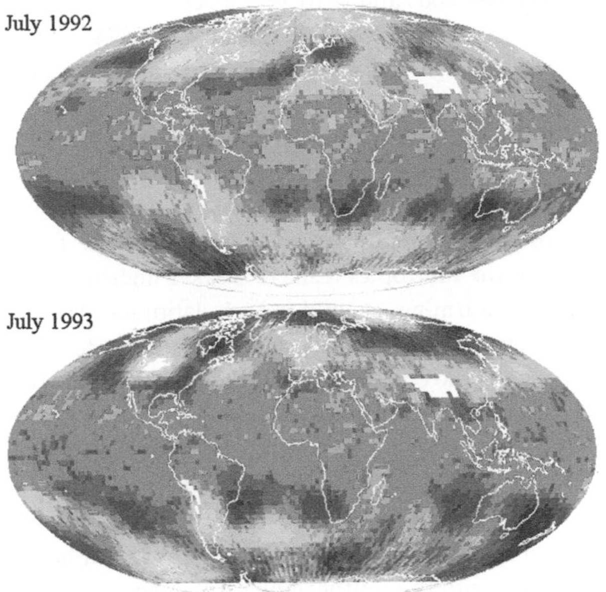

Figure 3.5 Temperature anomalies recorded by the Microwave Sounding Unit for the lower troposphere in the two (northern hemisphere) summers following the Pinatubo eruption. Note regions of boreal summer cooling over parts of North America, Europe and the Middle East. Scales are omitted but lighter tones represent anomalous cooling (the maximum amplitude of the temperature scale is about 8 °C). There are no data over Tibet. Data from Remote Sensing Systems (http://www.remss.com) via a program sponsored by the NOAA Climate and Global Change Program.

disturbing high-latitude atmospheric circulation. These temperature anomalies were associated with marked departures in sea-level pressure patterns in the first and second boreal (northern) winter. There was a poleward shift and strengthening of north Atlantic westerlies at around 60° N, associated with corresponding shifts in the positions and strengths of the Icelandic Low and Azores High atmospheric pressure centres. These effects have been modelled as a result of abrupt changes to the atmospheric circulation around the Arctic arising from differential heating effects of the volcanic aerosol in the middle atmosphere. In fact, there is a frequent flip-flop pattern in the North Atlantic jet stream known as the North Atlantic Oscillation (NAO) – it is positive when the difference in pressures between the Icelandic Low and Azores High is large. (The strongly negative NAO prevailing in the northern hemisphere winters of 2009 and 2010 were associated with some of the coldest temperatures on record in Europe for decades. Heavy snow falls led to widespread disruption of transport in the UK and other parts of Europe.)

Strong stratospheric heating in the tropics due to the presence of volcanic aerosol establishes a steeper north–south temperature gradient in the lower stratosphere, strengthening the circulation of the atmosphere around the Arctic (the polar vortex). At the same time, surface cooling in the subtropics leads to weaker high-altitude winds, and thus surface warming at mid-to-high latitudes. This reduces tropospheric temperature gradients, which also helps to strengthen the polar vortex, and the combined effects amplify the positive phase of the NAO, transporting warm, moist air to northern Eurasia and leading to winter warming. A significant contribution to this effect derives from mild night-time temperatures due to the formation of a thin but very low cloud deck. Any forcing that strengthens the polar vortex will have this effect. The stronger northern hemisphere response compared with the southern hemisphere once again reflects the greater landmass area in the north.

The temperature and precipitation patterns following the Pinatubo eruption have been fitted with some measure of success by Intergovernmental Panel on Climate Change (IPCC) models, though they do a better job of reproducing the summer-cooling effect – thanks to good observations of the aerosol distribution, size and properties – compared with the winter warming. Although the models reproduce the volcanically induced winter warming following the Pinatubo eruption, they predict only around 1 or 2 °C of warming in Siberia compared with the observed increase of up to 5 °C. The difficulties stem partly from poorer model performance for higher regions of the atmosphere.

3.2.3 Oceanic response

Oceans cover 72% of the Earth's surface and because they can store so much heat energy they play a very important role in climate. One quite recent finding is the effect that Pinatubo's radiative forcing had on the oceans. The reduced shortwave solar radiation falling on the sea surface resulted in heat loss from the surface ocean of 50,000 exajoules (equivalent to 100 times the total annual worldwide energy consumption) [25]. Globally averaged sea-surface temperatures dropped by around 0.4 °C. This effect is now recognised for numerous earlier eruptions based on studies of corals and tree rings (Section 4.3). Once mixed into the ocean, this cooling becomes a miniscule five thousandths of a degree but the overall effect has been sufficient to decrease global sea level by almost a centimetre (cooler water is denser and takes up less volume). The mixing of this cooled layer into the global ocean is a very slow process, so it means that Pinatubo's 1991 eruption likely still has a signature in the deep ocean. In the case of a much larger, sulphur-rich eruption such as that of Tambora in 1815 (Chapter 13), the ocean anomaly may have lasted well over a century. Just as remarkably, the impact of the 1883 Krakatau eruption is thought to have offset a large fraction of ocean warming and sea-level rise due to greenhouse gas emissions through the twentieth century [26].

3.2.4 Biological feedbacks

In the case of explosive eruptions such as Pinatubo's in 1991, there is evidence that despite the increase in the Earth's albedo and resulting global surface cooling, photosynthesis could be encouraged in some regions. This is a combined result of reduction in direct sunlight reaching the surface (many plants dislike very intense light and turn or close their leaves at midday to avoid damage) and increase of scattered light in the sky (which can penetrate more of a vegetation canopy than direct sunshine thus illuminating more leaves) [27]. Following Pinatubo, solar and sky measurements at the Mauna Loa Observatory, Hawai'i, indicated a decrease in the direct flux of up to 25% but at the same time a more than threefold increase in diffuse light. Increased photosynthesis helps to explain slowed growth of atmospheric carbon dioxide during the two years following the eruption [28] (its abundance still increased but not at such a rapid rate). By acting as a sink for atmospheric carbon dioxide, this effect can enhance the surface cooling brought about by radiative effects.

Meanwhile, cool air temperatures in the Red Sea region arising from the light-scattering effects of the Pinatubo aerosol are thought to have enhanced mixing in the water column, bringing nutrients to the surface and stimulating algal blooms. The usually clear blue waters of the Gulf of Eilat turned green with chlorophyll as algal mats spread across the sea floor, shading corals and hampering the flow of water. Stagnation of the water led to build up of hydrogen sulphide that, combined with the smothering effect, killed much of the coral.

3.2.5 Stratospheric ozone depletion

A few months after the Pinatubo eruption, global stratospheric ozone levels began to show a strong downturn. The clearest picture of the impacts was provided by the spaceborne TOMS. Ozone levels, integrated through the depth of the atmosphere, decreased 6–8% in the tropics in the first months after the eruption. These figures mask local depletions of up to 20% at altitudes of 24–25 kilometres. By mid 1992, ozone abundance in the stratosphere was lower than at any time in the preceding 12 years, reaching a low point in April 1993 when the global deficit was 6% compared with the average. Losses were greatest in the northern hemisphere. Total ozone above the USA, for example, dropped 10% below average.

The mechanisms for these dramatic decreases in ozone abundance in the stratosphere are complex but similar to those involved in the development of the Antarctic ozone hole. The key in Antarctica is the formation of 'polar stratospheric clouds', which are composed of ice crystals and condensed nitric acid. These provide surfaces on which ozone-destroying chemical reactions can occur. Following the Pinatubo eruption, ground-based lidar measurements as well as observations made from balloons and aircraft showed that the largest deficits in stratospheric ozone coincided with peaks in volcanic aerosol concentration. In effect, the sulphuric acid particles achieved on a global scale what polar stratospheric clouds routinely do for springtime Antarctic ozone.

The reactions themselves are several chemical cycles involving oxygen, nitrogen, hydrogen and chlorine. They act in such a way as to transform chlorine from stable compounds (such as hydrogen chloride and chlorine nitrate) into reactive forms (like hypochlorous acid, HOCl) that can destroy ozone. One point to bear in mind, however, is that the stratospheric chlorine involved in these reactions was sourced mostly from the chlorofluorocarbons manufactured as refrigerants, solvents and propellants. In other words, the ozone loss following Pinatubo was partly an anthropogenic effect. This raises an interesting question: can volcanic eruptions cause stratospheric ozone loss by themselves? Volcanoes emit copious quantities of chlorine as hydrogen chloride, and this could readily react with hydroxyl radicals to activate the chlorine for assault on ozone. However, hydrogen chloride is also very soluble in water, so it can be readily scrubbed out of a plume rising through a moist atmosphere. This may explain why enhanced stratospheric hydrogen chloride levels were reported after the

El Chichón eruption (up to 40% increases over background levels were observed within the cloud), while Pinatubo's estimated three megatonnes of chlorine emitted at the vent were not detected in the stratosphere. El Chichón erupted into a dry atmosphere while Pinatubo's eruption column rose through a particularly humid troposphere as a typhoon was crossing the Philippines at the time. The chemistry of the atmosphere is sufficiently complex that we still do not really know the circumstances that could lead to a volcanic eruption causing a major stratospheric ozone perturbation without help from humans.

3.3 RECIPE FOR A CLIMATE-FORCING ERUPTION

One of the foremost climatologists working on the atmospheric impacts of volcanic eruptions, Alan Robock at Rutgers University, has already prepared a press release predicting the climate response to the next large-magnitude, tropical, explosive, sulphur-rich eruption: he will only need to insert the name of the volcano! This may be somewhat tongue-in-cheek but it is true that we have come a long way in our understanding of the impacts of such eruptions through hundreds of studies of the Pinatubo effects. Nevertheless, there are many nuances to the picture and it is worth summarising those factors likely to distinguish an eruption capable of hemispheric- to global-scale climate impacts.

3.3.1 Sulphur content and eruption magnitude

We have seen that sulphur emission is crucial: it is the release of sulphur gases (principally sulphur dioxide or hydrogen sulphide) into the atmosphere during an eruption that leads to the formation of airborne sulphuric acid particles that may then perturb the Earth's heat budget. The old idea that suspended ash is responsible for the major impacts on the Earth's heat budget following eruptions has been convincingly overturned (although fine ash does have some significant local scale effects).

But what makes some magmas sulphur-rich and others sulphur-poor? As an example, consider that the climactic eruptions of Mt St Helens in 1980 and of El Chichón two years later were of similar size (both around M_e 5.1). However, the Mexican eruption released

7.5 megatonnes of sulphur dioxide into the atmosphere, seven times more than Mt St Helens. The reasons for such discrepancies are complicated and remain only partially understood. This is because sulphur abundance in magmas turns out to be controlled by several physical and chemical parameters among which temperature, pressure and melt composition are pre-eminent. The picture is also murky because of the multiple forms that sulphur may take in a magma – it may dissolve in the silicate melt as sulphate or sulphide ions, be incorporated into crystallising minerals such as pyrite and anhydrite or exsolve into a gas phase (as hydrogen sulphide or sulphur dioxide). Broadly, there is an inverse relationship between sulphur solubility in magma and silica contents – basaltic magmas typically contain up to 0.1% by mass of dissolved sulphur whereas silicic magmas may contain as little as 0.002%. This is important because larger explosive eruptions are almost invariably silicic in composition; thus, going up the scale in eruption magnitude need not correspond to linear increases in sulphur output.

Another important factor is the oxidising capacity of the magma. We may think of this in the same way that we consider the atmosphere an oxidising environment, and the bottom of a swamp a reducing environment. An oxidising agent is a chemical species that will accept electrons from another species. A familiar process in air is rusting in which iron changes its chemical valence (the number of electrons orbiting the atom) from Fe to Fe^{2+} to Fe^{3+} (the associated colour change and familiar reddish hue is due to the way that the redistributed electronic charge interacts with visible light). Magmas contain a miniscule fraction of reactive oxygen compared with the atmosphere and it is the much more abundant iron that plays a key role in controlling the oxidation state of magmas, shifting the following chemical equilibrium either to the left (more reducing) or right (more oxidising):

$$2Fe^{2+} + {}^{1/2}O_2 \leftrightarrow 2Fe^{3+} + O^{2-} \qquad (3.6)$$

This balance strongly controls the amount of sulphur that can be dissolved in silicate melt, as well as its valence (for example, S^{2-} or S^{6+}). Numerous laboratory experiments using kilns and high-pressure apparatus have confirmed that magmas can dissolve more sulphur when they are either highly oxidising or highly reducing. In between, the sulphur-carrying capacity is much lower.

The eruptions of El Chichón in 1985 and Pinatubo in 1991 were both notable in that their intermediate and silicic magmas, respectively, were uncommonly sulphur-rich. This is likely due in no small degree to their highly oxidising conditions. Taking the mass of Pinatubo's eruption as 13–18 gigatonnes and the estimated sulphur dioxide yield to the stratosphere, 17–20 megatonnes, suggests a sulphur fraction in the magma of between 0.05 and 0.075% by mass (accounting for the difference in the molecular weights of sulphur (32 grammes per mole) and sulphur dioxide (64 grammes per mole)). Such amounts are far higher than one would initially suspect for a silicic magma. What is more, they represent minimum values, since not all the sulphur gases released from the magma will have reached the stratosphere to be measured by the satellite sensors. Some will have been scavenged out by chemical and physical processes occurring in the ash- and ice-particle-rich plume during its ascent through the troposphere.

Other factors being equal, bigger eruptions (higher magnitudes) and more sulphur-rich magmas will cause stronger perturbations to atmosphere and climate. However, it is important to recognise that the climate response does not scale in any simple way with the sulphur yield of an eruption. Pumping more and more sulphur into the upper atmosphere ultimately leads to changes in the formation of sulphuric acid particles. In particular, it will result in growth of larger particles that are not only less effective in scattering sunlight but also faster to drop out of the stratosphere [29]. This leads to a saturation effect when mapping sulphur output of an eruption to its climatic impact. For sulphur injections more than a few times larger than Pinatubo's in 1991, there are diminishing returns in terms of temperature response at the Earth's surface. This is a particularly important concept when considering the climate forcing due to very large sulphur releases (Chapter 8 and Section 11.2).

3.3.2 Eruption intensity and style

The injection height of sulphur into the atmosphere is another critical determinant of climate impact. If an eruption column is confined to the troposphere then the atmospheric processing of the sulphur is speeded up, largely due to rapid deposition by rain. Sulphur from a typical explosive eruption needs to penetrate the much drier stratosphere if the aerosol it generates is to remain aloft for sufficient time to have a strong effect on radiation. So, eruption style (explosiveness and intensity) will influence the climatic outcome of a volcanic eruption. More intense eruptions, i.e. those with higher magma discharge rates, are more likely to loft reactive sulphur gases into the region of the atmosphere where they can generate climatically effective aerosol. Fumarole emissions and non-explosive degassing from lava ponds, lakes and domes simply cannot propel material to the stratosphere in the same way as major explosive eruptions. Nor can even quite violent fire fountains because of their inefficient conversion of thermal energy into buoyancy (Section 2.1).

Phoenix clouds are certainly capable of entraining material into the stratosphere, and have been responsible for some of the largest tephra fall deposits (Section 2.3). However, little is known about their efficiency, compared with Plinian eruption columns, in terms of injecting sulphur into the stratosphere. The two types of plumes differ in subtle ways. For example, there is no gas thrust region at the base of a phoenix cloud. It is conceivable that such differences and variations in thermal histories and pyroclast sizes in the two plume types may influence the fraction of sulphur gases that is removed from the ascending column before it reaches the stratosphere.

3.3.3 Eruption location

The location of a volcano, and thus where it might focus its sulphur injection into the stratosphere, strongly influences the geographic distribution of atmospheric heating and its interaction with vast atmospheric waves such as the jet stream. The latter are especially important in the northern hemisphere due to mountain ranges such as the Rockies. Thus, we can expect that two identical eruptions with identical sulphur outputs might result in different climate signals if they are located in different parts of the world. But there are other critical geographical factors to consider.

One is that the height of the tropopause varies with latitude – at the tropics it is around 16–17 kilometres above sea level but descends to 10–11 kilometres at high latitudes. In general terms, an explosive eruption requires a greater intensity (magma discharge rate) to cross the tropopause in the tropics than at mid to polar latitudes. However, there are two important factors that counteract this trend. The first is that since the primary mechanism by which volcanic-aerosol veils cool surface climate is by scattering sunlight back into space, a high-latitude eruption will have a more limited effect than a low-latitude one. This is simply because further from the tropics there is less solar energy to intercept in the first place.

Secondly, the atmospheric circulation works in a way to limit the effects of high-latitude eruptions. Very broadly, winds at altitude blow poleward from the tropics, and nearer to the surface in the reverse direction. In this regard, the atmosphere operates like a giant heat pump carrying equatorial solar warmth to higher latitudes (the polar regions would otherwise be significantly colder, and the tropics much hotter). A tropical eruption that pumps aerosol into the stratosphere results in localised heating. This, in turn, should increase the temperature difference in the middle atmosphere between the equator and high latitudes, enhancing north–south (meridional) air flows that spread the volcanic dust into both hemispheres, promoting climate forcing impacts at a worldwide scale. In contrast, volcanic aerosol injected into the stratosphere from high-latitude volcanoes will tend to have the opposite effect on the temperature gradient, acting to stagnate meridional air flow. Very little, if any, of the stratospheric aerosol formed as a result of eruption of a high-latitude volcano will reach the opposing hemisphere. The eruptions of Alaskan volcanoes Okmok and Kasatochi, in July and August 2008, respectively, did not noticeably perturb climate: Okmok's eruption was too small (100 kilotonnes of sulphur dioxide) and Kasatochi's, despite its substantial sulphur dioxide release (1.7 megatonnes, making it the largest pulse since 1991) was too late in the year to result in significant radiative forcing [30].

Nevertheless, both historical and modelling evidence suggests that high-latitude volcanic eruptions can have wide-reaching climatic effects at a hemispheric scale. Climate-model runs for the 1912 Katmai eruption (Alaska), and the same event scaled up three times in terms of sulphur release, confirmed pooling of the aerosol veil to high-boreal latitudes [31]. It also suggested that the sunlight scattering effects of the volcanic aerosol dominate over any impact on the NAO. But a consistent result that emerged was cooling over southern Asia during the northern hemisphere winter and a weakening of the Asian monsoon. However, the results of such modelling efforts are very sensitive to assumed sizes of aerosol and parameterisations of the microphysical processes that occur in clouds. A climate simulation to study the effects of a prehistoric eruption in Germany (Section 9.2) found a rather similar effect to Pinatubo's and a more significant role for atmospheric dynamics [32].

3.3.4 Eruption timing

As well as the latitudinal differences in tropopause height discussed in the preceding section, seasonal variations in the tropopause height will also influence the minimum eruption intensity necessary for an eruption column to reach the stratosphere. The seasons further influence tropospheric humidity. An eruption cloud rising through a humid atmosphere will tend to reach higher than one in a dry atmosphere since there will be an additional source of thermal energy to drive plume ascent, namely the latent heat released when the atmospheric water vapour entrained into the plume condenses out and then freezes at altitude. Physical models suggest this effect could account for several kilometres of extra plume height. A humid troposphere will also be more likely to strip out soluble plume gases (potentially including sulphur dioxide) into 'hydrometeors' (water droplets or ice crystals) before they reach the stratosphere.

Another important seasonal effect is the position of the Inter Tropical Convergence Zone (ITCZ). This imaginary circle shifts north of the equator in the northern-hemisphere summer, and south of the equator in winter. Because of the strong heating of the ground in the tropics, air in contact with the surface is heated and ascends due to its buoyancy. The ITCZ delineates the region in which these rising air currents diverge: to the north of the ITCZ, winds aloft head northwards, in opposition to the winds south of the ITCZ. The proximity of a stratospheric eruption cloud to the ITCZ can thereby determine the dispersal of the volcanic aerosol, and importantly whether it pollutes just one or both hemispheres.

In addition to the annual movement of the ITCZ there are other more complex circulation phenomena that can prove important for volcanic forcing of climate. These include the Quasi Biennial Oscillation in equatorial regions but, perhaps most importantly, the El Niño Southern Oscillation, which, depending on its phase at the time, could amplify or dampen the climate effect of an eruption. There has even been considerable scientific debate concerning the possibility that volcanic eruptions trigger El Niño Southern Oscillation events, though the evidence remains elusive.

Atmospheric circulation changes from week to week, with the seasons, and with alternating phases of planetary-scale atmospheric disturbances such as the North Atlantic Oscillation (Section 3.2.2). It is interesting to consider, for instance, how different Pinatubo's impact might have been had the eruption occurred in December rather than June of 1991. However, the largest seasonal effects are likely to be for high-latitude eruptions – for example, what if the 1912 Katmai eruption had taken place in December rather than in June? With a strong polar vortex in winter, emissions of sulphur dioxide could readily be trapped at high latitudes. It would take longer to oxidise the emissions to generate sulphuric acid aerosol because of the lack of sunlight and oxidising agents, and instead sulphur dioxide, a greenhouse gas, could have an effect on outgoing longwave infrared radiation. Furthermore, larger aerosol particles would form due to an increased aerosol density in the confined space of the polar atmosphere, contributing further to a greenhouse effect.

One recent modelling study has looked at the expected impacts of a five-megatonne sulphur dioxide injection into the lower stratosphere from a boreal volcano such as Katmai. It found that little sulphuric acid aerosol remained aloft in the spring following an August eruption due to widespread deposition during winter [33]. Another model of a very different kind of eruption (a prolonged effusive eruption in Iceland) also revealed contrasting lifetimes of volcanic aerosol according to eruption timing [34], again due to enhancement of particle sedimentation by winter precipitation. Combined with the reduced sunshine in winter, the radiative and photochemical consequences of volcanic emissions from an eruption late in the year should therefore be limited.

Another study has explored seasonal influences on a mid-latitude eruption. It looked at a much larger emission scenario, simulating a super-eruption of Yellowstone volcano (located around at latitude 44° N) with a 1700-megatonne release of sulphur dioxide (100 times that of Pinatubo) [35]. In the model, the aerosol veil of a summer eruption heads west and southward, driven by the Aleutian high-pressure system, compared with east and northward transport for a winter event, when westerlies dominate circulation. For the summertime case, more aerosol makes it into the southern hemisphere. This has a strong overall impact on the calculated radiative forcing at the Earth's surface.

While eruptions releasing much more sulphur than Pinatubo in 1991 eventually lead to limited payback in terms of climate response (Section 3.3.1), a string of Pinatubo-like eruptions (sulphur-rich; high intensity explosive; tropical) every 5–10 years or so might significantly extend and amplify volcanic forcing. Such pulses of volcanism are thought to have contributed significantly to the climate fluctuations of the Little Ice Age (Section 11.2.3). Between 1400 and 1850, up to around a half of the decadal variability seen in northern-hemisphere temperature reconstructions can be attributed to volcanism [36]. Given the longer response time of the oceans to radiative forcing compared with the land surface and atmosphere, recurrent sulphur-rich eruptions will likely have even more of a cumulative effect on deep-ocean temperature and circulation.

3.4 SUMMARY

The eruption of Pinatubo in 1991 provided a natural climate experiment and resulted in unambiguous evidence linking volcanism to changes in global temperature, wind and precipitation patterns. Around 18 megatonnes of sulphur were injected into the stratosphere, which generated up to 30 megatonnes of sulphuric acid particles that formed a veil around the whole planet. The globally and annually averaged surface temperature response was a cooling of 0.5 °C, driven largely by the radiative effects of the volcanic aerosol, specifically its scattering of incoming sunlight. Regionally and seasonally, there were more extreme patterns of storminess and drought, and heating and cooling.

Although Pinatubo has provided by far the clearest picture of these complex responses (thanks to its large sulphur release and the contemporaneous availability of sophisticated remote-sensing tools), similar eruptions in the past, such as those of El Chichón in 1982 and Krakatau in 1883, have been followed by comparable climatic change. The generalised findings are that the climate forcing following such eruptions lasts around three years, reflecting the time taken for most of the stratospheric aerosol to disperse. This results in surface and lower-atmosphere temperature anomalies that peak in the first year after eruption but which can still be discerned in sensitive records for up to seven years, by which time any remaining signal falls below the level of climate noise. The effects on sea-ice extent and global ice mass have a slightly longer response, lasting perhaps a decade. The longest response is seen in the oceans, where deep-sea temperatures, sea level, salinity and large-scale circulation can be perturbed for up to a century.

The magnitude of regional variability following large, sulphur-rich eruptions is sufficient to impact ecosystems and agriculture in many parts of the world. This is how an eruption on one side of the planet can result in major social and economic impacts on the other side.

4

Forensic volcanology

It is a capital mistake to theorize before you have all the evidence. It biases the judgment.

Sir Arthur Conan Doyle, *A Study in Scarlet*

Most historic eruptions with magnitudes of M_e 5 or more (Section 1.3.2) were the first on record for the responsible volcano. The deadliest eruption in history - that of Tambora in 1815, which directly or indirectly killed perhaps more than 100,000 people - is just one example. None of the volcanoes responsible for the four largest eruptions of the past century or so, Katmai 1912 (Alaska), El Chichón 1985 (Mexico), Mt Pinatubo 1991 (Philippines) or Chaitén 2008 (Chile), had previously erupted in recorded history, nor were they considered potentially hazardous. The message is clear - the largest eruptions in future are likely to come from previously little-known, even unheard-of, volcanoes.

To go about identifying these potentially dangerous volcanoes, it is necessary to deduce the timing, magnitude, intensity, style and gas yield of their past eruptions. And to understand their significance, we also need to determine and quantify the nature and extent of their impacts on the atmosphere, environment, climate and society. The latter endeavour is especially challenging since attributing cause and effect can be perilous. Association of climate change to a particular eruption requires careful compilation of evidence. For instance, short-term global climate change can have numerous explanations such as the El Niño Southern Oscillation, the North Atlantic Oscillation and variability in solar radiation.

The record of volcanism is written in diverse ways. Directly, it is recorded in deposits of ash and pumice found in rock sequences, lake and oceanic sediments and peat bogs, or in the sulphuric acid fallout that ends up sealed in glaciers and ice sheets. Information on past eruptions can also be gleaned from their indirect effects: for example climate changes inferred from tree-ring anomalies. For the prehistoric period, tephra records, ice cores, tree rings and archaeology yield the primary evidence for the timing, location and nature of volcanic events. Similarly, for the historic era, they supplement the often-sparse documentary record. Even for large explosive eruptions that take place today, direct observations during the event are intrinsically difficult and dangerous (except by means such as satellite remote sensing; Section 3.1), and thus much of our knowledge about them comes from field and laboratory studies of the deposits. An important application of the information obtained is in the evaluation of present-day volcanic hazards.

This chapter examines how data on the age, characteristics and consequences of eruptions can be elucidated from studies of rocks, ice and trees. Particular attention will be given to the reliability and accuracy of these different data sources, and the ways in which they can be integrated to reconstruct the nature and repercussions of past eruptions. The following chapter continues the forensic theme by reviewing what may be gathered from oral traditions and the archaeological record. These methodologies underpin many of the case studies elaborated on in this book.

4.1 READING THE ROCKS

The kinds of eruptions described in Chapters 1 and 2 yield a variety of characteristic materials and landforms. The rock record therefore represents a rich source of information on past volcanism (Figure 4.1). Its decipherment benefits from an understanding of the physics of eruptions and of sediment transport and depositional processes, as well as from empirical relationships drawn from meticulous descriptions of the deposits of modern eruptions such as those of Mt St Helens (1980, USA), El Chichón (1982, Mexico), Mt Pinatubo (1991, Philippines) and Soufrière Hills volcano (1995–present, Montserrat).

The relationship between an active lava flow or lava flow field and the final product that is literally set in stone is a rather close one. Lava landforms typically resemble, at least, the later stages of their parent flows. If the flow remains largely exposed at the surface, it is fairly straightforward to estimate its length and volume, and various techniques may be employed to date it (Section 4.1.3). With pyroclastic rocks, the correspondence between deposit and eruption style can be

Figure 4.1 Alternating bands of 'dry' and 'wet' pyroclastic current deposits on Lipari, Italy. Careful measurements of the thicknesses of layers, and the size, shape and composition of fragments in them, yields invaluable information on the magnitude, intensity and style of past eruptions.

more difficult to discern. To begin with, tephra may be deposited by two rather different physical processes - fallout and flow - often both occurring in the same eruption. They can also be dispersed over vast areas, especially in the case of ash fallout. For example, pumice from the 760,000-year-old Bishop Tuff eruption can be traced across an area of one million square kilometres in the USA. Because they can be regarded as having been deposited instantaneously, such widespread layers provide valuable time horizons that aid intercomparison of palaeoenvironmental records derived from a wide variety of sedimentary sequences, and also dating of archaeological and fossil finds.

This section sets out to highlight just how much information can be obtained on past volcanism from its direct solid products and their associated palaeoenvironmental contexts.

4.1.1 Characteristics of tephra deposits

Tephra deposits consist of fragments produced from fresh magma expelled by an eruption; and what are referred to as 'lithics' - pieces of old rock accidentally incorporated into the erupting mixture. Fresh materials include chunks of pumice, shards of volcanic glass, and mineral grains and crystals. Lithics are typically represented by ancient lavas or pyroclastic rocks that were wrenched from the volcanic conduit and vent by the erupting magma, or eroded from the ground surface beneath moving pyroclastic currents. Lithics can even include non-volcanic materials such as sedimentary or metamorphic rocks excavated from the volcano's basement. Mt Etna, for example, occasionally expels chunks of baked sandstone and limestone extracted from the piedmont on which the volcano rests. Densities of fragments in tephra deposits vary widely. A particularly frothy pumice may have a density of only 500 kilogrammes per cubic metre enabling it to float on water and, in some cases, travel great distances at sea. In contrast, bubble-free lumps of obsidian (volcanic glass) have typical densities of around 2600 kilogrammes per cubic metre.

Tephra that are carried through the air in drifting volcanic clouds settle out at rates according to their aerodynamic properties, namely density and drag, but also as a result of processes that cause particles to clump together while airborne. In general terms, smaller and/or lower-density fragments travel the furthest. This winnowing of ash and pumice is reflected in the thickness, range of particle sizes and the proportions of different components (e.g. lithics compared with crystals) found in the associated deposit. Systematic measurements of these properties, combined with knowledge of the aerodynamic properties of the particles, can be used to infer the rate at which they were blasted out of the eruption vent, the height to which they rose in the atmosphere, and the strength of the winds that transported them [37].

Key features of a tephra deposit thus include its thickness, the size of its constituent fragments, their physical nature (density, porosity and shape) and chemical composition, sedimentary structures such as the continuity and inclination of any layering found, and the orientation of particles. The 'grain-size distribution' of tephra deposits contains a wealth of information pertaining to the eruption itself and the dispersal of the plume. Grain-size analysis (granulometry) can be carried out in the field with a set of sieves, and in the laboratory with more sophisticated optical particle analysers (these are especially useful for very fine-grained ash). Numerous statistical treatments can be applied to such data. For example, a measure of how well sorted the grains are can help in discriminating between tephra fallout and pyroclastic current deposits. The winnowing action applied to tephra fallout results in a strong grading of particles at any particular location whereas pyroclastic currents tend to deposit more haphazardly, jumbling the finest ash with blocks of lava the size of a house. Collating such information constructs a picture of the volcanic history of an area – an endeavour known as 'tephrostratigraphy'.

Systematic observations of tephra deposits can be used to interpret the original conditions of an eruption (such as vent location, eruption magnitude, duration and intensity, column height, and wind speed and direction). The variations in thickness of a tephra deposit and their relationship to the underlying topography are among the leading diagnostic features. The fallout from volcanic clouds typically spreads a veneer of ash over an elliptical footprint stretching downwind from the vent. Within any smallish area on the ground, the thickness of ash tends to be roughly equal regardless of topography. This characteristic 'mantle bedding' is quite unlike the preferential accumulation of pyroclastic-current deposits in the valleys to which their parent flows were confined. Pyroclastic currents from very large eruptions can even iron out pre-eruption topography by burying valleys in pumice deposits hundreds of metres thick.

Hydrovolcanic eruptions are given away by a number of tell-tale features. One of the most curious of these are 'accretionary lapilli'. These are the tephra equivalent of a hailstone, consisting of concentric layers of ash that have usually agglomerated thanks to steam condensation. Other diagnostic characteristics of 'wet' eruptions are sediments that are strongly cemented, or plastered to steep slopes. 'Dry' deposits, on the other hand, tend to be depleted in very fine ash (because it has been winnowed out) and they often feel floury or sugary when scratched with the fingers at an outcrop.

4.1.2 Estimating eruption parameters

The classical way to chart thickness variations of pyroclastic rocks from a given eruption is to plot field measurements on a map, and draw contours of equal deposit thickness (Figure 4.2). The result is an 'isopach map'. In the case of modern eruptions, measurements can be made quickly before wind and water disturb the material (although a certain amount of erosion is actually helpful close to source, in order to cut down through fresh deposits and reveal their thickness). A basic capability of such maps is to identify the source of the eruption at the point of maximum deposit thickness. Though this might seem trivial, in the case of large eruptions in volcanic regions with many volcanoes, it need not be obvious which one was responsible for a given deposit. More usefully, isopach maps enable estimation of eruption sizes.

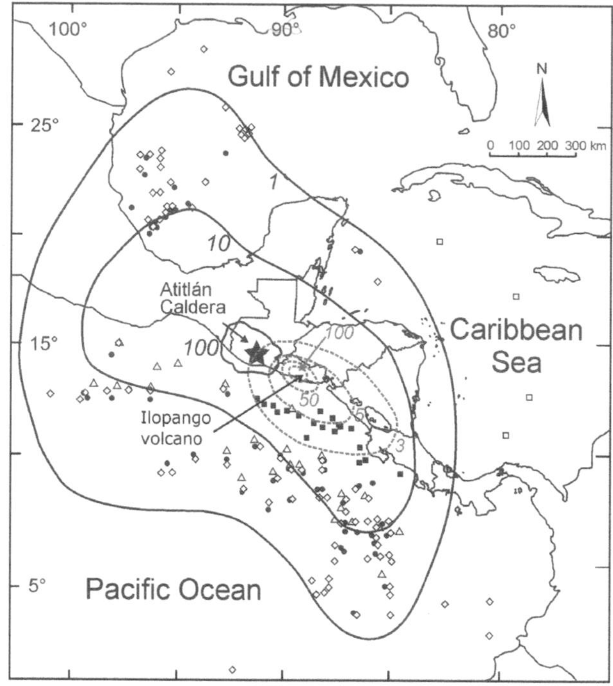

Figure 4.2 Isopachs for the (possibly) sixth century CE Tierra Blanca Joven eruption of Ilopango volcano in El Salvador (dashed lines) and 84,000-year-old Los Chocoyos eruption of Atitlán caldera, Guatemala (solid lines). Courtesy of Steffen Kutterolf.

Eruption magnitude

The minimum volume of a tephra deposit can be straightforwardly calculated by multiplying each isopach thickness by its corresponding area, and summing the products for all the isopachs. Unfortunately, the result will likely significantly underestimate the original volume because it will not account for far-flung fine ash (particles with a diameter less than about 0.1 millimetres) that was quickly eroded or just too fine to be recognised by eye and thus not captured in the field survey. In silicic eruptions, this component can easily represent half the total mass of tephra erupted. To gauge further the significance of this problem, consider that an ash layer only as thick as a human hair but spread over Asia would represent up to several cubic kilometres of rock!

One approach to calculating the missing volume is to fit a mathematical expression to the observations. The most common assumption is that the thickness of tephra fallout decays exponentially with distance from the source vent, with an expression of the following form [38]:

$$T = T_{max}\exp(-k\sqrt{A}) \quad [4.1]$$

where A is the area enclosed by the isopach contour representing the thickness T, T_{max} is the maximum deposit thickness, and k is a constant. Graphing the field data reveals the value of k and T_{max}, from which the volume, V, of the deposit can be determined from the expression:

$$V = 13.08T_{max}/k^2 \quad [4.2]$$

Given an estimate of the density of the tephra, the eruption magnitude (M_e) is then straightforwardly calculated. This mathematical approach enables calculation of deposit volume to an arbitrary thickness (one thousandth of a millimetre, for example), thereby modelling the volume of the entire deposit (including the portion that was not actually recorded in the field and quite possibly not even preserved in the rock record).

An alternative method for estimating deposit volume is to collect pumice fragments at various locations in the field and calculate the ratio of the mass of crystals to that of glass present in each, and then compare these values with the same ratio of crystals to glass for the ash found in association with each sample. If the ash is assumed to have been derived from fine fragmentation of the pumice then the two ratios should be equal. However, if fine glass is preferentially carried to the most distal parts of the deposit then there should be an enrichment of crystals with respect to glass closer to the volcano. The missing volume in the distal deposit can then be approximated by simple mass balance calculations [39].

It is important to note that many assumptions apply to both methods for calculating tephra volumes and the uncertainties they imply are difficult to quantify. The 'crystal concentration' method requires rather laborious sample preparation and has not been widely employed. Where the two techniques have been applied to the same deposit, divergent results were obtained. For example, for the second-century CE Taupo (New Zealand) eruption, the exponential decay method suggests a much lower volume.

Eruption intensity

Grain-size parameters can also be plotted on a map and contoured. One useful approach to modelling Plinian eruptions links field observations of a readily measured parameter - maximum particle size - to the physics of dispersal and sedimentation of pyroclasts of given size and density. The attraction of the technique lies in the relatively straightforward field-data collection involved and the surprisingly good results that can be obtained. Maximum pyroclast size can be observed in several ways according to taste. Some volcanologists arrive at an outcrop and extract the five largest clasts; others select only the three largest. Some record average dimensions to account for irregular shapes; others simply record the longest axis. Some will worry about the size of the search area (and spend ages trying to find the really large pyroclast they suspect to be hiding in several tonnes of tephra); others won't. Finally, rigorous field volcanologists with time on their hands will record clast dimensions as a function of stratigraphic position - that is, the height relative to the base of the deposit - since this can reveal changes in eruption vigour. An increasing grain size going up through a fallout deposit points to an eruption that intensified through time; decreasing grain size would indicate a waning eruption.

As with deposit thickness, pyroclast size tends to decrease exponentially away from the vent. And tephra-fall deposits typically become better sorted with distance. The exact patterns can be modelled to determine eruption-column height, wind velocity and magma discharge rate [40]. Such methods have yielded good agreement with independent observations of column height and wind speed for eruptions such as that of Mt St Helens in 1980. They can thus be applied with some confidence to ancient eruptions. For example, grain-size data suggest that the aforementioned outburst of Taupo was the most intense Plinian eruption yet identified in the geological record. Its eruption cloud apparently soared to a maximum height above the vent of more than 50 kilometres above sea level, driven by a phenomenal discharge of one million tonnes of magma per second (an intensity of 12).

The development of such models that relate characteristics of tephra deposits to the eruption conditions that produced them has been of tremendous importance in volcanology. Not only do the techniques help in evaluating the wider impacts of historic and ancient eruptions, they also contribute to volcanic hazard assessment. Building scenarios for the future activity of a given volcano relies strongly on understanding the vigour, magnitude, duration, style and frequency of its past eruptions. While some of these may have been observed and documented by eyewitnesses, many more are likely to be only written in the volcano's own log book - its stratigraphic record.

4.1.3 Dating eruptions

Timing is everything. When trying to establish cause and effect - an essential theme of this book - reliable estimates of eruption dates are crucial if we are to place volcanic events in geological, palaeoenvironmental, archaeological or historical context. Dates are also essential to establish the rates at which eruptions of different sizes occur - the frequency–magnitude pattern for individual volcanoes (Section 1.3.2) underpins assessments of long-term hazard probabilities.

The principal approach to dating a past eruption is to estimate the age of its products - solid or volatile. This can be achieved by finding the age of the products themselves or of the materials that they are encased in. In some cases, an eruption is dated by calculating the age of its effects, such as a climatic anomaly recorded in tree-ring growth (Section 4.3). In theory, tephra can be matched with the source volcano (and particular eruption) by fingerprinting its chemical composition. In the case of ice-core and tree-ring records, it can require much more detective work to identify the volcano responsible for a given acid layer or deficit in tree-ring width. Since chronometry is so important in resolving issues of cause and effect, this section reviews three of the common techniques used for dating volcanic rocks (Table 4.1). Before continuing though, it is helpful to review very briefly some nomenclature commonly used for specifying different time periods in the past. The 'Holocene' refers to the last 11,500 years since the end of the last glacial period; the final burst of bitter climate, the 'Last Glacial Maximum', peaked approximately 23,000 years ago. The 'Quaternary' era includes the Holocene but extends back to around 2.58 million years ago. The earlier part of the Quaternary is known as the 'Pleistocene'.

Potassium and argon

The decay of radioactive nuclides provides one of the most important clocks used for dating both geological and archaeological materials. One common chronometer is provided by the potassium–argon (K–Ar) system. There are three naturally occurring isotopes of potassium: ^{39}K, ^{40}K and ^{41}K (where the x in ^{x}K is the mass number, the sum of protons and neutrons in the nucleus of the potassium atom; all these forms have nineteen protons and hence different numbers of neutrons). The isotope ^{40}K decays radioactively in two ways: by capturing an electron and converting a proton to a neutron (electron capture decay) to become an argon isotope, ^{40}Ar; and by loss of an electron and

Table 4.1 *Principal methods for dating volcanic materials.*

Method	Principle	Suitable materials	Age range (years)
Potassium–argon (K–Ar)	Radioactive decay	Lava, tephra, K-rich minerals	1000–5,000,000
Argon–argon (Ar–Ar)	Radioactive decay	Lava, tephra, K-rich minerals	2000–5,000,000
Uranium series	Radioactive decay	Lava, tephra	50,000–500,000
Fission track	Radioactive decay	Volcanic glass, titanite and zircon crystals, obsidian flows, baked contacts and xenoliths	>1000
Thermoluminescence (TL)	Radioactive decay	Feldspar and quartz crystals, volcanic glass, baked soils	500–1,000,000
Radiocarbon dating (^{14}C, accelerator mass spectrometry (AMS))	Radioactive decay	Soils, charcoal	300–60,000
Obsidian hydration	Water absorption	Obsidian flows	<1,000,000
Palaeomagnetism	Changes in Earth's magnetic field	Lavas	2000–4000

conversion of a neutron to a proton (so-called β^- decay) to yield the calcium isotope ^{40}Ca. The half-lives of these two decay processes are 1.2 and 1.4 billion years, respectively. (Argon is one of the most abundant gases in the Earth's atmosphere, and nearly all of it derives from the decay of ^{40}K in rocks.) A rock containing ^{40}K will, through time, accumulate more argon as the ^{40}K decays. Knowing the half-life for this decay, the ratio of ^{40}K to ^{40}Ar can therefore be used to date the sample. The accuracy of this date will decrease if any ^{40}Ar existed in the rock when it was formed, or if argon has leaked out of the sample.

Lavas and tephra are particularly suitable for K–Ar dating because their initially very high temperatures reset the radioactivity clock, while rapid cooling on eruption locks the decay product ^{40}Ar into the crystalline structure of minerals. Potassium–argon dates therefore represent the time of eruption or crystallisation. Because the half-life is so immense and the abundance of argon in the atmosphere is high, estimating ages of young rocks (less than 100,000 years old) requires accurate determination of the minute quantities of radioactively produced argon present and careful correction for any contamination from the argon in air.

A popular variation on the potassium–argon technique is argon–argon dating. This is more complicated than conventional K–Ar dating since it requires taking the sample to a nuclear reactor where it can be bathed in energetic 'fast' neutrons. These are captured by ^{39}K in the sample yielding ^{39}Ar, which is not naturally present. Then the proportions of all three argon isotopes are measured in a mass spectrometer. The ratio of ^{40}Ar to ^{39}Ar provides a measure of the sample age, since ^{39}Ar is proportional to the potassium content of the sample. Argon–argon dates tend to be much more precise than K–Ar ages and have revolutionised understanding of the largest volcanic eruptions on Earth, the flood basalts (Section 6.2).

Radiocarbon

Radiocarbon dating is one of the most widely known techniques for estimating the ages of ancient materials. There are three principal isotopes of carbon: ^{12}C, ^{13}C and ^{14}C. Carbon-12 and carbon-13 are both stable, whereas carbon-14 decays with a half-life of 5730 years. The source of the radioactive isotope is neutron capture by atmospheric nitrogen at altitudes of 12–15 kilometres (nitrogen is the most abundant component of air):

$$^{14}N + n \Rightarrow {}^{14}C + p \qquad [4.3]$$

where n indicates a neutron and p a proton. Impacts of cosmic rays (which are mostly protons) with atmospheric oxygen provides the source of the neutrons. The ^{14}C produced is rapidly oxidised to $^{14}CO_2$, which is taken up by plants and animals via metabolism. The natural abundance of ^{14}C in living tissue is extremely low – only 1 in a trillion carbon atoms will be radioactive – but their decay is matched by continuous replenishment from the air and water. Once the plant or animal dies, however, equilibrium is disturbed, and the ^{14}C undergoes β^- decay back to nitrogen:

$$^{14}C \Rightarrow {}^{14}N + \beta^- \qquad [4.4]$$

To estimate the age of an organic sample requires measurement of either the remaining concentration of ^{14}C, or the residual radioactive emission of β^- particles. From the calculated half-life and measurements of modern activity levels of radiocarbon, the ^{14}C age of the sample can be derived. By convention, dates are referred to years before 1950 CE (and written as years BP for 'before present').

The particular relevance of radiocarbon dating for volcanology is the availability of organic materials engulfed or covered by tephra or lava. These include organic materials in palaeosols (ancient soil layers) charred by a lava flow or pyroclastic deposit, as well as vegetation consumed by hot pyroclastic currents. Many a field volcanologist has spent hours sifting through crumbly deposits of pumice trying to find the fragment of charcoal that will date the responsible eruption. Such determinations are generally less susceptible to the 'old wood' problem encountered in archaeology in which the estimated age of wooden artifacts may considerably predate the time of its human use.

If these radiocarbon ages are to represent true calendrical ages (CE or BCE) then ^{14}C production would have to have been constant through time. In the 1950s and 1960s, however, discrepancies emerged between dates assigned by radiocarbon and independent methods, suggesting that this was not the case, and that the radiocarbon method would, therefore, require calibration for it to yield calendrical dates. The obvious way to achieve this was to determine the radiocarbon ages of precisely dated tree rings (Section 4.3) and derive a calibration curve. Further back in time, calibration has been based on analysis of many hundreds of radiocarbon dates for remarkable annually layered sequences of lake sediment. One of the most exceptional sequences of such 'varves' is found below lake Suigetsu in Japan. These deposits also contain a rich tephra record that includes fallout from one of Japan's largest Quaternary eruptions – the Akahoya ash from a caldera-forming eruption of Kikai volcano that took place towards the end of the sixth millennium BCE.

Fission tracks

The radiometric methods just described require the measurement of concentrations of daughter and/or parent nuclide concentrations. Another approach to exploiting radioactive decay as a clock is dosimetry, in which the radiation itself or its effects are recorded. Some minerals and volcanic glasses contain small amounts of the uranium isotope ^{238}U, which undergoes spontaneous nuclear fission with a half-life of 4.5 billion years. The ejected particles are highly energetic and leave a trail of radiation damage. These fission tracks can be preserved in certain minerals and glass, and accumulate with age. Measurement of the total number of tracks per unit volume along with the concentration of ^{238}U can be used to estimate the age of the material since it was formed or last re-heated (re-heating can fuse pre-existing tracks resetting the radiometric clock). Zircon is the most commonly used mineral since it has a naturally high concentration of uranium. Dating of rocks as young as a few thousand years requires such uranium-rich materials, otherwise the long half-life of ^{238}U would result in too few tracks to measure reliably. Many tephra layers and some basalt lavas have been dated by fission-track analyses of zircon crystals. The glassy silicic lava known as obsidian has also been dated by this method.

4.1.4 Tephrochronology

There are certain environments where even very thin layers of ash have been rapidly buried and preserved for aeons. The requirements include fast background sedimentation rates to seal in the ash before it is blown or washed away, and limited disturbance of the layer after burial (for instance by plant roots or burrowing creatures). These conditions can be met in very diverse settings on land and at sea, with the result that tephra can be found in deposits from lakes, peat bogs, caves, floodplains, shallow seas and deep oceans, as well as in windblown sediment (loess), palaeosols and ice cores. Importantly, such sedimentary sequences can also represent a rich source of information on past environmental conditions and human occupation in the form of diagnostic structures, chemical and isotopic composition, pollen, fossils, charcoal, stone and bone tools, ceramics, ornaments and so on.

Once a tephrostratigraphy in a particular region has been established and especially when its component tephra layers have been attributed to dated eruptions, the sequence can provide a very useful chronometer. In Iceland, for instance, where there have been dozens of recorded eruptions in the historic period (the last millennium or so), a hole dug in the ground just about anywhere is likely to reveal a set of dark tephra beds, some thick some thin, separated by brownish palaeosols. The appearance is reminiscent of a bar code and, with experience, it is possible to recognise the pattern and bracket the ages of all the palaeosols according to the known eruption dates. Indeed, it was Iceland where the science of 'tephrochronology' was first established by the pioneering volcanologist, Sigurður Þórarinsson. By deciphering the bar code, he was able to study climate change in Iceland and its effects on agriculture and ecology since the time of settlement. The ash horizons from larger eruptions can sometimes be traced from land into the ocean basins and, thus, tephrochronology also provides one of the most effective tools for linking terrestrial and marine (and sometimes ice core) records and events.

As we have seen, with distance from the source, tephra-fall deposits tend to thin out, and are increasingly composed of very small particles – including grains less than one hundredth of a millimetre across. Once incorporated into the sedimentary record – for instance in a peat bog, a lake bed or the seafloor – it may well be impossible to see such a tephra layer because of the abundance of other sediment. However, with painstaking work to separate out the volcanic glass shards, and powerful microscopic and analytical techniques to classify their morphologies and composition, it is possible to identify otherwise hidden volcanic deposits. These are referred to as cryptotephra or microtephra [41]. One laboratory technique, in particular, has revolutionised this kind of work – laser ablation-inductively coupled plasma-mass spectrometry (LA-ICP-MS for short). It is not only extremely sensitive (it can measure element abundances as low as one part per trillion) but it can measure simultaneously pretty much everything in the periodic table from lithium to uranium. Importantly, in the context of cryptotephra, it can do all this with minute samples. By comparing the major, minor and trace element abundances of one deposit with others, it is possible to make firm correlations between them [42,43], and potentially to identify the volcano and specific eruption responsible (if its deposits near the volcano are well characterised) (Figure 4.3).

One exceptional site for such work is the Lago Grande di Monticchio in southern Italy [44] (Figure 4.4). It is in fact a maar – a lake in an ancient volcanic crater – but because it has only minor inflow and no outflow its varved sediments have piled up sequentially with very little disturbance for about 130,000 years (the deposit is about 100 metres thick). The core can be dated with a high degree of accuracy by counting the annual layering of alternating organic-rich and mineral-rich horizons. Thanks to the lake's strategic location in range of volcanoes of the Roman and Campanian provinces as well as the Aeolian Islands and Pantelleria, it has trapped tephra from more than 350 discrete eruptions. The thicknesses of the layers vary from less than a tenth of a millimetre up to two metres. Although the record of volcanism preserved in the core is immensely rich, it is incomplete. It does not include, for instance, fallout from the 79 CE eruption of Vesuvius that destroyed Pompeii and Herculaneum. Clearly, the

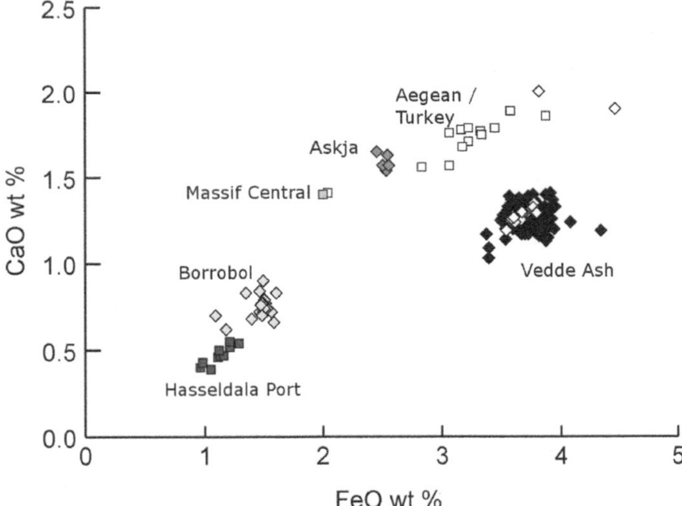

Figure 4.3 Correlation of a far-flung tephra layer extracted from two lake sediment cores (from Rotmeer, in southern Germany, and Soppensee, on the Swiss Plateau). The iron oxide (FeO) versus calcium oxide (CaO) composition of this tephra (represented by the open diamond symbols) is compared with available data for intermediate and silicic tephra horizons from known European and Icelandic sources [42]. A perfect match is found for the Vedde Ash (black diamonds) produced by a large Icelandic eruption about 12,000 years ago. The distance between Iceland and the sites in Germany and Switzerland is around 2500 kilometres demonstrating the wide dispersal of very fine ash. Courtesy of Christine Lane.

prevailing atmospheric circulation at the time of an eruption plays a key role in the transport of volcanic clouds and thus where ash will end up falling out.

Cryptotephra finds are revealing much wider geographic ranges of fallout for a number of ancient eruptions than were previously mapped from identification of visible tephra layers. They are also being increasingly sought for in archaeological and palaeoenvironmental sequences, since they provide valuable time markers to connect records from one site to the next. The construction of such a time lattice between sites, potentially separated thousands of kilometres apart, is enabling new investigations of the impacts of abrupt environmental change on human evolution and behaviour.

Naturally there are some difficulties with the technique. An important assumption is that the cryptic grains were incorporated in

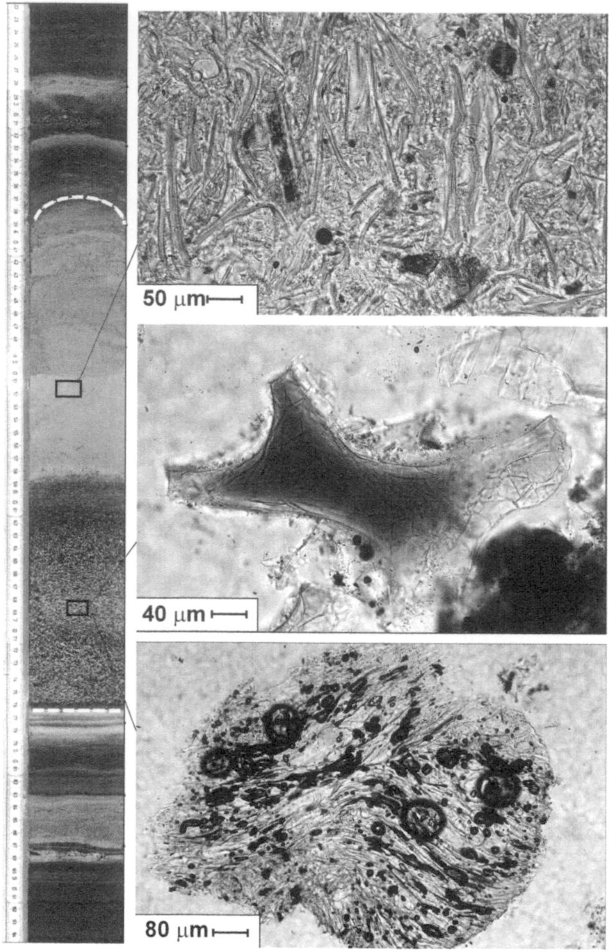

Figure 4.4 Fallout from the Campanian Ignimbrite eruption (between the white dashed lines) located at a depth of 25 metres in the core from Lago Grande di Monticchio [44]. Deposits above and below the dashed lines include other tephra fallout and reworked tephra. Insets show photomicrographs of the tephra from different layers. This section of core is approximately 70 centimetres in length. Courtesy of Sabine Wulf.

their host sediment soon after eruption. This might not be the case since ash can be redistributed and reworked by air and water currents, and even plant and animal disturbances, in more complex ways for long periods of time (Section 2.9). Another problem is that geochemical 'fingerprints' are not as unique as the expression might suggest. Some volcanoes produce tephra of very similar composition in consecutive eruptions spanning many thousands of years. This can make it very difficult to discriminate the particular event that yielded a given tephra layer (and thus identify its age) or to say which of several similar but discrete layers in an ice core correlates with a single compatible layer found in marine sediment cores. A related issue is that some eruptions tap chemically zoned magma chambers and thus produce tephra with a chemical composition spectrum rather than a simple fingerprint. Also, volcanic glass (especially of basaltic composition) can alter after deposition (especially in more acidic or alkaline environments), shifting its chemical composition. Lastly, the erosion or burial of deposits that takes place on and near a volcano can obliterate or obscure long periods of its eruptive history. Thus, it may not always be possible to match a far-flung ash horizon to a particular lava or tephra outcrop near the vent.

4.1.5 Volatile yields

Chapter 3 examined in detail the impact of the gas emissions, principally of sulphur, during volcanic eruptions on the atmosphere and climate. Much of what we have learned is a result of studies relating to the 1991 eruption of Pinatubo. If we want to evaluate the potential hemispheric- to global-scale impacts of a past eruption it is thus vital to estimate its contemporaneous yield of sulphur and other gases to the atmosphere. For an eruption that occurred thousands, even tens of millions, of years ago this is quite a challenge. No trace of the emitted gases will be left to measure in the atmosphere. So, what to do? One solution is to be found in ice cores (Section 4.2) but first, we examine clues remaining in the erupted rocks.

During the long period before a magma chamber fails and feeds an eruption, it has been sitting in the crust, probably cooling and evolving through fractional crystallisation (Section 1.2). Some of the crystals that form may sink to the base of the chamber or clump against its walls but others will remain suspended in the silicate melt. As they grow, they often trap minute blobs of the melt – a little like insects trapped in amber. A typical microscope section of a volcanic rock will reveal a variety of crystals, some up to a few millimetres across, surrounded by a groundmass composed of much smaller crystals, and glass that cooled too rapidly to crystallise after eruption. The larger crystals, such as olivine, which form early in the evolution of a magma body, are often peppered with so-called melt inclusions. These have, of course, long since cooled down, and are generally also glassy. Their importance is that, in effect, they represent fossils of the pre-eruptive magma. Specifically, they have the potential to preserve the complements of volatiles that the magma contained prior to eruption. Using laboratory techniques, including LA-ICP-MS and infrared spectroscopy, it is possible to measure the abundances of sulphur, chlorine and fluorine, as well as water and carbon dioxide, in the minute inclusions.

The next step is to use the same techniques to measure the volatile contents of the silicate glass that forms a matrix surrounding the crystals in tephra samples (and referred to as matrix glass), or forming the rinds of lava flows that chilled rapidly on eruption. In theory, since this was the melt from which gas separated to drive the eruption, there should only be small quantities of volatiles remaining. This is typically found to be the case, indicating how efficiently magmas can lose their dissolved gases during eruption. Focusing on sulphur - the key element for volcanic forcing of climate - we now have two quantities: the pre-eruptive and post-eruptive concentrations. These can be reported as mass fractions: we shall refer to them as S_{initial} and S_{final}, respectively. For example, if S_{initial} is 0.1% (i.e. one thousandth of the mass of the pre-eruptive melt consisted of sulphur) and S_{final} is 0.025%, then, for every tonne of melt erupted, 0.075% of a tonne, i.e. 750 grammes, of sulphur was released into the atmosphere.

With this information, all that is needed to estimate the total quantity of sulphur expelled, M_S, is the magnitude of the eruption in question - for instance, the total mass of magma erupted, m (in kilogrammes), obtained from isopach maps (Section 4.1.2):

$$M_S = (S_{\text{initial}} - S_{\text{final}}) \times m/100 \qquad [4.5]$$

The '100' simply scales the result appropriately if the S_{initial} and S_{final} values are given in per cent (a correction should also be applied to account for the presence of crystals). These so-called 'petrological estimates' of eruption yields of sulphur and, to a lesser extent, halogens, have received much attention, partly because they offer a means to assess the possible climatic impacts of historic and ancient eruptions [45].

The search for suitable host crystals requires careful examination of microscopic sections of volcanic rocks, and even more careful analysis of the inclusions themselves, which can measure only a few thousandths of a millimetre across. While the method appears to work well for some systems, it is not universally applicable. For example, the crystals may not always be leak-proof containers for their high-pressure melt samples, and inclusions can interact with their host minerals. Most problematic, however, is the possibility of volatile exsolution prior to crystal growth and entrapment of the melt, leading to the inclusion recording a lower volatile content than that of the original melt. This very likely explains why the petrological technique singularly fails to explain the sulphur release during Mt Pinatubo's eruption in 1991 (Section 3.1): the sulphur contents in the matrix glass and melt inclusions are more or less the same, which would suggest a sulphur-free eruption plume. Instead, the eruption is known to have released about 17 megatonnes of sulphur dioxide (equivalent to 8.5 megatonnes of sulphur).

Several geologists have explored the reasons for early, pre-eruptive gas separation in magmas, and some have advanced modelling approaches to estimate the volatile contents of pre-eruptive bubbles [46]. But another problem with petrological estimates of volatile yields is that they depend on knowledge of the eruption magnitude, which is often only poorly constrained from the rock record (Section 4.1.2). This is especially true in cases where tephra dispersal is very widespread, perhaps largely at sea, and where substantial burial, erosion or re-deposition limit efforts to identify original thicknesses of sediment in the field.

4.2 ICE CORES

A further invaluable archive of past atmospheric and climatic conditions on the Earth is kept within the frozen wastes of the polar regions and some lower-latitude ice caps and glaciers (e.g. in the central Andes, the Himalayas and Tibet). Needless to say, the technical challenges of extracting cores from some of the most inhospitable regions of the Earth are formidable. However, the rewards include arguably the most detailed and fascinating palaeoenvironmental records available for the last 800,000 years or so [47] (Figure 7.2). In addition to abundant information on temperatures, precipitation, and air chemistry and circulation, the ice cores contain a unique record of volcanism resulting from the fallout of minute sulphurous particles formed in the atmosphere after eruptions (Section 3.1).

Ice sheets grow by seasonal and storm accumulation of snow. In very cold regions where there is no melting, each new snowfall buries the previous layer, trapping air and any materials that were deposited on the surface. These might include fallout from volcanic clouds. In effect, glacial ice represents the accumulation of atmospheric sediment. Over decades, as the layers are buried deeper and deeper, the firn (old snow) transforms into increasingly compacted ice: what starts out as a layer getting on for a metre in thickness ends up being compressed to a few centimetres. Despite this densification process, the ice sheets on Greenland and Antarctica have reached thicknesses of about three kilometres. Ice cores provide a particularly good record of past atmospheres because there is no disturbance of annual layers due to the action of plants or animals (as is the case for many other kinds of terrestrial and marine deposits).

The trick to finding a good site for ice coring is to identify the ice divide where ice accumulates rather than flows towards the margins of the ice sheet. A key then is to elaborate a depth–age timescale for the core that is extracted. This can be achieved by counting the annual layers (picked up chemically, isotopically, or by oscillations in light reflectance and electrical conductivity). Many measurements are now made 'on-line' as soon as the core is removed from the borehole (Figure 4.5). The onsite laboratory for the NEEM (North Greenland Eemian) project in Greenland, which is dug into the ice, is better equipped than most university facilities – its only downside being that the 'indoor' temperature is minus 20 °C!

Once the timescale is established, serious work can begin. Ice cores contain a deposition signal (representing the fallout of atmospheric gases and particles to the surface) and a post-depositional signal (compaction, chemical diffusion, wind deflation and re-deposition). Oxygen and hydrogen isotopes and the calcium ion record are essential indicators of climatic variability, while peaks in sulphate ions (SO_4^{2-}) and in electrical conductivity of the ice indicate volcanic fallout. Claus Hammer from the Glaciology Group at the University of Copenhagen pioneered work on the volcanic acid layers preserved in ice cores [48]. Subsequently, further cores have been drilled in the Arctic and Antarctic and studied by international research groups. Over the last decade, the time resolution of the records has improved greatly thanks to a technique known as continuous flow analysis in which a square-sided rod of ice is placed end-on atop a melting stage. The melt forming at the stage, and the trapped air bubbles that are released, are continuously extracted and fed to a panoply of spectrometers for compositional and isotopic measurements.

One initial challenge in piecing together the glaciochemical record of volcanism is that eruptions are not the only source of sulphate deposition to the ice. There are other sources, notably sea salt. Fortunately, volcanic fallout (sulphuric acid) is readily distinguished from the other sources which correlate with sodium or calcium ion

Figure 4.5 An electrical conductivity measurement (ECM) being made on an ice core in the field at North GRIP (Greenland) by Anders Svensson. The ECM records acidity or H^+ concentration in the ice, and this is by far the fastest way to detect volcanic sulphuric acid fallout. Between the electrodes is a 1250 V DC potential. The conductivity versus depth profile is obtained by pulling the electrodes along the core. Photograph by Dorthe Dahl-Jensen. Inset (top left) shows a visible ash layer from an eruption of Katla (Iceland) about 29,000 years ago in a freshly drilled North GRIP ice core. Photograph by Sigfús Johnsen. Both images courtesy of Jørgen Peder Steffensen.

levels. Following correction for sea salt, the glaciochemical timeline bristles with anomalies, the majority of which reflect volcanism (Figure 4.6). A further caveat to the approach is that although the dating of the ice core by counting of seasonal layers is fairly robust, it is not fail-safe. The greater the depth from which the core is retrieved, the more likely it is to have suffered deformation. This could remove slices of time, or even double them up. Ice-core dates invariably come, therefore, with an uncertainty range attached. Nevertheless, while absolute dates may have uncertainties of hundreds of years for cores representing the last glacial period, the accuracy of relative ages between events

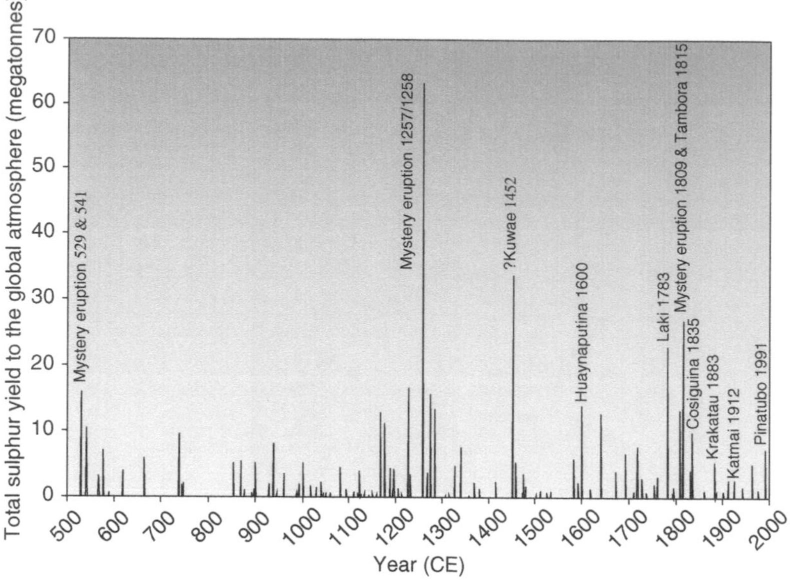

Figure 4.6 Volcanic sulphur emissions to the stratosphere from 501 to 2000 CE calculated from the spatial pattern of sulphate fallout found in ice cores from Greenland and Antarctica [56]. The original data on total global aerosol loadings have been converted to equivalent sulphur release assuming a mix of 75% sulphuric acid and 25% water by mass. Data from http://climate.envsci.rutgers.edu/IVI2.

can be as little as a few years. This is very important in discerning time lags and causation of observed changes [49].

One of the people closely associated with past work on the glaciochemical record of volcanism is Greg Zielinski who, with his colleagues, analysed the Greenland Ice Sheet Project 2 (GISP2) core to identify historic and prehistoric eruptions. Their first efforts took the record back to 7000 BCE with the core measured at biennial resolution [50]. The error of the depth–age scale was estimated to be ±180 years at 7000 BCE swamping the ±2 year sampling error for individual eruptions, and the time lag between an eruption and its fallout reaching Greenland. Generally, particles from high-northern-latitude eruptions arrive quickly, within a few months, on the Arctic ice sheets, while acid fallout from tropical eruptions peaks within 1–2 years depending on prevailing circulation of the upper atmosphere. Fallout from eruptions south of the tropics is unlikely to make it to the Arctic (and similarly, fallout from boreal eruptions is unlikely to reach the Antarctic).

Having identified volcanic signals in the GISP2 core, Zielinski and colleagues compared their eruption time series of events with the known record of volcanism for the past millennium and found that 57 of the 69 sulphate anomalies could be attributed to known eruptions. Of the 57 eruptions, 45 were of equatorial or mid-latitude northern-hemisphere volcanoes, and the remainder were Icelandic. These correlations need to be viewed somewhat cautiously given the 2% error in the depth–age timescale, and there are some spurious selections in the original work. Furthermore, they had overlooked the fact that many of the published eruption ages were uncalibrated radiocarbon ages (Section 4.1.3). By calibrating the dates, a better correlation between known eruptions and ice-core volcanic markers emerged. Nevertheless, the record of volcanism in ice cores has proved extremely important for our understanding of rates and climatic impacts of volcanism. The GISP2 team subsequently pushed the ice core record of volcanism back to 110,000 years ago [51]. More recently, sulphate records from the European Project for Ice Coring in Antarctica (EPICA) ice core (drilled to a depth of over three kilometres) have provided a usable sulphate record back to around 430,000 years ago [52].

One of the most striking revelations of the ice-core record is just how many very large eruptions have occurred, even in the comparatively recent past, for which we have no individual volcano to point the finger at. For instance, just over two hundred years ago, in 1809 (Figure 4.7), an eruption around the size of Krakatau's infamous explosions in 1883 occurred but we have no idea where, except that it was very likely in the tropics given that its sulphate fallout reached both polar regions [53]. Considering that globalisation was well advanced by the early nineteenth century, it is challenging to think where in the world such an event could have gone (apparently) unrecorded. Further back in the ice-core archive there are hundreds of other sulphate anomalies attributable to volcanism, which are as yet anonymous. Such 'mystery eruptions' are the subject of Chapter 11.

4.2.1 Geochemical fingerprinting

As well as the uncertainty in ice-core dating, an obvious problem with the ice-core record of volcanism is that a sulphate anomaly does not identify the responsible volcano. Ice-core records are, thus, especially useful when ash grains can be found associated with a given sulphate layer (Figure 4.5 (inset)). The glass shards can be filtered out of the core by melting part of it and then chemically analysed. The geochemical

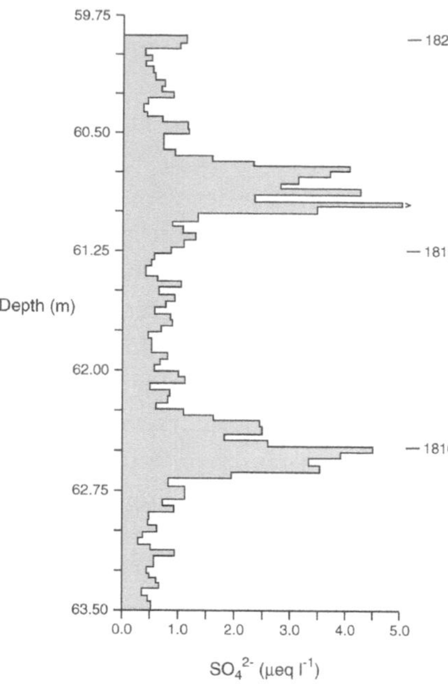

Figure 4.7 Sulphate concentration in an ice core from central Greenland, accurately dated by counting seasonal oscillations in oxygen isotope ratios. The later spike represents fallout from the 1815 eruption of Tambora (Chapter 13). The earlier spike points to a major equatorial eruption in 1809, which injected around 25–30 megatonnes of sulphur into the stratosphere. The protagonist in this case has yet to be identified. Based on data in reference 53.

fingerprint of the particles can be matched against tephra analyses for known eruptions as introduced in Section 4.1.4, though care must be taken to avoid or account for biases that might arise from the measurement techniques, from different eruption phases, or from chemical sorting of ash during transport. In some cases, this provides strong corroboration of the identity of a given layer, for example in the case of the Huaynaputina eruption (Perú) in 1600 [54] whose ash and acid fallout are found in the Greenland ice core. In other cases, for instance the Bronze Age 'Minoan' eruption of Santorini, the geochemical data have proved much more controversial. It is likely that some sulphate layers have been attributed to particular known eruptions simply because the given event was the most renowned for the period concerned. Even where tephra cannot be assigned to a known eruption or even to a particular volcanic field, if they are sufficiently distinctive and widespread they can still provide tie points to link ice-core, marine and terrestrial records. Some of the larger Icelandic eruptions and their associated tephra deposits have proved particularly important in establishing such connection points between Greenland, the north Atlantic and Europe [55].

While the most productive cores for identifying volcanic markers have come from Greenland and Antarctica, including the GISP2 and, more recently, the North GRIP (Greenland) and EPICA (Antarctica) projects, useful results have also emerged from lower-latitude cores, especially from Perú and Bolivia. Some large explosive eruptions in the tropics have sourced fallout identified in cores from both polar regions [56]. However, all ice-core records of volcanism are affected to a greater or lesser extent by the proximity of the core sites to active volcanoes. For instance, the Greenland records are biased towards preserving fallout from eruptions in Iceland, Alaska, the Aleutians and Kamchatka.

4.2.2 Volatile yields

One further important application of the ice-core record is in providing independent estimates of the sulphur output of large eruptions. Considering that volcanic events are being detected in the cores from their sulphate content it makes sense that there might be some correlation between the amount of sulphate present - measured in units of mass per unit surface area of ice - and the quantity injected into the atmosphere by the eruption responsible. The trick is to find a suitable calibration.

The most widely used calibration is based on measurements of the radioactivity in ice cores due to the hundreds of nuclear weapons tests carried out in the atmosphere in the 1950s and 1960s. Because the yield of each bomb is known, a correlation can be established between the quantity of radionuclides released into the atmosphere and the concentration of the corresponding radioactive fallout in the ice (usually measured by counting β^- particle emission). This same correlation can then be applied to the volcanic markers if it can be assumed that the volcanic sulphate was delivered to the ice by comparable atmospheric circulation, starting from similar latitude. In reality, these conditions are difficult to confirm for ancient eruptions since usually the responsible volcano is unidentified and, during certain periods of glacial and interglacial history, atmospheric circulation patterns have been very different from those found today.

Very low snow deposition rates, and post-depositional effects that affect the ice, such as densification, diffusion and wind deflation or re-deposition, as well as other potential sources of sulphur (including marine sources), also pose difficulties in interpreting sulphate layers in the ice cores. Nevertheless, multiple estimates of stratospheric aerosol loading of eruptions such as Tambora 1815 (Chapter 13), whose sulphate marker is found in both Arctic and

Antarctic cores, are reasonably consistent with each other, and with estimates obtained by other methods (e.g. based on the petrological method outlined in Section 4.1.5), providing some confidence in the approach. Recent integration and modelling of data from multiple ice cores in both Greenland and Antarctica (Figure 4.6) suggest the total sulphuric acid aerosol released from Tambora in 1815 was 108 megatonnes, compared with 30 megatonnes for Pinatubo (1991) and 22 megatonnes for Krakatau (1883).

4.3 TREE RINGS

Trees grow by accreting a new layer of woody tissue each year. The pattern of growth and its state of preservation depend on several environmental variables including temperature, length of growing season, precipitation and environmental shocks such as fires or unseasonal frosts. Thus, by studying these patterns, it is possible to reconstruct key aspects of a tree's environmental history. It is even possible with careful inspection of tree rings to resolve the stage in the growing season when particular events occurred. The most useful trees for palaeoclimate research are often those living near the limits of their environmental tolerance, since they tend to be most sensitive to abnormal conditions. Seasonal ring patterns are most pronounced in trees living at higher latitudes but even some tropical species develop annual rings.

While the record of a single tree will tend to reflect localised environmental influences that it experienced during its lifetime, by correlating and combining growth patterns of many trees sampled across a wide area, it is possible to build up a picture of regional- and even continental-scale climate trends. Importantly, and in contrast to ice cores, tree rings can be dated, without error, to the year of growth.

The oldest known tree is a bristlecone pine (*Pinus longaeva*) that goes by the name of Methuselah, growing in the thin air of the White Mountains of east central California. The venerable pine was discovered and dated by the dendrochronologist Edmund Schulman in the 1950s, and turned 4776 in 2010. The trees have crooked trunks worn smooth as driftwood, and exposed roots in which drifts of pine cones accumulate (Figure 4.8). Growing at the upper forest border at more than 3000 metres above sea level, they are particularly sensitive to temperature fluctuations during the growing season. Narrower rings reflect cooler, shorter growing seasons. Trees can also record unusual weather events such as summertime frosts. If two successive nights with temperatures below −5 °C and an intervening day at around

Figure 4.8 Bristlecone pines (*Pinus longaeva*) in the White Mountains of California. Some trees are almost 5000 years old but this youngster is probably only a couple of thousand years or so old. The US Forest Service's Ancient Bristlecone Pine Forest reserve and information centre is generally open from mid May through the end of October, and is well worth visiting.

freezing occur during the growing season, the weaker cells in the accreting tree ring are crushed and dehydrated, leaving a distinctive trace known as a frost ring. Bristlecones can remain upright for millennia after their death, and by correlating ring patterns between living and dead trees it is possible to push the palaeoclimate record back as far as 10,000 years.

In one of the early studies looking for a link with volcanism, bristlecone frost rings were identified, dated and matched across seven localities in the western USA to establish a regional picture [57]. This revealed a close correspondence between the occurrence of frost rings and eruptions known from historical and ice-core sources. It also led the authors to suggest an early date for the Minoan eruption of Santorini, fuelling a long-running archaeological controversy (Section 9.3.3). Many of the frost rings, as well as low-growth rings indicating cold summers, were later found to coincide with the timing of extremely narrow rings in partly fossilised oak trees recovered from peat bogs in Ireland [58], as well as tree-ring anomalies from Scandinavia [59], pointing to hemispheric-scale climate change.

Palaeoclimatologists have exploited this correspondence between tree-ring chronologies collected from many parts of Eurasia and North America to reconstruct regional to global temperature histories. Temperate trees are particularly sensitive to variations in precipitation, which tend to hide any temperature response. Thus, these compositing approaches tend to focus on sensitive conifers such as spruce and larch, growing close to their environmental limits in the sub-Arctic or at high elevation in lower latitudes. Patterns of cells accreted later in the growing season (forming so-called latewood) most closely reflect summer conditions, and are particularly worthwhile to focus on given the known correspondence between sulphur-rich explosive volcanism and summer cooling (Figure 4.9; Section 3.2.2). In order to convert the records to a useful climate

Figure 4.9 Example of a volcanic forcing of climate revealed in tree rings. This is an oak tree sample (Q68A) from Castlecoole, County Fermanagh (Northern Ireland), showing the narrow growth rings for 1816 and 1817 (1816 has the white pointer). Chalk has been rubbed into the surface to highlight the rings. The older rings are at the bottom of the picture; time runs from bottom to top. These two narrow rings appear to be the tree's response to the 1816 'year without a summer' following the eruption of Tambora in Indonesia in 1815 (Chapter 13). Most Irish oaks show this anomaly. Courtesy of Mike Baillie.

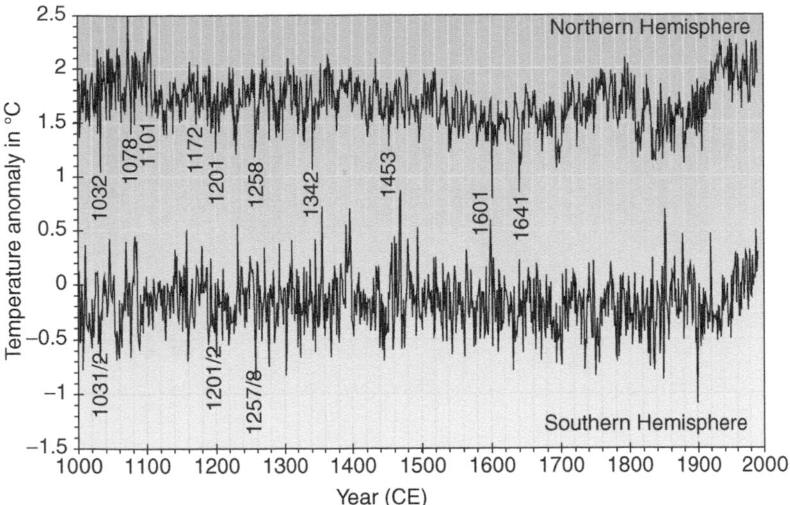

Figure 4.10 Summer temperature anomalies (with respect to the 1961–1990 mean) largely based on tree-ring measurements for the northern and southern hemisphere. Several of the cold summers correspond to climate forcing following known large, sulphur-rich volcanic eruptions, for instance that of Huaynaputina (Perú) in 1600. It is likely that several other anomalies are due to otherwise unidentified eruptions (note the prominent mid-thirteenth century cool summers in both north and south hemisphere records, which correlate with the most striking sulphuric acid spike in the GISP2 ice core; Figure 4.6 and Section 11.2). Data from P. D. Jones, K. R. Briffa, T. P. Barnett & S. F. B. Tett, 1998, IGBP PAGES/ WDC-A for Paleoclimatology Data Contribution Series #1998-039, NOAA/NGDC.

variable, ring patterns for the last century are calibrated against weather-station measurements of average summertime temperature [60]. Figure 4.10 illustrates two of the many available chronologies and reveals several cold summers that immediately follow known large eruptions. This is a remarkable and consistent result considering the natural variability in climate from one year to the next, and it demonstrates the value of tree rings in volcanological research in providing accurate dates and insights into the response of climate to past eruptions. As well as providing such timelines of average temperature, there are now tree-ring data from many parts of the world, enabling maps of temperature anomalies to be reconstructed [61] (Figure 4.11).

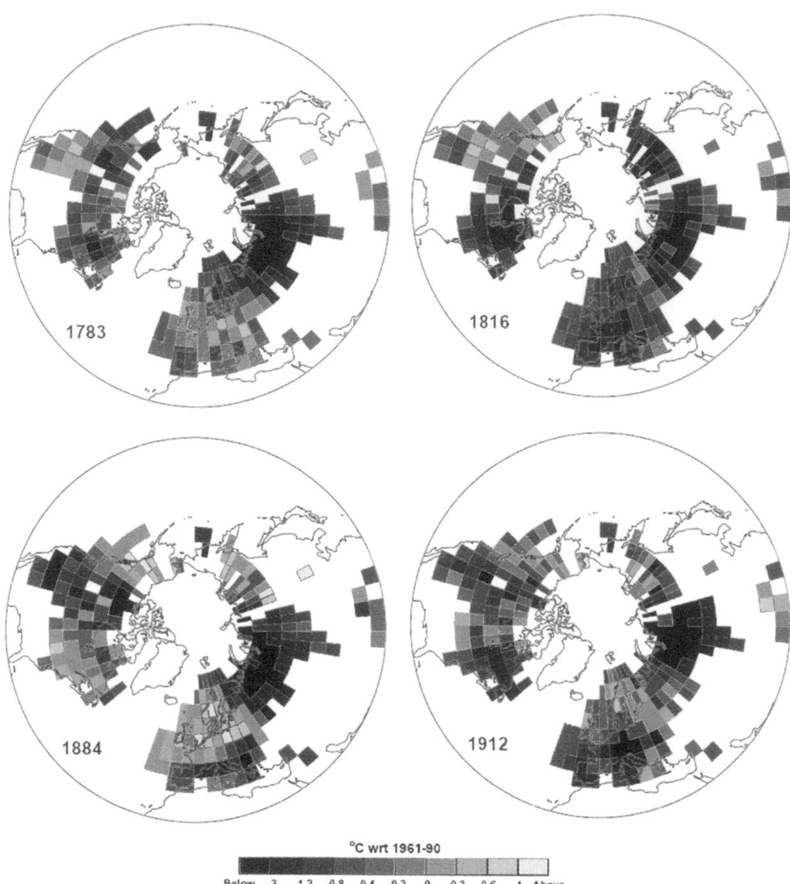

Figure 4.11 Hemisphere temperature anomaly reconstructions for 1783 (during the Laki eruption), 1816 (post-Tambora), 1884 (post-Krakatau) and 1912 (post-Katmai) based on tree-ring data. Courtesy of Keith Briffa (http://www.cru.uea.ac.uk/cru/people/briffa/temmaps).

4.4 SUMMARY

The rock record around a volcano represents a tremendous, albeit imperfect, archive of eruptive history. The forensic examination of tephra sequences can reveal the vent location, size and intensity of eruptions, as well as the height reached by their ash columns in the atmosphere, and also the quantities of gases such as sulphur dioxide released. Various analytical techniques, including radiocarbon dating, allow estimation of the ages of the eruptions. Combined with reconstructions of eruption size, identification of eruption chronologies underpins efforts to characterise the frequency–magnitude curves for individual volcanoes as well as for global volcanism.

A large explosive eruption can cover millions of square kilometres with tephra, which, if it can be identified in the sedimentary record, provides an outstanding marker bed to correlate rock sequences from one place to another. Combined with studies of plant and animal fossils (pollen are especially useful) and detailed sedimentological and geochemical analysis, it is possible to resolve thorny problems in archaeological dating, to investigate human origins and migrations and to understand the causes, magnitudes and extents of past environmental change (which could even be related to the impacts of the same eruptions yielding the tephra).

Large explosive eruptions will generally leave their imprint in the ice sheets of the Arctic and/or Antarctic, and sometimes in glaciers at lower latitudes (though many are now contracting rapidly). This is primarily in the form of sulphate derived from the sulphuric acid particles generated in the middle atmosphere following powerful eruptions. In this case, while there may be unambiguous evidence that an eruption has occurred, it may be very difficult to identify the responsible volcano. However, on occasion, there is also very fine ash incorporated in the sulphate-enriched layer, which can provide geochemical clues to the source. (The acidic and trace-metal fallout from explosive volcanism can also find its way into the drip-water that forms stalactites and stalagmites in caves, providing another potential repository of information on ancient eruptions.)

Lastly, because eruptions can change climate on regional to global scales, the aggregation of climate proxy records obtained from tree rings, corals and ice cores (among other sources) can also pinpoint past occurrences of major volcanism. Again, this will not reveal the identity of an otherwise unknown eruption but it can provide corroborating evidence that a large eruption has taken place, and extend the evidence for the climate impacts of volcanism beyond what we have learnt from the Pinatubo experience.

When considering any given past eruption, we want to be able to answer the following six questions: When did it happen? How large was it? What kind of eruption was it? How much sulphur, chlorine and fluorine were released? What impacts did the eruption have on the environment and climate? What were the immediate and lasting impacts on people? This chapter has introduced the various records that help in answering the first five of these questions. The methods for finding answers to the last question are explored in the next chapter.

5

Relics, myths and chronicles

> . . . many believe that the terrible monster is already dead; but I think that he is just resting after his exertions, and that someday he will surely come out of his hiding place again . . .
>
> Ayta folklore

Since large eruptions are rare, to understand the wider spectrum of volcanic activity requires delving into the past. The preceding chapter outlined how the rock record, combined with assorted proxies for climatic and environmental conditions such as tree rings and ice cores, yields tremendous information on the nature and atmospheric impacts of past eruptions. This chapter extends the theme of 'forensic volcanology' by examining what can be learnt of the complex and long-term interactions and intersections between people, volcanoes and volcanic activity. In this regard, both archaeology and memories and records of eruptions in oral or inscribed forms offer a mine of information since they offer insights into both the physical aspects of past eruptions and social responses to living on volcanoes. The investigation and scope of these sources is introduced here.

Not only can impacts of some of the very largest eruptions be examined - events more extreme than anything seen in the historic period - but it is possible to evaluate how different societies have flourished, coped, declined or collapsed in the face of eruptions of different style and scale. It is possible, too, to take account of additional environmental factors that may have acted to amplify or suppress human consequences of damaging eruptions, and to infer the psychological impacts responsible for catalysing major political, economic and social change. Often, the creative responses to both the resources provided and threats posed by volcanism have led to beneficial and positive developments and change in human society and culture.

Sometimes there appear to be no changes at all. A clearer understanding of the range of inter-relationships between volcanism and human society can ultimately contribute to better preparedness for future disasters, especially those of a scale or impact rarely, if at all, seen in the historic record.

5.1 ARCHAEOLOGICAL PERSPECTIVES

Anyone who has visited Pompeii or Herculaneum will know that volcanic ash is one of the ultimate preservatives (Figure 5.1). Such sites, instantaneously and simultaneously destroyed and frozen in time, have much to tell us. Not only do they convey viscerally, for instance in the tortured postures of the Pompeii and Herculaneum victims, the direct physical effects of volcanic eruptions (Figure 2.7), they also yield tremendous insights into the relationships between society, culture and disaster. One of the most fascinating aspects of the 79 CE Vesuvius eruption, yet one which is seldom discussed in any detail in the countless books on the subject, is that the devastation sent barely a ripple through the Roman world. Emperor Titus may have been moved sufficiently to visit the region (he made a dedication at the inauguration of a new sundial at Sorrento), and he struck some commemorative coins, but there is virtually no further reference to the disaster

Figure 5.1 Street scene at Pompeii (with ever-present Vesuvius) showing stepping stones for crossing the road and the grooves worn into the stone by cart wheels.

in the extant literature of the day. Perhaps it was decided to exercise a collective lapse of memory as a way to cope with the trauma but, in any case, Rome clearly did not fall (not for a good while longer, in any case). The Campanian region lost its prestige as Rome's favourite playground but there was very little dent in its exports of agricultural products. Before long the fate of Pompeii and Herculaneum was just a 'vague memory in folklore' [62].

Since the story of the 79 CE eruption of Vesuvius has been recounted so profusely and elaborately elsewhere, this remarkable human story does not feature prominently in this book (but see the recommended reading for this chapter). Instead, I begin by reviewing an exceptional archaeological site in El Salvador, not because the eruption that entombed it 'shook the world' but because it illustrates the breadth of detail on ancient human activities that can be gathered through meticulous archaeological survey in the context of Pompeii-style preservation. The broader implications for understanding disaster response are then explored in two contrasting archaeological case studies representing very different geographical and cultural contexts.

But before 'digging in', it is worth emphasising the importance of one of the most prized of rocks: obsidian. Of course, being a volcanic product, it is not so surprising that it crops up again and again in the contexts of 'Pompeii' style burial of ancient sites. As a material of immense aesthetic appeal - it comes in a variety of hues from pink-grey through black, sometimes with intricate banding, and always vitreous - it is loved both by volcanologists and archaeologists. Furthermore, obsidian can be geochemically fingerprinted in the same way described for tephra in Section 4.1.4, making it possible to match archaeological obsidian to geological outcrops and thereby map ancient trade routes [63].

5.1.1 El Salvador's 'Pompeii'

Mesoamerica is home to dozens of volcanoes and some of the world's most extraordinary archaeological sites (Figure 5.2). One of the most significant intersections between volcanism and human societies in the region was the very large eruption of Ilopango volcano in El Salvador, possibly in the first half of the sixth century (Section 10.2). But, within a century (the relative dating is poorly constrained), the Zapotitán valley in the centre of El Salvador had recovered demographically, with an estimated population of 40,000–100,000. Around

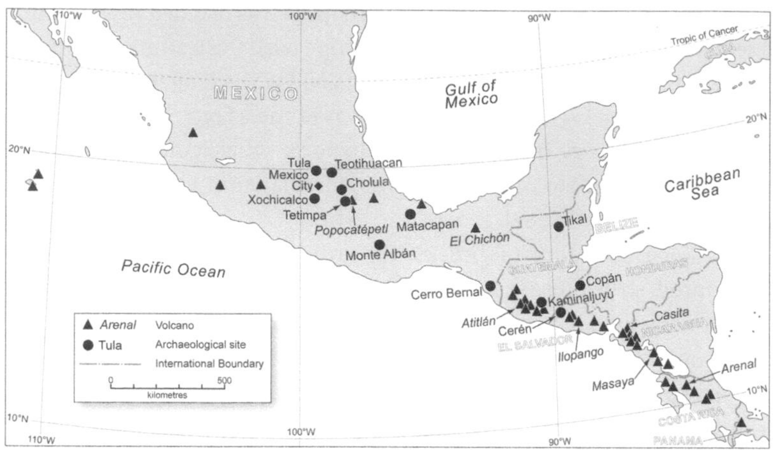

Figure 5.2 Map of Mesoamerica showing major volcanoes and archaeological sites.

600 CE, however, nearby Loma caldera on the northwest flank of San Salvador volcano exploded violently, burying a thriving rural community under as much as seven metres of tephra. The site, named Cerén, has been excavated and researched by a team co-ordinated by Payson Sheets from the University of Colorado. Cerén was just five kilometres away from the Zapotitán valley's largest regional centre and economic hub, the site of San Andres, to which the villagers would have travelled to barter for obsidian tools, settle disputes over land tenure or access, and to attend important religious or ceremonial events.

Cerén was discovered by chance in 1976 and has been excavated and studied now for over 30 years. The deposits that entomb it were laid down as a result of a series of volcanic explosions of Loma caldera. First, as magma came into contact with river water, steam explosions blasted wet pyroclastic currents into the village (Figure 5.3). Once the external water was used up, the eruptive plume turned dry and tephra fallout gradually buried Cerén, with some larger bombs punching through the thatch roofs and igniting fires. Although no victims of the eruption have been found, the event does seem to have caught the inhabitants by surprise because many valuable items were abandoned in the dwellings. This makes the site particularly illuminating of the lifestyles of its prehistoric inhabitants. So rich is the preservation that the thatch roofs still contain the mummified remains of their resident rodent population: six mice in one! Foodstuffs left in serving bowls even show the furrows left by the diners' fingers.

Figure 5.3 One of the buildings at Cerén (El Salvador). The striping on the walls corresponds to the undulating layers of tephra from pyroclastic currents that entombed the village.

The best-known household at Cerén illustrates general aspects of the lifestyle of its former occupants. It consists of four distinct structures - domicile, workshop, storehouse and kitchen - in a plot containing a small potager and cornfield. The buildings were constructed on a platform composed of fired soil that helped to drain rainwater away. Four solid adobe posts, head-height, were sunk at the corners of the platform and a meshwork of sticks affixed between them. The vertical sticks in the meshwork poked up above the adobe walls providing an opening below the palm thatch roof that would have admitted both light and air. A mixture of clay and grass was packed into this framework producing thick walls known locally as *bajareque*. Such walls are better able to resist earthquake shaking because of the reinforcing lattice of sticks; purely earthen walls are more prone to disintegrate during tremors producing large and potentially lethal chunks. Cerén's use of *bajareque* reveals an early instance of anti-seismic construction.

Inside, the family room measured just four square metres and contained two storage jars resting against the back wall and a number of smaller pots. One contained a spindle whorl, a miniature *metate*, cylinders of red paint and pieces of seashell. Next door was the bedroom, identified by rolled-up sleeping mats stowed in the rafters below the roof thatch (suggesting that the eruption occurred by day). The room seems also to have served as a dining area because two pots full of beans and bunches of chilli peppers were found there. As well as stowing bedding away in the roof thatch, the occupants kept their obsidian blades there, probably to help preserve the fine cutting edges and to keep them out of reach from children. Such tools would have been obtained through trading in San Andres and considered valuable.

The storehouse was larger than the family room and contained a wealth of artefacts including pots full of seeds, an incense burner, hammer stones and obsidian tools. A grindstone for grain was found raised half a metre off the floor by thick forked sticks and had been narrowly missed by a filing cabinet-sized lava bomb that had pierced the roof, smashed a large storage jar, and embedded itself in a corner wall. Another victim of the eruption was discovered by the doorway: a duck that had been tethered to the wall.

When they discovered the household kitchen, next to the storehouse, the archaeologists were particularly excited. They had been looking for a kitchen for years having failed to find even a single sooty cooking pot. Unlike the other buildings, its floor was composed of older volcanic ash (from the Tierra Blanca Joven eruption, see Section 10.2), which would have usefully absorbed kitchen spills. Decorated gourds were used for food preparation and presentation, and another grindstone was found on a low shelf below which were several more vessels, baskets, an incense burner and a pile of beans placed on a leaf mat. Obsidian tools were again found inserted into the roof thatch along with another implement made from bone. The cooking pot was removed from the hearth, the fire had been allowed to burn out, and food (*tamales* made from maize) had been consumed but the dishes not washed up. Sheets speculated that the eruption had therefore taken place between dinnertime and bedtime.

A stone walkway led from the domicile to the workshop, which consisted of an un-walled platform with a thatch roof. Its function was identified from the quantities of obsidian debris found outside, suggesting use as an area for re-sharpening stone tools. South of the storehouse there was a small well-tended garden with manioc (cassava) and other plants and, beyond, a maize field with parallel ridges occupied by clusters of young corn plants spaced about a pace apart. The plot was not irrigated, and the maize ears were mature, suggesting that the eruption had occurred during the rainy season, probably August.

The largest structure excavated at Cerén initially baffled the team – footwear abrasion of the approach paths and porch suggested considerable traffic but only a few large pots and bowls were found inside. A long bench in the front room indicating communal use, and the size of the vessels, which pointed to storage and dispensation of large quantities of a liquid, gave the game away, however. The archaeologists had stumbled upon the village alehouse! Its patrons presumably drank a beer made from maize.

Cerén revealed how its inhabitants prepared and preserved foods and how they constructed their buildings. They likely exchanged painted gourds, cacao, cotton thread or garments or perhaps surplus food for the valuable imported items they required (such as obsidian and jade from Guatemala and salt from the Pacific coast). They enjoyed a good standard of living (notwithstanding the co-habiting rodents) and displayed a well-developed aesthetic sense. The buildings were designed with sensitivity to the seismic climate of the region and the residents found uses for the local ash deposits as an absorbent.

But what happened to the villagers themselves? The only possible human victim discovered at Cerén was found on a hillock overlooking the river, in the form of a cavity in the tephra that contained three molars. However, there is a building nearby, and a fragment of thatching very close to the teeth. It is a custom still in traditional El Salvador for someone who sheds a tooth to toss it up on the roof for good luck. Sheets speculated that the residents fled to the river for relief from burns inflicted by the initial pyroclastic clouds. Probably, though, most people escaped while Loma caldera ramped up its precursory activity.

Cerén provides a spectacular window into the lifestyles of prehistoric Classic Mayan rural society. Sheets concluded 'we had no idea that common people lived so well in Southern Mesoamerica some fourteen centuries ago . . . the richness of life . . . puts into stark contrast the desperation of the lives of many Salvadorans today'.

5.1.2 Arenal volcano, Costa Rica

The interpretation of archaeological materials is not without its challenges and controversies but, broadly speaking, ancient cultures are reconstructed and identified by their tools, ceramics, decorative arts, construction and burial practices and so on. Thus, when material cultures vary little over time or space, it suggests a continuity of social organisation. Sheets and his colleagues have dug up numerous other sites in Central America, revealing further the interplay between human ecology and volcanism. The most dramatic case is reserved for Chapter 10 but instructive in its own way is the story that emerged from archaeological investigations around Arenal volcano in Costa Rica. Here, despite repeated explosive eruptions over the past 7000 years, the archaeological record suggests that people achieved a remarkably stable pattern of human occupation [64].

Figure 5.4 Arenal volcano in Costa Rica has been erupting continuously since it burst to life in 1968. It makes a popular tourist attraction as when the cloud clears it is possible to observe the summit pyrotechnics from nearby hot springs.

Arenal assumes the form that any volcano should aspire to – a near-perfect cone that rises just over a kilometre above its base (Figure 5.4). It also behaves in exemplary fashion, having been in almost continuous eruption since re-awakening from centuries of slumber in 1968. Around 20 violent explosive eruptions have been recognised in the rock record stretching back to the mid-fifth millennium BCE. Though of modest size – around M_e 4 – these events would certainly have disrupted and displaced the local populations.

Palaeoindian occupation in the Arenal neighbourhood dates back to 10,000–7000 BCE and is recognised from discoveries of Clovis points – magnificent, flaked blades that are found throughout North America. Population gradually increased and, by the fourth millennium BCE, maize was being grown. By 2000 BCE people were living in small villages composed of clusters of circular huts with thatch roofs. They ground corn using a stone *metate* and *mano* (a kind of mortar and pestle), added squash and beans to their crops, and ate from decorated ceramic bowls. Apart from some minor embellishments in the decoration of the pottery around 600 CE, and relocation of cemeteries further away from the villages, this is how people lived in the area for 3500 years despite recurrent volcanic devastation. Although people would certainly have suffered as a result of the eruptions, and probably abandoned the worst-affected area for decades until it recovered naturally, the society as a whole did not change in an abrupt manner.

Sheets suggests that this stability reflects various cultural factors that made Arenal's settlers especially resilient. Firstly, although they were sedentary insofar as they were settled in villages and practised mixed agriculture, they were self-reliant in terms of food security and, based on analysis of carbon isotopic composition in skeletal remains, just 12% of their diet consisted of maize – a larger part of their diet consisted of wild foods (unlike the residents of Cerén whose diet consisted almost exclusively of domesticated crops including maize and manioc). As well as being able to source locally all their nutritional needs, they could also obtain raw materials for building, tools, pottery and pigment from their environs, without having to rely on external trade.

Also in their favour were a fully egalitarian social organisation, independent economic and political structures and rather low population density in the region. Thus, when survivors became refugees they did not end up competing unduly with their new neighbours. Nor did they find themselves resettling amidst an unfamiliar culture that might have led to conflict. Once the tephra deposits weathered and dispersed, and soils became re-established, the abandoned area was resettled by people of a very similar culture. An intriguing possibility is that the very fact that complex societies did not emerge in the Arenal area was, in itself, an adaptation to the volcanism, which was frequent enough to be appreciated through oral traditions (Section 5.2). However, it is more likely that the region was simply less attractive for settlement due to rather marginal terrain, soils, climate and ecology, reflected in the low population density of just five people per square kilometre (compared with around 200 per square kilometre in the region of Cerén).

5.1.3 Papua New Guinea

Papua New Guinea is home to dozens of volcanoes, among them Blup Blup, Bam, Bagabag and Billy Mitchell. Someday, I want to submit a research grant proposal to conduct fieldwork on all four just to see if it can get past the reviewers. An eruption of Long Island at the western end of the Bismarck volcanic arc around 400 years ago is one of the largest known in history and it is widely recorded in oral traditions. Among the more recent eruptions of note is that of Rabaul volcano in 1994, at the other end of the arc on New Britain Island, which resulted in considerable damage to the nearby town. Most of its inhabitants had evacuated spontaneously ahead of the volcanic

Figure 5.5 Part of a decorated Lapita pot. Photograph courtesy of Stuart Bedford.

paroxysm, likely sparing thousands of lives and providing a contemporary example of the threat that inhabitants of the Bismarck Archipelago have faced for millennia. The region is also key to understanding human migrations across the western Pacific. While humans have occupied New Guinea for at least 40,000 years, it was only a little over 3000 years ago that they spread beyond the Solomon Islands in what would be a sequence of rapid colonisation of all the remote islands as far as Samoa, Tonga and New Caledonia. The ancestral home of these Lapita people (named after the site where their distinctive pottery style was first described; Figure 5.5) was in the Bismarck Archipelago [65]. One of the prized commodities it offered them was obsidian, most versatile of materials for lithic technology. A measure of the value of the region's obsidian is that it has been found in Lapita archaeological contexts as far as 3000 kilometres away in Fiji, demonstrating extensive seafaring trade and exchange networks.

The Lapita kept pigs and dogs, and tended 'gardens' of tree and root crops in the forests. They manufactured distinctive ceramics with dentate patterns and, though they became more sedentary over time, they remained accomplished seafarers. What led them to migrate from their heartland is uncertain but perhaps it was a result of accumulating population pressure owing to the fruits of their horticultural skills. Who knows, perhaps some of the volcanic eruptions encouraged a few pioneering Lapita to cross the seas in search of a less dusty environment?

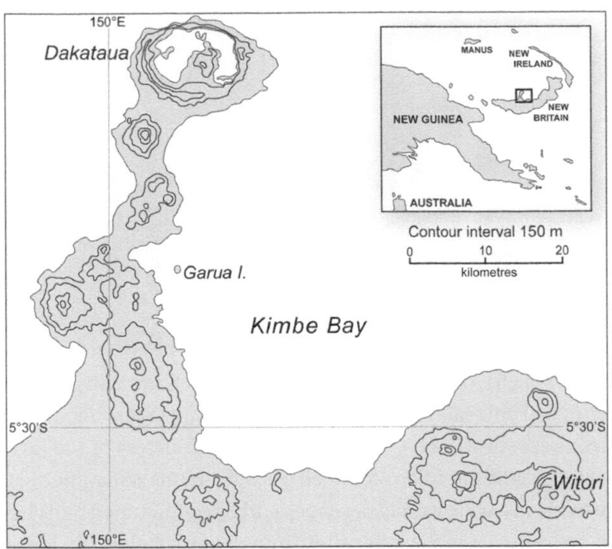

Figure 5.6 Map of the Willaumez Peninsula with Dakataua volcano at the end of the peninsula and Witori volcano not far away. Contour intervals are 150 metres. Based on reference 65.

The long-term picture of human adaptation and migration in the region, and its complex relationships with volcanic disturbances of varying magnitude and frequency, has been the focus of multidisciplinary research in western New Britain for more than 20 years, led and inspired by the archaeologist Robin Torrence from the Australian Museum in Sydney [66,67]. This archaeological and palaeoenvironmental work, which is centred on and around the Willaumez Peninsula (Figure 5.6), is underpinned by a strong tephrochronological framework: a pit dug just about anywhere reveals a 'to-die-for' (as an archaeologist friend put it) layer-cake of ancient soils interspersed by tephra layers (Figure 5.7, left). Radiocarbon dating provides a chronology for the sequences and, by now, most of the tephra fallout layers have been associated to eruptions of particular volcanoes in the region.

Six major episodes of volcanism affected western New Britain in the past six millennia: five at Witori volcano (dated roughly 4200 BCE, 1400 BCE, 300 CE, seventh or eighth century CE and another shortly thereafter) and one at Dakataua, located at the tip of the peninsula, around the mid-seventh century. The magnitudes have been roughly calculated and range from an M_e of about 5.8 up to 6.5 for the mid-second millennium BCE Witori eruption [68]. The first of these was catastrophic for the local environment, with up to 70 centimetres of

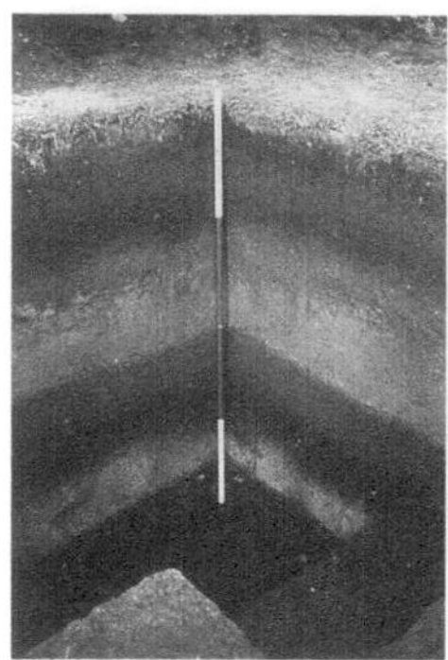

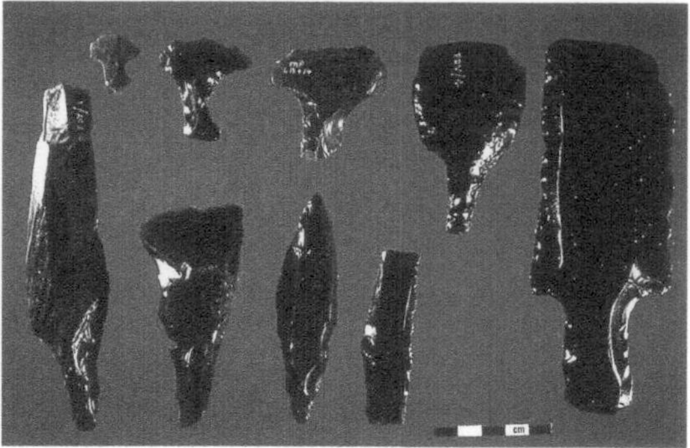

Figure 5.7 (Left) Archaeological excavation at Bitokara Mission on the Willaumez Peninsula. Pale grey bands are tephra layers and the darker horizons are soils. Photograph by Jim Specht; courtesy of the Australian Museum. (Right) Obsidian stemmed tools from the Willaumez Peninsula were made in a wide variety of shapes and sizes. Stemmed tools like these were recovered at many other sites from deposits below the lowermost white tephra layer, which is derived from the 'W-K2' eruption of Witori volcano. Photograph by Georgia Brittain; courtesy of the Australian Museum.

tephra fallout accumulating at one of the study sites, 50 kilometres from the crater. Here, the pace of forest and soil recovery was gradual and the area was abandoned for up to 2000 years. Twenty-five kilometres away on Garua Island, just offshore from the Willaumez Peninsula (Figure 5.6), much less tephra accumulated but the island was still abandoned for as long as two centuries.

The people who returned were mobile but retained cultural links with their past. Among them was the specialist knapping of magnificent 'stemmed tools' from obsidian, the same variety of volcanic glass found in Ceren's rafters (Figure 5.7, right). Their function has been widely debated without reaching much consensus but it is possible that some had more ideological and ceremonial significance, while others were more utilitarian. The distribution of stemmed tools from western Papua to Bougainville in the Solomon Islands demonstrates long-range social and cultural links between mainlanders and islanders. Torrence suggests that the ritual and cosmological connotations of stemmed tools point to a social system aligned around the acquisition of such valuable items reinforced by knowledge of songs, dances, costumes and ceremonies learned via exchange with the chiefs of neighbouring clans and even overseas communities. Society became increasingly complex as it experimented with different systems of social exchange, and the construction of symbols to identify status and structure social links. The increasing evidence of land-use management is evident from palaeoenvironmental analysis of pollen, phytoliths (plant structures made from silica) and charcoal abundance. These reveal not only the ecological succession following the eruption – from grasses, followed by gingers, palms and then shrubs and trees – but also the increasing practice of slash-and-burn agriculture [69].

The mid-second-millennium BCE eruption of Witori was the big one, covering much of the region in up to a metre of tephra; an area of at least 10,000 square kilometres was buried to a depth of 20 centimetres or more. The fallout clogged waterways and coastal lagoons, and surely choked coral reefs as ash was drained off the land surface for decades afterwards. Despite the devastation, the area was reoccupied within two centuries (it is difficult to pin down precisely the chronologies). The palaeobotanical evidence suggests again that reforestation was slow and that people lived in a more disturbed and managed open environment. This time, there was a pronounced change in social and cultural patterns following the event and it is plausible that the eruption played some part in it. Stemmed tools disappeared from the lithic repertoire but a new and highly distinctive variety of ceramic, so-called Lapita pottery, emerged in abundance.

Torrence believes that these archaeological signatures testify to profound social change. She suggests that effects of physical devastation of the landscape and its resources, combined with the social and political consequences of thousands of refugees having to find new homes in potentially hostile territories, undermined the old social fabric built around the procurement of stemmed obsidian tools and acquisition of status. For starters, the obsidian sources on and around the Willaumez Peninsula were buried and quarrying ceased, so it is unsurprising that manufacture of stemmed tools declined abruptly. This immense physical and social disruption likely challenged accepted beliefs (perhaps the stemmed tools were supposed to protect people from volcanic disasters). This provided a window of opportunity for innovation and entrepreneurism. New technologies and food-production practices emerged; perhaps the patterns on Lapita pottery reflect a new cosmology intended to supersede the failed symbolism of the stemmed tools. A different kind of society was forged amidst the chaos and confusion of disaster.

The next significant eruption of Witori took place around the beginning of the fourth century CE. It was lesser in magnitude (around

M_e 5.8) but still damaging, and another period of up to a century or two of abandonment ensued. The society that returned was similar to its predecessor except that it no longer made any pottery locally, nor was the Lapita style seen again. Finally, around the seventh to eighth centuries, there were Plinian eruptions of both Witori and Dakataua [70]. By now, society was well rehearsed in coping and adaptation strategies; abandonment was short-lived. These eruptions may, however, have had global climatic repercussions - the M_e 6 Dakataua eruption has been linked tentatively to one of the most prominent sulphuric layers in the Greenland ice-core record for the past two millennia.

What emerges from this exceptional, long-timescale perspective on the human prehistory of a specific region is a social trajectory of increasing adaptation to an unstable and unpredictable landscape. Forest management and agriculture intensified, leading over time to an increasingly sedentary lifestyle. Larger disturbances from greater eruptions correlate to some degree with the scale of human response and adaptation; certainly, during the course of a few hundred generations, communities in New Britain evolved patterns of lifestyles adapted to the volcanic landscapes and the frequent eruptions. It is very likely, for instance, that the persistent re-occupation of devastated lands partly reflects the lure of obsidian lodes on the Willaumez Peninsula and to the southwest of Witori. In reality, people probably returned very soon after eruptions to survey impacts and, perhaps, to collect obsidian from the surface. After the powerful 1994 eruption of Rabaul, people erected territorial markers atop the fresh tephra to claim land even though they couldn't actually live on it! Another significant factor in this part of the world is that the pace of ecosystem recovery following each eruption would have been quickened by the humid tropical climate. The size of New Britain and its proximity to numerous other islands including Papua itself would have promoted rapid re-colonisation and the return of the species of flora and fauna so useful to hunters and gatherers and a people practising shifting agriculture.

Also important to consider, and easy to lose sight of in such a geographically focused study, are the implications of migrations of refugees from the disaster zone. Being a refugee is not easy; there are many modern histories where people have returned to their homelands as soon as they can. To be accepted in safer territories and given access to food, shelter and other essential resources requires pre-existing social organisations beyond the clan. Even if survivors migrated into unoccupied territories, they would still rely on lifelines back to kindred communities elsewhere. The general trend since the mid-fifth millennium BCE towards shorter hiatuses in human occupation following eruptions, rather than a correlation between eruption size and period of abandonment, suggests a strengthening and widening of social organisation through time and experience.

5.2 ORAL TRADITIONS

Around the world, myths and legends bring people in touch with their ancestors' experiences of geological disasters. These stories, usually passed down through oral tradition, can be difficult to decipher because of the culturally based metaphor and symbolism that have shaped and infused them over centuries and millennia. However, they can provide valuable information for modern scientific understanding of past volcanic activity, as well as inform present-day management of volcanic risks. They also condition the mind-set and perception of people who live with risk (whether it be from tsunami, flood, earthquake or eruption) and thus play an important function in preparing societies for future disruption. In the long term, this might inform where people avoid settlement and, in the short term, provide explanation and understanding of events (for instance seismic shaking) and knowledge of what to do in the event of an eruption.

Examples abound from many parts of the world, but I focus here on remarkable oral traditions from three very different volcanoes: Pinatubo (Chapter 3), Kīlauea on Hawai`i and Crater Lake in Oregon.

5.2.1 Mt Pinatubo 1991: an eruption foretold

Given that the volcanological world knew nothing of Pinatubo before 1991 (Chapter 3), it is humbling to consider the indigenous knowledge of the volcano held by the Ayta people, of whom some 10,000 were living on the volcano and its environs up to the time of the volcano's re-awakening. The following Ayta legend strongly suggests a folk memory of an ancient eruption of Pinatubo [71]. The story was transcribed in 1915 (i.e. long before the 1991 eruption) and survives as a microfilm of a typewritten transcript held at the University of the Philippines.

The two main protagonists are the 'king of the spirit hunters', Aglao, and 'the terrible spirit of the sea', Bacobaco, who would metamorphose into a huge turtle and feast on deer in the hunting grounds of the spirit hunters. Aglao resented this but could not confront

Bacobaco 'who carried his thick shield on his back, and who threw fire from his mouth'. However, Aglao was joined by Blit, brother of the spirit of the wind, and together with the other spirit hunters they attacked Bacobaco with arrows. Bacobaco fled to 'the lake at the foot of Mount Pinatubu' and jumped into the lake:

> '... but the water was so clear, that Blit could see him at the bottom. Finding the lake a useless place of refuge, he climbed the Mount Pinatubu in exactly twenty-one tremendous leaps. When he had reached the top, he at once began to dig a big hole into the mountain.... pieces of rock, mud, dust, and other things began to fall in the showers around the mountain.... he howled and howled so loudly that the earth shook ... The fire that escaped from his mouth became so thick and so hot that the pursuing party had to turn away. For three days the turtle continued to burrow itself, throwing rocks, mud, ashes, and thundering away all the time in [a] deafening roar. At the end ... all was quiet ... But the lake, with its clear water was now filled with rocks, and mud covered everything. On the summit of the Pinatubu was the great hole, through which Bacobaco had passed, and from which smoke could be seen constantly coming out. This showed that although he was already quiet he was still full of anger, since fire continued to come from his mouth ...'

The conclusion of the narrative seems extraordinarily prescient of the 1991 cataclysm:

> 'But now, you do not see smoke coming out of the Pinatubou mountain ... and many believe that the terrible monster is already dead; but I think that he is just resting after his exertions, and that someday he will surely come out of his hiding place again ...'

5.2.2 Kīlauea

In Hawai`i, it is the goddess Pele who, from her home in the summit crater of Kīlauea volcano, controls eruptions (Figure 2.8). Legends of Pele doubtless record individual eruptions experienced by the Hawaiians, who first arrived in the islands around 800–1000 CE. One of the most important narratives that has been handed down the generations through chants and storytelling is that of her first arrival at Kīlauea with her sisters [72]. En route she had fallen in love with Lohi`au on Kaua`i Island and, after settling into her new home, she asked her sisters to fetch him. Only the youngest, Hi`iaka`aikapoliopele (Hi`iaka for short), accepted, on the condition that Pele would not destroy her favourite forest of `ōhi`a lehua trees to the east of the summit (an area known as Puna). But, while she managed to bring Lohi`au back, suffering many hardships on the journey, she took longer than the 40 days she had been granted for the task. Pele presumed her little sister had run off with Lohi`au and, needless to say, Hi`iaka's first sight of home on return was of her beloved forest ablaze! In retaliation, Hi`iaka walked up to the summit of Kīlauea with Lohi`au and made love to him, right in front of Pele. Enraged, the goddess killed Lohi`au and flung him into the crater. Hi`iaka then dug fervently to recover his body, excavating deeper and deeper. Rocks were flying and she was warned not to dig too far or else water would come in and extinguish the fires of Pele. At last, Hi`iaka reached Lohi`au and the two were reunited in spirit.

The volcanologist Don Swanson, who served for years as the Director of Hawai`i Volcano Observatory, suggests there is a 'remarkable correspondence between the oral record and the results of modern research' [73]. Specifically, the Pele story fits with what is now known, through tephrostratigraphy and radiocarbon dating, of the two largest volcanic events to have taken place in Hawai`i since it was settled, namely the 60-year-long `Ailā`au eruption in the fifteenth century and the subsequent collapse to form Kīlauea's summit caldera. The burning of Hi`iaka's forest suggests a lava flow and fits with the massive destruction that must have been caused by the `Ailā`au flow, erupted between 1410 and 1470. This flow travelled more than 30 kilometres, reaching the coast on the eastern tip of the island. With an estimated magnitude, M_e, of 6.1 (around 5.2 cubic kilometres of magma) the effusive eruption poured basalt over a total area of 430 square kilometres, devastating the native forest. One of the popular tourist attractions on Hawai`i today is the magnificent Nāhuku (Thurston) lava tube, which insulated the flowing `Ailā`au lava from heat loss and thereby helped to convey it over such a great distance. Another tube from the same eruption is among the longest known on Earth – its sinuous track extends over 65 kilometres! As the largest eruption since Polynesian settlement, the `Ailā`au eruption would have affected the lives of the Hawaiians in many ways.

Swanson goes on to associate Hi`iaka's digging for Lohi`au with the initial formation of Kīlauea's caldera. The 'flying rocks' suggest explosions accompanying the collapse, and the warning about the water putting out Pele's fire might indicate that the crater reached a substantial depth. He also re-interpreted the written account of the Reverend William Ellis, who led a party of missionaries that were the first Europeans to visit Kīlauea. Ellis reached the summit of the volcano on 1 August 1823 and was informed by his Hawaiian guides that Kīlauea:

> 'had been burning from time immemorial . . . and had overflowed some part of the country during the reign of every king that had governed Hawaii: that in earlier ages it used to boil up, overflow its banks, and inundate the adjacent country; but that, for many kings' reigns past, it had kept below the level of the surrounding plain, continually extending its surface and increasing its depth, and occasionally throwing up, with violent explosion, huge rocks or red-hot stones'.

Swanson interpreted the reference to a period of 'many kings' reigns' as indicating around 10–15 reigns. From evidence that a king's period of reign was in the region of 20–25 years, this would suggest a change in activity sometime between the mid-fifteenth and early-seventeenth centuries, consistent with the `Ailā`au eruption and caldera collapse, followed by intermittent explosive eruptions and caldera enlargement up to the late-eighteenth century. Previously, volcanologists had believed that the caldera formed during a violent explosive eruption in 1790 that killed many soldiers in Keōua's army, which was on its way to battle for control of the island. They had also thought that the ash deposits found around Kīlauea had all derived from this single blast. However, the interpretation of the oral traditions is consistent with new radiocarbon dates that reveal around three centuries of sporadic explosive activity following initial caldera formation in the period 1470–1500 CE.

Swanson concluded that: 'There is a lesson here, plain to see. But, it is difficult to interpret anecdotes, particularly those cloaked in thick poetic metaphor. We are used to thinking scientifically, not metaphorically, when we tackle volcanic problems'. In understanding volcanic hazards on Hawai`i 'it is important to know that Kīlauea erupts explosively more often than once thought . . . [and] that the caldera formed, with only minor explosion, following a long-lasting eruption'.

5.2.3 Mt Mazama

One autumn, around 5700 BCE, Mt Mazama (one of the major Cascades volcanoes, located 200 kilometres inland from the Oregon coast) erupted, crumpling its profile to leave a roughly nine-kilometre-diameter caldera, known as Crater Lake (Figure 5.8). With an astounding maximum depth of 589 metres, it is the deepest lake in the USA.

Figure 5.8 Crater Lake, Oregon, formed by an eruption around 5700 BCE. Wizard Island is a cone built up since the caldera-forming event and its crater is more than 800 metres above the deepest part of the lake. Llao Rock dominates the caldera rim (centre of photograph).

The age estimates are based on radiocarbon dates for charcoal found in the pyroclastic deposits and also from the age of a layer in the Greenland ice core containing sulphate fallout and fine ash geochemically fingerprinted to the Mazama eruption. The seasonal clues come from pollen records.

The paroxysmal eruption took place in two phases: the first was a Plinian eruption from a single vent to the northeast of Mazama's summit, which propelled ash to an estimated height of 50 kilometres above sea level. A vast area was covered in tephra. The vent widened and the eruption intensity increased, eventually limiting air entrainment at the core of the plume, which subsequently collapsed, fountaining down on to the valleys on the northern and eastern slopes of Mazama. The second phase saw the mountain founder into the emptying chamber, some five kilometres down, with magma disgorging from the ring-like fracture forming around the incipient caldera walls. Pyroclastic currents gushed down all sides of Mazama, travelling up to 70 kilometres and devastating a vast area. The deposits from this stage of the eruption attained thicknesses of 100 metres, and ash fell over a large part of the Pacific Northwest. In total, an estimated 50 cubic kilometres of magma were blasted out from a silicic magma chamber in the M_e 7.1 eruption. This was accompanied by emission into the atmosphere of at least 34 megatonnes of sulphur and 210 megatonnes of chlorine [74].

Astonishingly, considering the remove in time, evidence of the eruption can be found in the oral traditions of the present day Klamath

Tribe, descendants of the ancient Makalak people, who once lived to the southeast of Crater Lake [75]. In one version of the legend, the eruption is linked to the battle of two great spirits, the Chief of the Below World, called Llao, and the Chief of the Above World, known as Skell, who gives laws and destroys evil beings. From time to time, Llao would come up from his home inside the Earth to stand atop Mt Mazama, then one of the highest volcanoes of the Cascades. On one of these excursions, he spied the Makalak chief's daughter and fell in love with her. He promised her immortality if she would only join him beneath the mountain. But she refused to go, and an angered Llao promised to destroy her people with fire. Enraged, he burst through the entrance to Mazama, bestrode the summit and carried out his threat.

> 'Mountains shook and crumbled. Red hot rocks as large as hills hurtled through the skies. Burning ashes fell like rain. The Chief of the Below World spewed fire from his mouth. Like an ocean of flame it devoured the forests on the mountains and in the valleys. On and on the Curse of Fire swept until it reached the homes of the people.'

Skell took pity on the people and defended them from the neighbouring volcano, Mount Shasta. The two spirits battled furiously, throwing incandescent rocks the size of hills, unleashing landslides of fire (pyroclastic currents) and causing the earth to tremble. The terrified people fled to the waters of Klamath Lake. Two seers offered to sacrifice their lives by leaping into the fiery pit atop Llao's mountain. Skell, moved by their bravery, advanced, and Llao retreated into Mount Mazama with a terrific explosion. At sunrise on the following day, the mountain had disappeared having fallen in on Llao, leaving a large hole. This final event surely marks the caldera collapse that formed Crater Lake. Llao dwells today in Crater Lake 'rather octopoidal and of a dirty white color'.

Various archaeological finds (including abundant obsidian tools) help to connect us between the modern volcanological interpretation of the Mazama eruption and this vivid story, among them a pair of Makalak sandals found entombed beneath the pumice.

5.3 CREPUSCULAR LIGHTS, CANNONADES AND CHRONICLES

The effects of volcanic particles on sunsets were first studied carefully following the explosive eruption of Krakatau (aka Krakatoa) in

Indonesia on 26 and 27 August 1883. Indeed, the first and still the most complete compilation of the optical effects of volcanic aerosol was published in 1888 in a major treatise commissioned by the Royal Society of London [76]. It provides a prolific source of information on the nature and interpretation of these atmospheric phenomena and was based on reports in ships' log books, and the eyewitness observations of scientists and lay-observers worldwide. One of the more striking apparitions noted was a reddish-brown halo around the Sun known as Bishop's rings, after the Reverend S. E. Bishop, who observed them from Honolulu on 5 September 1883:

> 'Permit me to call special attention to the very peculiar corona or halo extending from 20° to 30° from the Sun, which has been visible every day with us, and all day, of whitish haze with pinkish tint, shading off into lilac or purple against the blue'.

The mathematician Robert von Helmholtz described his observations of twilight glows, which he made in Berlin in late November 1883 to the journal *Nature* [77]:

> 'A greenish sunset at 3.50, an unusually bright red sky with flashes of light starting from south-west. An interesting physiological phenomenon, which we call 'Contrast-Farben', was there beautifully illustrated by some clouds, no longer reached by direct sunlight; they looked intensely green on the red sky. At 4.30 the streets were lighted by a peculiarly pale glare, as if seen through a yellow glass. Then darkness followed and the stars became visible. But half an hour afterwards . . . the western sky was again coloured by a pink or crimson glow. Persons . . . mistook it for a Polar aurora; others spoke of a great fire in the neighbourhood . . . At 6 o'clock all was over'.

It was assumed at the time that these and other reported optical effects resulted from scattering of light by fine ash particles suspended in the atmosphere, a view that prevailed for nearly a century. The Honourable F.A. Rollo Russell and E. Douglas Archibald, writing in the Royal Society of London's tome, concluded that: 'a stratum of matter was present in the upper atmosphere of a great part of the globe during the autumn and winter of 1883–4'. On the basis of reported fallout, Russell suggested that the layer was composed of 'glass-like laminae or very thin fragments and spherules [which constituted a] mass of fine dust . . . of fragments of pumice blown into very thin transparent plates [forming] . . . an immense volume of glassy pumice dust of microscopic and ultramicroscopic minuteness'. With startling correspondence to modern estimates, Russell estimated the dimensions of these particles to have been around 0.1–1 thousandths of a millimetre (Section 3.2.1), and that the 'dust' cloud travelled westwards at around 100 kilometres per hour based on the first appearance of optical phenomena at different locations. Several observers estimated the altitude of the cloud with varying success. One observer in France erroneously estimated the cloud height to be 500 kilometres. Archibald synthesised various data, such as the duration of crepuscular glows, and concluded that the cloud had occupied a layer between 20 and 35 kilometres above sea level, again entirely consistent with modern measurements of stratospheric volcanic clouds (Section 3.1).

It was not until the 1963 eruption of Mt Agung volcano on Bali that the stratospheric aerosol layer (observed only a few years previously by high-altitude balloon experiments conducted by the pioneering atmospheric scientist Carl Junge) was linked to volcanic eruptions. The climatologist Hubert H. Lamb subsequently noted that cold northern-hemisphere summers were often preceded by large volcanic eruptions. In 1970, he published the first comprehensive historical database in which he classified volcanic aerosol layers according to their Dust Veil Index (DVI) [78]. The DVIs were largely based on observational and empirical studies and did not explicitly consider the sulphur yield of eruptions, which is now known to be critical to the atmospheric and climate-changing potential of stratospheric plumes.

Worldwide eye-catching sunsets, Bishop's Rings and other rare atmospheric optical effects only require a keen eye to observe them. They have been sufficiently striking for astronomers, diarists and chroniclers through the ages to take note of. In this way, textual mining of the literary record provides a means to identify occurrences of very large eruptions back through the historic period, and even into ancient times. For some regions in particular, the manuscripts of scholars, scribes and ecclesiastical historians yield abundant but accidental information on past volcanic events by way of sightings of reddening or dimming of the Sun and other stars, sunspots seen with the naked eye, reddish haloes around the Sun, dark total lunar eclipses and red or purple twilight glows.

Richard Stothers, of NASA's Goddard Institute for Space Studies, and Michael Rampino, of New York University, were the first to sift through a large body of the ancient Latin, Classical and Byzantine Greek, Syriac and Armenian literature specifically on the lookout for such references [79]. They identified major eruptions of Vesuvius in 217 BCE and 472 CE, Mount Etna in 44 BCE, and two mystery eruptions around 536 CE and 626 CE. In the case of 44 BCE (the year Julius Caesar was assassinated), there is the following passage in *Natural History* by Pliny the Elder (who became one of the most famous victims of the 79 CE Vesuvius eruption):

> 'Portentous and protracted eclipses of the sun occur, such as the one after the murder of Caesar the dictator and during the Antonine war which caused almost a whole year's continuous gloom'.

Referring also to the time of Caesar's murder, a passage in Plutarch's *Parallel Lives* appears to corroborate the phenomenon:

> '... among events of divine ordering, there was ... the obscuration of the sun's rays. For during all that year its orb rose pale and without radiance, while the heat that came down from it was slight and ineffectual, so that the air in its circulation was dark and heavy owing to the feebleness of the warmth that penetrated it, and the fruits, imperfect and half ripe, withered away and shrivelled up on account of the coldness of the atmosphere.'

Stothers and Rampino concluded that, for the interval between 700 BCE and 630 CE, 'in the case of the most explosive eruptions, having ... widespread and obvious atmospheric effects, the record is probably very nearly complete ... at least in Europe'. Stothers went on to pore over translations of Babylonian astronomical and meteorological diaries. These provide near daily observations from 61 BCE to the fourth century CE and identified several more likely volcanic events [80]. Chinese records have also proved a rich source of information on atmospheric disturbances that may have arisen from the presence of high-altitude volcanic clouds. I shall return to the testimony of mediaeval historians in Chapter 11.

Returning to the case of Krakatau's cataclysm in 1883, the scientific report commissioned by the Royal Society of London documented not only optical phenomena but also auditory ones. The most violent detonations occurred on the morning of 27 August 1883 and were heard across nearly 10% of the Earth's surface, as far as Diego Garcia and Rodrigues in the Indian Ocean, 3650 and 4800 kilometres distant, respectively. The chief of police on Rodrigues, James Wallés, recorded that 'several times during the night of the 26th–27th reports were heard coming from the eastward, like the distant roars of heavy guns'. Indeed, volcanic explosions have often been mistaken for the cannonades of ships in distress. In some cases, the sounds have been interpreted in a more sinister way. In Aceh (Sumatra, 1650 kilometres from Krakatau) it was thought that a fort was being attacked, and the troops were put under arms.

Rafts of pumice drifting across the Indian Ocean were another far-reaching manifestation of the Krakatau eruption. The descriptions in the Royal Society report offer a telling comparison of lyricism of seafarers from different nations. For instance, while Captain Williams of the *Shah Jehan* scribbled thus: '[we] passed a lot of pumice', his French counterpart, Captain Hogon of the barque *Mary Queen*, was moved to describe 'une infinité de parcelles de rôche brulée sur l'eau'.

5.3.1 Visual arts

An unprepossessing mound situated on the Konya Plain in Central Anatolia contains one of the most important Neolithic sites unearthed to date. In its heyday, between 8000 and 7000 BCE, Çatalhöyük was one of the largest communities on Earth, with a population of up to 10,000 living in closely spaced, box-like dwellings. Excavations began in the 1960s and uncovered magnificent murals, which are said to include the earliest known iconographic representation of a volcanic eruption [81] (Figure 5.9). The volcano in question looms above clusters of buildings and appears to be caught in the act of showering its flanks with lava bombs. One candidate volcano put forward is 3250-metre-high Hasan Dağı at the eastern end of the Konya Plain, which provided the Neolithic town with sources of obsidian used in the manufacture of tools, weapons and jewellery. However, the mural's claim to depict the earliest eyewitness account of an eruption (and one of the oldest maps in existence) is not beyond doubt. A more recent interpretation suggests that it shows instead a leopard skin and a geometric pattern [82].

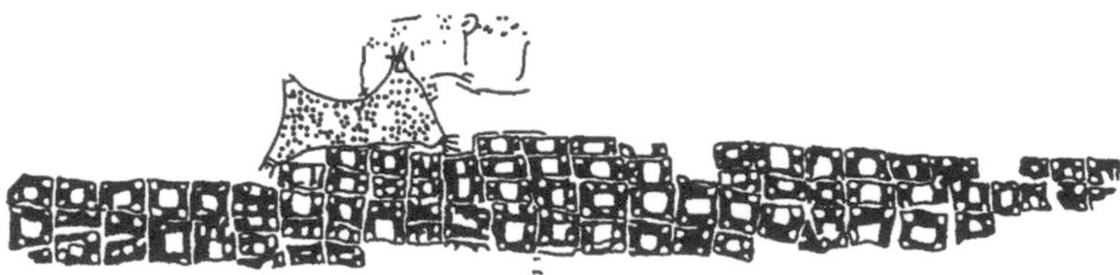

Figure 5.9 Reproduction of wall painting from Çatalhöyük purportedly depicting the Neolithic town and an erupting volcano, possibly Hasan Dağı, 50 kilometres away. Or is it a leopard skin above a geometric pattern? Based on a drawing by Grace Huxtable.

While the interpretation of the Çatalhöyük mural may be open to debate, there is no question about the many depictions of volcanoes in innumerable paintings from the historic period. And one particularly ingenious study focuses on paintings depicting volcanically enhanced sunsets! To appreciate its basis, a quick digression into atmospheric optics may be helpful. It is clear that volcanic aerosol can have a dramatic impact on visibility through the atmosphere (Section 3.2.1). A measure of this is the optical depth, τ, defined by the relationship between the radiation, I_0, entering the top of the atmospheric layer containing the aerosol, and the radiation leaving the bottom of the layer, I:

$$I = I_0 e^{(-\tau/\cos\theta)} \qquad [5.1]$$

where θ is the angle between the vertical and the Sun (the solar zenith angle). The optical depth depends on the depth of the layer and the concentration and optical properties of the particles contained in it. The value of I is smaller than that of I_0 because some of the original energy will have been scattered, and some may have been absorbed by the layer. If $\tau = 0$ then the layer is transparent (i.e. $I/I_0 = 1$). For values of θ of 55° and $\tau = 0.05$, then the ratio $I/I_0 = 0.92$. In other words, 92% of incident radiation is transmitted through to the base of the layer. For the same solar zenith angle, and $\tau = 1$, then $I/I_0 = 0.18$, and 82% of the incoming light has been intercepted or deflected. Optical depth can be given for a specified wavelength or wave band in the electromagnetic spectrum but, typically, reported measurements refer to visible light.

Christos Zerefos from the National Observatory of Athens and his colleagues found a way to estimate volcanic aerosol optical depths in canvases by well-known painters including Claude Lorrain, Caspar Friedrich, J. M. W. Turner, Edgar Degas and Gustav Klimt [83]. In total, they analysed θ and the proportion of red versus green in the skies of more than 500 paintings portraying sunsets, spanning the period from 1500–1900. They then plugged the measurements into a model for the propagation of light through the atmosphere in order to estimate the optical depths (τ) represented in the paintings! They found an astonishingly close correspondence between their calculations and Lamb's DVI. Notable volcano sunsets in the data set include Turner's *The Decline of the Carthaginian Empire* ($\tau = 0.46$) painted in 1817 (two years after Tambora's eruption) and Degas' *Race Horses*, which was painted two years after the 1883 eruption of Krakatau ($\tau = 0.36$). The sunset hues in Edvard Munch's famous *The Scream* have also been attributed to the

Krakatau eruption, though there is some debate over the year it was painted.

5.4 VOLCANO FORENSICS: A CASE STUDY

To provide an example of forensic volcanology, this section reviews the detective work that uncovered a remarkable eruption, in 1861, of a little known volcano in Eritrea, called Dubbi. One reason for including this casework is that it is based on personal experience but Dubbi assumes significance, also, since it turns out to have been responsible for the largest known documented eruption in Africa.

5.4.1 The 1861 eruption of Dubbi volcano

During the mid 1990s, I spent much of my research time studying satellite images of volcanoes. I was particularly interested in a remote Ethiopian volcano called Erta 'Ale in the arid, tribal region known as the Danakil Depression. It was while poring over the satellite images that I noticed another volcano, called Dubbi, 120 kilometres to the east and near the Red Sea coast of Eritrea. What drew attention to Dubbi was a bright apron of pumice covering the flanks of the volcano and the plain beyond. Even more conspicuous were several dark lava flows, superimposed on the pumice, and stretching up to 22 kilometres from the summit craters. The tephra and lava looked very pristine and piqued my curiosity.

Reference to the Smithsonian Institution's *Volcanoes of the World* database quickly identified an eruption in 1861 that had been reported in the *Times of London* newspaper, and subsequent venting in 1863 and 1900. The events had come to the attention of the outside world largely because of disruption to maritime traffic in the Red Sea. The most complete description of the 1861 eruption was provided by Captain R. L. Playfair, the British Resident at Aden, and reported in the *Times*.

> 'On the night of the 7th or the morning of the 8th of May, the people of Edd [a coastal village 40 kilometres north of Dubbi] were awakened by the shock of an earthquake followed by others which continued with little intermission for about an hour. At sunrise, a quantity of fine white dust fell over the village. ... About noon, the character of this dust ... resembled red earth. Shortly afterwards, it increased to such an extent that the air was perfectly darkened and we had to light lamps in our houses. It was darker than the darkest night, and the whole place was covered with dust, nearly knee-deep. At night

> [on May 9], we saw fire and dense smoke issuing from a mountain called Djebel Dubbeh . . .'

At Perim Island (at the mouth of the Red Sea, 200 kilometres to the southeast), explosions were heard from 02:00 local time on 8 May and were initially attributed to a military bombardment. They were sustained for up to ten hours but continued intermittently for two days. Two steamers, the *Candia* and *Ottawa*, encountered ash fallout 'like a London fog' on 10 May in the southern Red Sea. Travelling south to Aden, the boats were delayed for ten days in the Dhalak Islands. Some accounts suggested that an 'ash-rain' fell on the Ethiopian plateau 400 kilometres west of Dubbi on 18 July 1861, which might be attributable to further eruptions of Dubbi. Activity certainly resumed in September 1861, with explosions again audible, this time in the coastal town of Massawa, 340 kilometres northwest. An eyewitness account made by a Somali officer who climbed the volcano, probably in October 1861, was relayed by Captain Playfair and published in several periodicals of the day:

> 'The top of the mountain appeared as if it had been white, but was blackened by the fire. We dug in the ashes about a foot and a half before reaching the earth. On the mountain we saw 19 craters; 18 of which smoke in the daytime and at night give light like a lamp . . .'

Further reports also indicated that there were victims of the eruption:

> 'The names of the villages burned, which were located near the mountain, are Moobda and Ramlo. One hundred and six men and women were killed and their bodies were not found. The number of animals killed is unknown.'

According to Captain Playfair a total of 175 people were killed by the eruption. This represents a high toll given how sparsely populated the region is. Only one later account reported that 'large lava flows' had been erupted in September 1861 and had spread towards the sea. These were very likely to be the same flows that were so prominent in the satellite imagery. However, it was unlikely they could have claimed so many lives.

Intrigued by these historical accounts, I organised fieldwork in the area with my PhD student, Pierre Wiart, and Eritrean colleagues. We flew into Asmara and, after a few days making preparations, set off overland towards the Red Sea coast. Our first stop was at the port city of Massawa, achromatised in desert sunlight, and still showing the signs of the heavy aerial bombardment inflicted during Ethiopia's civil war.

Figure 5.10 Dubbi volcano in Eritrea is little known and yet produced one of the largest known eruptions in Africa in history. (Top) The volcano seen from the north; prominent lava flows descending from the summit were erupted in 1861. (Bottom) A section of pyroclastic deposits from the first phases of the 1861 eruption. The narrow white layer is a band of ash that fell out from phoenix clouds associated with underlying pyroclastic-current deposits. These were likely responsible for over 175 deaths recorded during the eruption. The thick layer occupying the rest of the photograph above the white layer consists of pumice fallout from a Plinian eruption cloud.

This would be the last town we would pass through, so we picked up as much fuel and water as we could carry on our 4WD vehicle and set off southwards, following the coast around the Gulf of Zula, later cutting inland through the Danakil Alps. En route, every single one of the plastic jerry cans on the roof rack sprung a leak and our petrol and even more precious water dripped out!

By late afternoon on the following day, we caught our first glimpse of Dubbi (Figure 5.10, top), rising above a blinding white plain of pumice, and its neighbouring volcanoes, including the much larger and wider caldera-topped Nabro. In the next days we scoured the terrain, searching for any remnants of 1861 deposits. We clambered over the lava flows, hiked to the summit craters, which were still steaming in places, and looked in gullies for tephra sections (Figure 5.10, bottom). At each key site, we collected samples for geochemical analysis and, for each tephra layer, used a balance and set of sieves to calculate the distributions of particle sizes.

Integrating all the information – from the newspaper columns, the satellite-image interpretation, fieldwork and laboratory analysis – we were able to reconstruct a clear picture of the eruption [84]. Around two in the morning on 8 May 1861, strong earthquakes – measuring up to 5.9 on the Richter scale – awoke people in Edd. Magma was on the move. By noon that day, Edd was plunged into a darkness that lasted several hours. We found a cliff in a small gully 12 kilometres from Dubbi that cut perfectly through the deposits from this phase of the eruption. It exposed a 4.5-metre-thick layer of pumice fallout from the main explosions. But we also recognised the deposits of pyroclastic currents beneath the fallout and were able to trace them up to 25 kilometres from the volcano. There was no evidence for any soil development between these contrasting pyroclastic deposits suggesting they belonged to the same eruptive sequence. This finding at once explained the heavy human losses reported during the eruption; such mobile pyroclastic currents could readily have overwhelmed a village within their radius. A further explosive eruption may have occurred in mid July, yielding a light ash fall 400 kilometres to the west on the Ethiopian plateau.

In September–October 1861, the eruption radically changed its character, disgorging basaltic lava flows from a chain of small craters at the summit. There was still enough gas in the magma to cause the mild explosive activity witnessed by the Somali officer. We already knew from satellite images that the lava flows had travelled as far as 22 kilometres, but we were astonished, on the ground, to discover how thick they were: we climbed up several lobes of the flows and measured thicknesses of over 20 metres. Combined with the lava-flow areas measured from the space imagery, we calculated that at least 3.5 cubic kilometres of basalt had been erupted. Unfortunately, most of the original pumice deposit had been blown away by desert winds, and we could only find remnants in protected gullies. It was not possible to estimate its original volume, therefore, but it had clearly been substantial. The lava flow volume alone, though, places Dubbi's 1861 eruption amongst the largest in history.

This finding encouraged us to examine the summertime temperature reconstructions for the northern hemisphere derived from tree-ring methods (Section 4.3). Intriguingly, this revealed a cooling of

0.33 °C between 1861 and 1862, ranking as the tenth largest anomaly in the six-century-long record. If this is genuinely a volcanic cooling signature, then the 1861 Dubbi eruption rates as one of the most significant climate-changing eruptions in history. It also has important hazard implications. In 1861, it would still be another eight years before the Suez Canal would open but, nevertheless, numerous vessels were assailed by Dubbi's fallout. Ash clouds can cut visibility to zero. Any future explosive eruptions of Dubbi could, therefore, pose a significant collision risk to the much larger volume of maritime traffic that now sails the Red Sea.

5.5 SUMMARY

By examining the spatial and temporal distribution of archaeological materials and identifying the nature of human subsistence patterns prior to and following eruptions, archaeology can reveal the long-term (decades, centuries, millennia) effects of volcanism. Sometimes it can expose the immediate impacts (Figure 5.11) [85]. This chapter has only scratched the surface (both geographically and temporally) of this immensely rich field of enquiry. What transpires is that some prehistoric events had negligible long-term impacts on society even when the short-term effects were disastrous, while others are known to have had prolonged repercussions. Such different responses can arise from a

Figure 5.11 The footprints of Acahualinca (huellas de Acahualinca) found in the suburbs of Managua, Nicaragua are thought to reveal the tracks of 15 or 16 people fleeing through muddy tephra deposits during a powerful hydrovolcanic eruption of Masaya volcano around 2100 years ago.

wide range of factors including the scale and physical consequences of the eruption, the contemporary climatic conditions, land use and, crucially, the support and adjustment mechanisms of the disaster survivors. Where social, economic and kin networks extended beyond the devastated area, refugees could maintain cultural traditions, preserving cultural continuity when the impacted zone was re-colonised. In this case, the archaeological record will not indicate dramatic long-term effects. Although such models for vulnerability and impact are simplistic they can help to consider the range of potential impacts of both past and future eruptions on different kinds of society.

Another deep time perspective can be found in the oral traditions, memories and legends of living communities around the world. This is a potentially rich area for scientific research into volcanism and its impacts. Had volcanologists heard sooner of the legends of the Ayta people, it is possible that Pinatubo could have been recognised as an active volcano before it sprang to life in 1991, and better preparations could have been made for the largest eruption on Earth in the past century. Both archaeology and anthropology have much to offer contemporary disaster management.

Large eruptions pumping megatonnes of sulphur gases into the atmosphere, especially the middle atmosphere, lead to generation of atmospheric hazes that can wrap around the whole Earth. The fine particles of which they are composed interact with light in such a way as to dim the night-time stars, change the appearance of lunar eclipses, cast haloes around the Sun and colour sunsets and sunrises spectacularly. These kinds of phenomena did not go unnoticed to astute astronomers, scientists and historians of the past, and their documentation represents another rich source for volcanological and climatological data mining. Volcanically tinted sunsets even appear to have inspired great art! Of course, not every manifestation of a livid Sun, frigid summer, weird storm or stellar extinction will point to a volcanic eruption thousands of kilometres away, but where consistent patterns emerge across a wide geographical span, and other potential explanations can be discounted, the evidence builds.

6

Killer plumes

> . . . the long geologic time scale virtually guarantees that potential catastrophes such as large-body impacts and flood basalt volcanism will happen from time to time . . ., and the results of these very energetic events should be an important aspect of the geologic and biologic records.
> M. R. Rampino, *Proceedings of the National Academy of Sciences USA* (2010) [86]

Mass extinctions punctuate the palaeontological record. These overturns of species and groups of species are so monumental they help to define the framework of the geological timescale, with many smaller extinction events providing finer subdivisions. Perhaps the most famous mass extinction is that which occurred at the boundary between the Cretaceous Period (of the Mesozoic Era) and the Palaeogene Period (of the Cenozoic Era), 65.5 million years ago when, among countless other creatures, the dinosaurs finally went to the wall. The cause of these biological mega-crises has been sought in several directions but there are two main competing mechanisms - bolide (meteorite or comet) impacts versus volcanism on a massive scale. The extra-terrestrial theory argues that especially large bolides - perhaps 10 kilometres or more across - strike Earth from time to time and the collisions would generate sufficiently vast dust clouds to chill the surface, killing off organisms in the oceans and on land. For a long time, the Cretaceous–Palaeogene boundary and associated extinction have been linked to the impact that formed a 170-km-diameter crater at Chicxulub in México.

However, it is a startling fact that most of the recognised mass extinction events coincide with vast outpourings of (predominantly) basalt lavas known as large igneous provinces (LIPs) associated with mantle plumes. The associated gas emissions would likely initiate dramatic climate change, providing the kill mechanism. After a couple of decades during which the bolide argument has held sway, the causes of the Cretaceous–Palaeogene mass extinction look increasingly likely to involve volcanism.

Some of the leading proponents of the idea that volcanism caused mass extinctions now suggest that cyclic processes occurring in the Earth's core may be the ultimate cause of LIPs. Others have suggested that massive bolide impacts triggered them, and that the combination of effects was responsible for mass extinctions. This chapter investigates the coincidences and arguments - given the weight of circumstantial and theoretical evidence, it seems very likely that some mantle plumes are indeed 'killer plumes'.

6.1 MASS EXTINCTIONS

Mass extinctions are geologically brief 'bottlenecks' in genetic diversity that pinch out the palaeontological record. It is uncertain how brief is 'brief' because of the limited resolution of dating techniques but the timescales of mass extinctions are probably much less than 0.5–1 million years. The extinctions act worldwide and involve countless varieties of flora and fauna representing a broad range of habitats on land and in the sea. Such drastic loss of biodiversity leads to irreversible changes in ecological structure; entire ecosystems on land and in the oceans are re-organised. Mass extinctions represent drastic pruning of life's family tree in which entire branches were lopped off. The recovering shoots often represented novel environmental adaptations and solutions leading to new ecologies.

6.2 MORE ABOUT LIPS

Volcanologists do benefit from a helping hand from time to time. For example, it is thanks largely to the British Broadcasting Corporation that the expressions 'super-volcano' and 'super-eruption' have infiltrated our conversations at conferences. Despite its amusing acronym, the term 'large igneous provinces' seems in equal need of sprucing up for modern times, not least since 'large' hardly conveys the actual, gargantuan dimensions of these phenomena. These provinces typically cover areas of hundreds of thousands of square kilometres, and their volumes run into millions of cubic kilometres of erupted lavas and tephra (as well as comparable quantities of intrusive rocks). In terms of rock composition they are predominantly basaltic, though some LIPs contain significant quantities of silicic pyroclastic rocks and, as we shall see, one class almost exclusively consists of silicic ignimbrites and associated tephra-fall deposits. There are many important scientific questions concerning LIPs: what were their origins and role in the Earth's internal and surface evolution, and what have been their impacts on global climate, environment and the biosphere?

Large igneous provinces are found worldwide in three kinds of geographic settings: in the oceans they form plateaux such as Kerguelen and Ontong Java in the Southern and Western Pacific oceans, respectively; at the margins of continents that were once joined; and within the continents (Figure 6.1). A dozen LIPs are known to have formed over the last 250 million years, of which nine are within the margins of continents and the remainder are oceanic plateaux. We shall focus on the continental LIPs since these have received the most scientific attention (their rocks are most accessible) and they had the most direct impacts on the atmosphere.

A typical continental LIP – usually referred to as a 'continental flood basalt' province – is composed of thousands of individual lava flows (Figure 6.2). Some flows have been traced over distances of 1500 kilometres. One important and comparatively recent finding concerning LIPs is that they formed over much shorter timescales than used to be thought. Thanks to the accuracy of argon–argon dating of igneous rocks (Section 4.1.3), we now know that the majority of lavas erupted in LIPs were emplaced in periods of one million years or less (previous potassium–argon dates suggested time spans of millions or even tens of millions of years). This has huge ramifications for understanding the impacts of LIPs on the environment and life.

Since we have not witnessed a LIP eruption, we don't really know what they were like and some of our inferences are drawn from lava flood eruptions of minute comparative scale (such as the 1783–4 Laki fissure eruption in Iceland; Section 12.1). What we can say with confidence is that, over the million or fewer years during which typical LIPs have formed, eruptions would have been discontinuous but extremely intense. Lava eruption rates would have been extraordinarily high – much higher than the time-averaged rate of around one million cubic kilometres per million years, which is equivalent to one cubic kilometre per year. A more intense lava eruption will generate longer flows and higher fire fountains at the vent, and release gases such as carbon dioxide and sulphur dioxide into the atmosphere with much greater vigour.

Large igneous provinces are generally considered to be related to new mantle plumes (Section 1.1) [87]. They are discontinuous in space

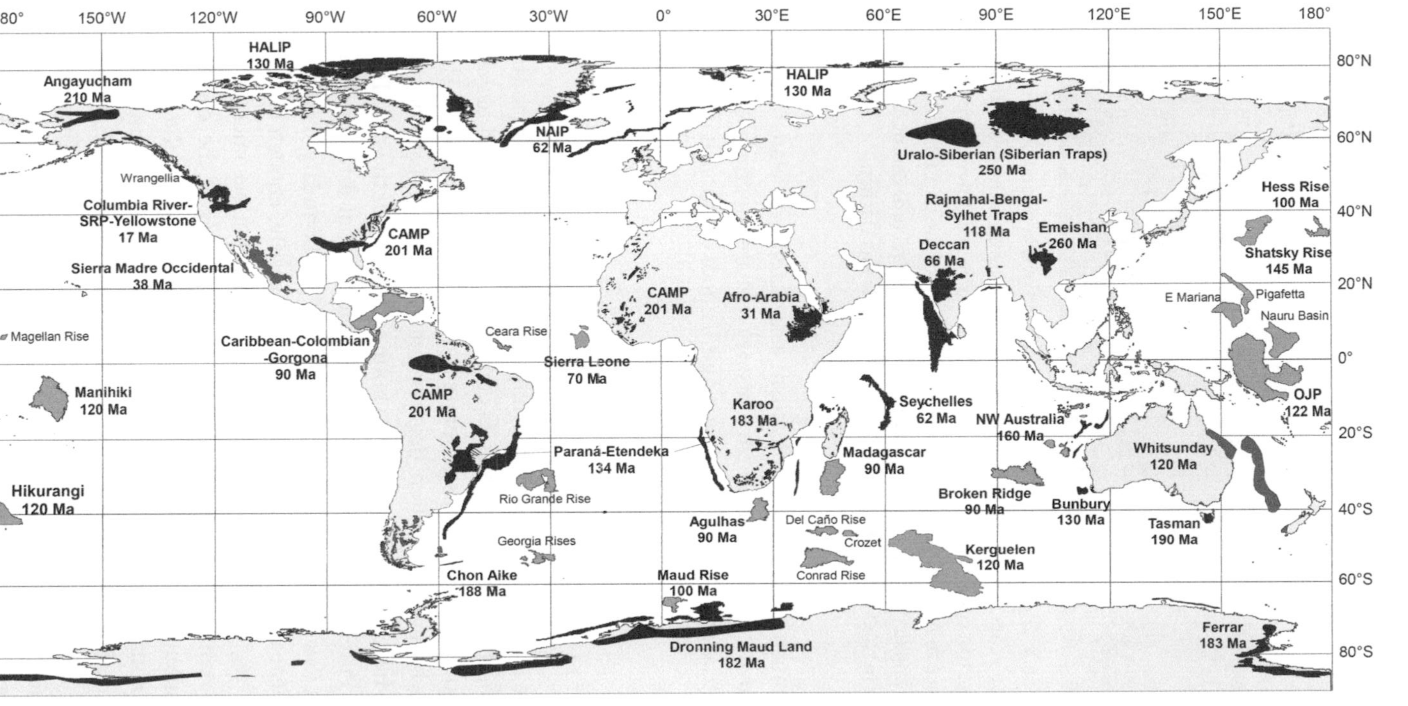

Figure 6.1 Global distribution of LIPs of the past 275 million years. Abbreviations: CAMP, Central Atlantic Magmatic Province; HALIP, High Arctic Large Igneous Province; NAIP, North Atlantic Igneous Province; OJP, Ontong Java Plateau; SRP, Snake River Plain. Courtesy of Richard Ernst [88] and used with permission of Elsevier.

Figure 6.2 The Ethiopian flood basalts erupted over a period of one million years, about 30 million years ago. This timing coincides with a change to a colder and drier global climate, a major continental ice-sheet advance in Antarctica, the largest sea-level drop of the past 60 million years and extinction events apparent in the fossil record. Photograph taken in the Simien Mountains looking towards Ras Dashen.

and time, much more so than volcanism associated with plate tectonics. The provinces punctuate Earth's history and are superimposed on the more steady production of magma and seafloor creation at oceanic ridges [88].

6.3 LIP ORIGINS

The mantle is the Earth's main internal shell: it stretches down to the boundary of the outer core, which is composed of an iron- and nickel-bearing liquid, whose motions generate the Earth's magnetic field. The outer core is several hundred degrees hotter than the base of the mantle, so heat must pass across the boundary between the two. This heat flow represents one of the sources of energy that drives mantle convection (Section 1.1). Mantle rock close to the boundary with the core will heat up and its density will fall. Eventually, enough of this lowermost region of mantle will be heated up that it becomes sufficiently buoyant to counter the strong viscous forces of the cooler, overlying mantle (remember this is all happening in the solid state) and rise as a mantle plume, as much as 300 °C hotter than the ambient temperature. Because of this incubation time, mantle plumes have big heads and skinny tails: they look a bit like mushrooms. The head continues to grow as the plume ascends towards the top of the mantle, partly through a snowballing effect as it assimilates the rock through which it is rising, and partly because the tail rises faster and catches up with the head, feeding it. When the plume head approaches the top of the mantle, a journey that has taken it around 20 million years, it flattens out (like an opened mushroom), reaching a diameter of up to 2500 kilometres, with a stalk (the tail) that is about 100–200 kilometres wide (Figure 1.2).

The high temperature of the plume head, coupled with its huge diameter and volume, ensures extensive melting as it decompresses towards the top of the mantle. This yields enormous quantities of magma that then rise through the crust to erupt at the surface. The volcanic rocks end up covering a roughly circular area as much as 2500 kilometres across. It takes several hundred thousand to a few million years to consume all the plume head but that still leaves the tail, which continues to rise. If the heat source is maintained at the core–mantle boundary then the tail continues to be fed. This can sustain plume volcanism at the surface for tens of millions of years, though these hotspots are much more localised and erupt at rates that are a fraction of those associated with the plume head. One of the best current examples of a hotspot volcano is the Big Island of Hawai`i, where the plume tail feeds several volcanoes intermittently (including Kīlauea, Mauna Loa and Lō`ihi seamount). However, because the Pacific Plate has been coasting across the hotspot, there is a long trail of other volcanic islands running northwards, and then seamounts, of ever increasing age, that can be traced nearly 6000 kilometres. These are all constructed from the Hawaiian plume magmas over the last 70 million years. Because the rocks from these various islands and seamounts have been dated (by radiometric techniques, Section 4.1.3) we can calculate the speed of the Pacific Plate motion from the pairs of age and distance of each island (or seamount) from Hawai`i. This reveals that the Pacific Plate is racing at around seven centimetres per year. The fact that oceanic LIPs tend to be obliterated by plate recycling is important since it means that the further back in time one goes the less complete is the record of LIPs. The oldest parts of the oceans formed about 250 million years ago, so we can only hope to find oceanic LIPs from earlier times if they failed to subduct (which has occurred).

But is there evidence for the massive volcanism that we would expect at the end of the Hawaiian chain corresponding to the time that the plume head bled its magmas to the surface? Actually, no – in this case, the trail of seamounts leads us to the subduction zone at the top of the Pacific Plate. Almost certainly, there was a LIP formed when the Hawaiian hotspot first formed more than 70 million years ago, but it has since been carried down the subduction zone, where much of it will have been recycled back into the mantle.

Good examples of present-day hotspot volcanoes that can be linked to their original plume-head products include Iceland (associated with the opening of the North Atlantic Ocean about 56 million years ago) and Yellowstone (at the end of the Snake River Plain and associated with the Columbia River Flood Basalt, which is the world's youngest – about 16 million years old – and smallest LIP, with a volume of around 0.17 million cubic kilometres). Another present-day hotspot volcano, which frequently erupts, is Réunion Island in the Indian Ocean. Its activity and lava products are reminiscent of Hawai`i's – not surprising, given its similar origin and oceanic setting. But, in the case of Réunion, we can trace the hotspot back in time 65 million years via submarine volcanic plateaux and ridges to arrive on land, in central India, where we find the more than 1.5 million cubic kilometres of lavas covering an area of around half a million square kilometres – the Deccan Traps (Figure 6.3). The total volume of igneous rocks (including intrusions) in the Traps may exceed a staggering eight million cubic kilometres, much of it erupted within half a million years. One of the individual lava flows – the Rajahmundry Trap – has been traced over an astonishing distance of 1500 kilometres!

Let's make a few 'back-of-the-envelope' calculations to imagine how this massive flood basalt might have been built up. If the individual lava flows of which it is constituted average 500 kilometres in length, 20 kilometres in width and 100 metres in thickness, then each flow unit has a volume of 1000 cubic kilometres. That would mean there must be 1500 component flows to add up to the total volume of Deccan lavas of 1.5 million cubic kilometres. If these were all erupted in half a million years, then the average interval between individual eruptions is 333 years. In other words, we could construct our continental LIP by erupting a 1000-cubic-kilometre lava flow every 333 years for half a million years.

In 2008, Steve Self (then at the Open University in the UK) and his co-workers caught the attention of the volcanological world by reporting analyses of Deccan melt inclusions (Section 4.1.5). Few would have imagined that such old samples could still retain their trapped volatiles. Self and colleagues estimated that for each cubic kilometre of lava erupted, at least three megatonnes of sulphur dioxide would have been released into the atmosphere (plus a comparable amount of hydrogen chloride) [89]. For an individual pulse of 1000 cubic kilometres of magma, that would mean a release of four to five gigatonnes of sulphur

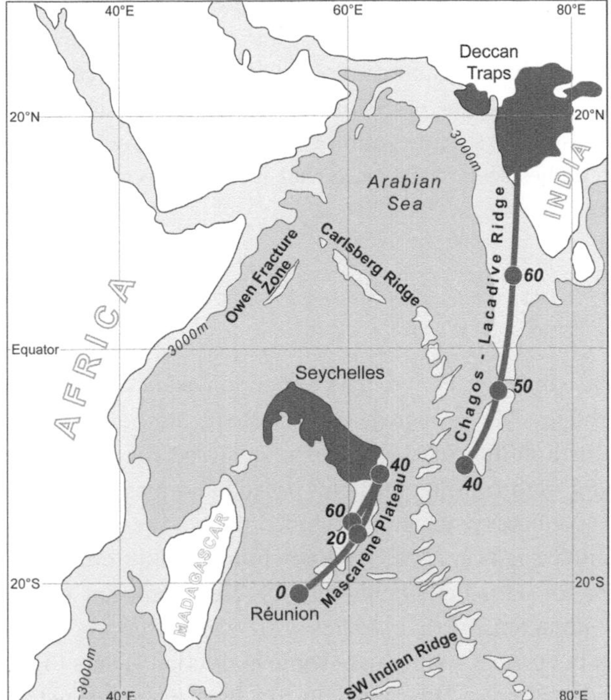

Figure 6.3 Map of the western Indian Ocean showing volcanism associated with the Réunion–Deccan plume (the plume is today beneath Réunion). The Seychelles were part of the Deccan Traps but were separated by spreading of the Carlsberg–Central Indian Ocean Ridge. The 200–300-kilometre-wide Chagos–Lacadive Ridge connects them, across the Carlsberg Ridge, to the Mascarene Plateau and Réunion Island. Dates shown in millions of years. Modified from [87] and used with kind permission of the Mineralogical Society of America.

dioxide (at least 200 times as much as Pinatubo's eruption in 1991). We have very little idea how long each flow took to erupt but estimates vary from years to a few decades. If we assume ten years, that would imply an average eruption rate of 100 cubic kilometres of lava per year - enough to top up around forty Olympic swimming pools every second. (For comparison, the longest-lived lava eruption today - that of Kīlauea on Hawai'i - erupts at a rate of around 0.08 cubic kilometres per year, which is comparable to the rate at which magma is supplied to the volcano from depth.) Such massive eruption intensities along with the expanse of active lava covering the surface are likely to have generated especially energetic, towering fire fountains and vigorous atmospheric updrafts, enabling a good fraction of the released sulphur gases to reach the stratosphere. Here, they would have oxidised to form a sulphuric acid haze capable of cooling the Earth's surface. Along with the sulphur and chlorine emissions, massive amounts of carbon dioxide would also have streamed out of both the erupting magma as well as magma bodies deep in the crust that were connected to the surface via magma plumbing systems.

6.4 LIPS, BOLIDES AND EXTINCTIONS: THE COINCIDENCES

The evidence linking LIPs and mass extinction events is circumstantial but it is impossible to disregard it. Richard Stothers of the NASA Goddard Institute for Space Studies re-analysed the best dates then available for both continental flood basalts and mass extinctions for the past 250 million years and concluded that all the major extinction events could be associated with a known flood basalt eruption with a high degree of statistical likelihood [90]. This result has been reconfirmed by numerous further studies with ever better precision and is illustrated in Figure 6.4. There are some extinction events that do not correspond to a known LIP, some LIPs that do not coincide with a mass extinction, and some evidence that extinctions are more frequent when bolide impacts and LIPs coincide, but none of this detracts from the extraordinary coincidence of large continental flood basalts with some of the most severe mass extinction events. Some LIPs are killers, it seems. Note, also, that because the preserved record of LIPs deteriorates going back in geological time (especially for oceanic LIPs beyond 200 million years or so) it is quite likely that some mass extinction events recognised from fossil data have simply lost their corresponding LIP after it was flushed down a subduction zone.

This might seem to present an open-and-shut case to blame continental flood lava eruptions for several major mass extinctions, including the most severe of all which took place at the end of the Permian Period, 251 million years ago, and the most famous of all, at the end of the Cretaceous Period, 65.5 million years ago, when the dinosaurs (amongst innumerable other fauna) perished. There is a complication, however, in the form of bolides. Meteorites and comets do strike the Earth from time to time – the larger ones thankfully rather infrequently. Meteor Crater in Arizona is one of the most famous holes in the ground left by a meteorite strike. Formed around 50,000 years ago, it is 1.2 kilometres across and was created by an iron meteorite of

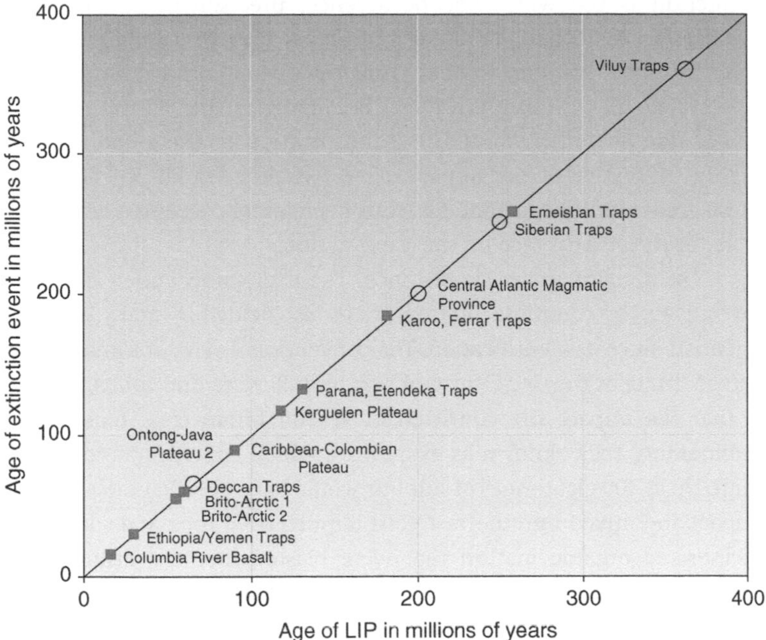

Figure 6.4 Ages of the 15 major flood basalts of the past 370 million years compared with mass extinction events. Larger circles denote four of the 'big five' extinctions. Data from reference 100.

around 30 metres diameter. In the geological record there is evidence for much larger direct hits from extra-terrestrial objects, whose impacts are hard to imagine (except by Hollywood directors). One of them has become more (in)famous than any other in the scientific world: the Chicxulub crater on the Yucatán peninsula of México, also known as 'the crater of doom'.

6.4.1 The end of the dinosaurs

Chicxulub's claim to fame is its great size – with a diameter of 180–200 kilometres the meteorite responsible must have been ten kilometres or more across – and it dates very close (some say precisely but this is contentious) to the very end of the Cretaceous Period (65.5 million years ago). The impact itself likely resulted in an ephemeral hole 30 kilometres deep [91], accompanied by massive ejections of crushed and shocked rocks, magnitude 11 earthquakes and mega tsunami. The 'smoking gun' said to link Chicxulub unambiguously to mass extinction is the worldwide occurrence, precisely situated at the Cretaceous–Palaeogene boundary, of a very bizarre geochemical feature. This is the 'iridium anomaly' and it has been found at hundreds of sites worldwide. Iridium is a metal belonging to the platinum family and is rare in the Earth, and its concentration in the global layer is highly suggestive of the impact of an extra-terrestrial object. Given these credentials, Chicxulub has thus been widely accepted as responsible for the end-Cretaceous mass extinction.

On account of its significance, the Chicxulub crater has been investigated by countless different methods wielded by many teams of scientists since its identification. The conventional view, summarised by Peter Schulte of the Geo-Centre of Northern Bavaria and colleagues [92] is that the impact site consisted of several kilometres thickness of sedimentary rocks known as evaporites, which contain abundant sulphur. Up to 500 gigatonnes of sulphur would have been vaporised by the impact and mixed into the vast palls of pulverised rock and soot (from incinerated organic matter) that were blasted into the atmosphere. The sulphurous clouds and soot are argued to have provoked massive climate cooling and acid rain that devastated photosynthetic life, initiating an extinction wave that travelled up the food chain [93].

However, the impact hypothesis has recently been losing ground and a surprisingly acrimonious debate has surfaced in the scientific literature between rival camps. The starting point for the counter-argument is, in essence, the striking correlation between LIP and extinction dates illustrated in Figure 6.4. All extinction events in the past 300 million years coincide with LIPs whereas only one impact – Chicxulub – unambiguously dates to the time of an extinction event despite extensive searches. Furthermore, there are many other known large impacts in the geological record that do not coincide with extinctions. Most intriguingly, recent stratigraphic evidence reported by Gerta Keller of Princeton University and her colleagues [94] reveals that the Chicxulub impact actually occurred some 300,000 years before the Cretaceous–Palaeogene boundary. Meanwhile, new dating of the Deccan Traps lavas suggests the zenith of their eruptions coincides precisely (within the uncertainties of the dates) with the peak in extinction [95]. Thus, there has been an increasingly vocal argument that volcanism played a significant role in the catastrophe.

Schulte and colleagues responded to these claims thus: 'the data and interpretations presented to warrant the staggering conclusions of this paper are insufficient, contradictory, and in part erroneous . . . A much more extensive discussion of the data . . . would be required to seriously contradict the general outcome of the earlier studies' [96]. To which Keller and colleagues replied: '[Schulte and colleagues] made their case by repeatedly resorting to factual misrepresentations, misinterpretations, out of context quotes, selective use of references, ignoring critical studies and bogus arguments. Amazingly, this was done in the most strident tone and accusations of misuse of biostratigraphy, geochemistry, mineralogy and sequence stratigraphy' [97].

Such ill-tempered geological spats do arise from time to time, and this one is by no means over. The 2010 review article by Schulte and colleagues provoked further published comment [98]. The authors of one critical letter described the article's extinction scenario as 'simplistic' and inconsistent with 'countless studies' of the fate of terrestrial and marine organisms at the end of the Cretaceous. They added that the extinction and survival patterns witnessed in the palaeontological record across the Cretaceous–Palaeogene boundary are so varied that they must point to multiple causes for extinction, including bolide impact, volcanism, sea-level change and associated climatic change. They concluded that 'The general importance of impacts to extinction is called into question, as well as the importance of the Cretaceous–Palaeogene impact as a single cause'.

The crux of the matter boils down to the stratigraphic relationships between the impact (and its associated debris), the iridium anomaly and the Cretaceous–Palaeogene boundary (and extinction event). According to Keller and colleagues, no evidence supports the assertion that the iridium layer, crater and boundary event are of the same age, and no iridium excess has yet been found in an undisputed Chicxulub impact deposit. On the other hand, Schulte and colleagues point out that the Deccan Traps were erupted over a time period that exceeds the duration of the extinction event. Why, they ask, should just one of the flood lavas have precipitated catastrophe when previous effusions of comparable magnitude did not? They also stress the immensity of the potential sulphur release from the vaporised salt deposits at Chicxulub's ground zero.

But they are overlooking a potentially major issue: the non-linear relationship between magnitude of sulphur release to the atmosphere and climate impact that we have already encountered in the context of volcanic explosions (Section 3.3.1). As the atmosphere is loaded with more and more sulphur, the additional effects on radiation are slight. This is due to generation of larger sulphuric acid particles that have less backscattering effect on incoming sunlight and shorter periods of suspension in air. It could well be argued that the repetitive thumping of the atmosphere with massive hits of sulphur and carbon dioxide from Trap lavas would have had the more devastating impact for life on Earth. Thus, the very fact that the Deccan volcanism *was* more prolonged and pulsatory than an instantaneous asteroid impact may have made it more deadly overall.

If Keller and colleagues are right about the timing of the Chicxulub impact, then one outstanding puzzle remains - what caused the global iridium anomaly? One suggestion is that there was a second, much larger and as yet unidentified bolide impact but this is hardly a parsimonious explanation. Presently, all that is certain is that the astonishing associations and coincidences between bolide impacts, iridium anomalies, massive extinction rates and the Deccan volcanism will be debated for many years to come.

6.4.2 Is the Earth's mantle a serial killer?

One idea proposed to link bolides, LIPs and extinctions is that some impacts actually initiate massive eruptions by triggering mantle plumes [99]. It has even been suggested that the craters from very large impacts are 'auto-obliterated' by near-instantaneous burial in lava floods, conveniently hiding the key evidence. That could explain the presence of a global iridium anomaly at the Cretaceous–Tertiary boundary without, it seems, a crater to link it to. But this hypothesis has been rejected by Vincent Courtillot and Peter Olson from the Institut de Physique du Globe de Paris and Johns Hopkins University in the USA, respectively. In fact, they identify an even more extraordinary coincidence in the Earth's behaviour that points to an entirely terrestrial, deep-Earth mechanism for extinguishing life at the surface [100]. But to explain it, we need first to consider why the Earth has a magnetic field.

In truth, there remain many mysteries surrounding the nature and behaviour of the Earth's magnetism but it is nevertheless widely accepted that it originates from the motions of the Earth's outer (molten) core. Being composed of iron and nickel, and therefore highly electrically conductive, its fluid motions act like a dynamo to generate the magnetic field. The effect is called the geodynamo and the resulting magnetism can be reasonably well approximated by imagining that there is a bar magnet at the Earth's centre. The nature of the Earth's rotation ensures that the magnet's opposing poles more or less line up with the geographic north and south poles. The equations required to explain the detail represent an exceptionally complex field of geophysics and remain vigorously researched but there is one other truly remarkable feature of the Earth's magnetic field: its polarity reverses from time to time. The key evidence for this derives from studies of the weak magnetism of igneous rocks erupted and intruded over geological history. On eruption, when lavas are very hot, their iron-rich constituents align, like tiny compass needles, with the prevailing magnetic field of the Earth. Once the rock cools below 570 °C (the so-called Curie temperature), the 'needles' (better described as 'magnetic domains') are literally set in stone. The direction they now point in can be measured using a magnetometer, and one way to operate such a device is by towing it from the back of a ship. This enables unambiguous measurements of the magnetic polarity recorded in submarine lavas of the oceanic crust.

Imagine starting a geophysical cruise directly above the Mid Atlantic Ridge, somewhere south of Iceland, and now travelling east (above the Eurasian Plate) dragging the magnetometer in your wake (Figure 1.1). At first, the instrument shows that the igneous compass needles in the seafloor are pointing more or less towards the north magnetic pole. But after you have travelled about ten kilometres the device reveals an unambiguous flip of the hypothetical compass, to point to the south magnetic pole! You continue your voyage in the same direction and find the switching pattern repeats over and over. If you now return to your starting point and instead sail west (across the North American Plate), you find the identical sequence of magnetic stripes. Recalling that the ocean crust is created at the ridge and spreads either side, our voyages represent journeys back in time – the further the distance from the ridge, the older are the lavas of the seabed below.

This amazing picture of seafloor spreading, first recognised in the 1950s, was, in fact, one of the most potent pieces of evidence that led to the theory of plate tectonics (which heralded a genuine revolution in thinking in the Earth sciences at the time). More significantly, for our purposes, it reveals the reversals of Earth's magnetic field. For the period of the last 158 million years, for which excellent seafloor data are available, just under 300 reversals have been identified. That equates to an average of roughly two every million years, though the actual rate is unsteady. Closer inspection of the longer records available shows that, every 200 million years or so, the Earth's magnetic field stops reversing for 30–50 million years. These intervals of constant magnetic polarity are called 'superchrons' – the last one began around 110 million years ago. Following a superchron, the rate of magnetic flips increases, reaching a peak of around 4–6 reversals per million years. What Courtillot and Olsen noticed is that LIPs (and associated extinction events) seem to occur 10–20 million years after each known superchron (Figure 6.5). For example, the 'Cretaceous Long Normal Superchron' ended about 18 million years before the Deccan Traps formed, and the 'Kiaman Long Reversed Superchron' ended ten million years ahead of the largest continental flood basalt, the 250-million-year old Siberian Traps.

The long interval between superchrons – 200 million years – suggests they are controlled not by processes occurring in the Earth's core (since these operate over much shorter timescales) but rather the convection of the much larger mantle. Further, the characteristic stirring time of the mantle is rather similar to the 30–50 million year lifespan of a superchron. Courtillot and Olsen propose that mantle convection itself influences the rate of heat flow across the boundary between the core and the mantle, and that this regulates the geodynamo. High heat flow excites turbulence in the outer core that, in turn, induces magnetic polarity reversals; whereas low heat flow eventually stalls reversals leading to a superchron. They also believe that the patterns of heat flow around the core–mantle boundary, and in particular its geographic distribution with respect to the Earth's rotational axis, affect the frequency of magnetic reversals.

The timing of superchrons is comparable to another large-scale geophysical feature of the Earth known as the 'Wilson cycle', which

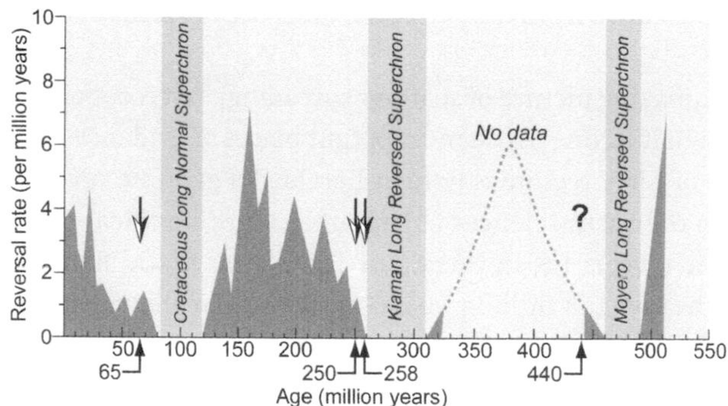

Figure 6.5 Changes in the rate of magnetic reversals over the past half-billion years compared with major extinction events (up arrows) and associated trap eruptions (down arrows). Polarity superchrons shown by light shading. Redrafted from reference 100 and used with permission of Elsevier.

describes the alternating episodes in which continents collided to build super-continents, only to disaggregate again via rifting and drifting. Courtillot and Olsen wonder if a consequence of this cycle is to accumulate dense, sinking mantle rock just above the core–mantle boundary during the peak of mantle convection in the rifting phase. This would tend to insulate the core, reducing heat flow out of it and preventing magnetic reversals. But over a few tens of millions of years, the dense basal layer of mantle heats up by conduction and starts to become buoyant. The core–mantle boundary becomes a mantle plume nursery! Once ripened sufficiently to hatch, the nascent plumes disturb the core–mantle boundary enough for it to return to high heat flow conditions. The superchron is over and magnetic switches resume. It takes another 10–20 million years for the lavas to erupt (and perhaps trigger a mass extinction) suggesting plume rise rates in the mantle of a few tenths of a metre per year (several times faster than present-day plate-tectonic motions).

This hypothesis needs to pass thorough testing and evaluation before it can be accepted but, unlike the bolide impact theory, it does explain why superchrons precede LIPs. It is hard to imagine that a large comet or asteroid would intercept Earth's trajectory through space just because 20 million years had elapsed since its last superchron. On the other hand, it remains possible that the apparent timing of superchrons and LIPs simply arises through chance.

6.5 KILL MECHANISMS

Countless hypotheses have been advanced to explain the processes by which volcanism or bolide impacts could lead to mass killing. Early ideas for magmatically induced mortality included poisoning by metals such as lead, copper and cadmium released into the environment by volcanic gases. Other ideas included dramatic global cooling – even glaciations – arising from sulphate aerosol veils formed in the stratosphere from all the volcanic activity associated with LIPs. A further hypothesis is that all mass extinctions begin with a collapse in biomass production at the base of the food chain – for instance through killing off photosynthetic bacteria in the oceans – that leads to global starvation. Andrew Knoll from Harvard University and his colleagues have approached the debate from the perspective of physiology and pathology, asking whether there is evidence in the fossil record that points to the physiological processes that caused death [101].

'Palaeophysiology' is fraught with difficulties because, by definition, extinct organisms are unavailable for modern studies to test their environmental tolerances. However, although we cannot, for example, experiment with the long extinct ammonites to explore how they might have coped with variations in water temperature, salinity, oxygenation and so on, we can observe their relatives, the coleoids (which include beasts such as the nautilus and cuttlefish), which thrive today in various oceans of the world, and make useful inferences. Thus, experiments using living taxa can suggest the physiological impacts that would be expected for different kinds of kill mechanism. With that information in mind, we can compare varieties of organism that survived a mass extinction with those that succumbed, and consider if patterns emerge that fit with certain physiological pathways to the graveyard. Once we know the kill mechanism, we can test if it is compatible with our supposed trigger mechanism (e.g. eruption of a LIP). One additional point to make is that a kill mechanism does not have to annihilate a community instantaneously to extinguish it. In the case of many marine species, even a modest decrease in a population of 1% per generation would finish them off in around a century. This is important because, in thinking about the triggers of mass extinctions, it means that lower amounts of environmental 'stress' can still be sufficient to bring a population down, fast.

As we have alluded to, one plausible scenario by which a LIP might trigger a mass extinction is from the gases and particles emitted into the atmosphere during the eruptions. Suspects include toxic substances such as trace metals, sulphur dioxide (which can lead to surface cooling if it reaches the stratosphere) and carbon dioxide (which leads to global warming). In the case of carbon dioxide, for instance, it has been estimated that the Siberian Traps may have pumped up to 40,000 gigatonnes of the gas into the atmosphere – as much as 1000 times the amount that was already there. (Much of the carbon dioxide was sourced directly from the fizzing magma but it is thought that additional carbon emissions were sourced from combustion of coal deposits in the crust.) Such an amount of carbon dioxide released into the atmosphere would lead to a powerful greenhouse climate and corresponding warming, with many knock-on effects. One consequence of biological significance is that as the oceans warm up their capacity to hold oxygen drops. This alone could asphyxiate many organisms living in already poorly oxygenated parts of the sea. Bacteria capable of metabolising sulphates in seawater would have thrived in the low oxygen conditions and would have generated further carbon dioxide as well as hydrogen sulphide, which would fizz through the surface waters and into the atmosphere. Many plants and animals would have been subjected to a combination of asphyxiation, carbon dioxide and hydrogen sulphide poisoning, heat stress, and even increased ultraviolet radiation due to loss of ozone in the atmosphere (Section 3.2.5).

Studies of these different biological threats are abundant thanks, in large part, to concerns surrounding the effects of present and future climate change. If we look at the oceans today, one basic observation is that oxygenated waters support more biomass composed of larger and heavier organisms representing a greater diversity compared with oxygen-poor waters. Furthermore, where carbon dioxide concentrations are high in the oceans, organisms grow slower, live shorter lives and have lower reproductive rates. This is because high carbon dioxide limits the oxygenation of tissues and upsets the organism's pH balance. One especially damaging consequence of this, as we shall see below, is for the growth of skeletons.

Hydrogen sulphide is bad news for nearly all animals, plants and fungi because it inhibits the functioning of the structures in cells called mitochondria, which are vital for metabolic processes (Section 7.3.2). For humans, just 0.07–0.1% of hydrogen sulphide in air is lethal. From time to time people are poisoned by the gas in geothermal areas such as Rotorua in New Zealand, usually when it has leaked through the soil and accumulated in an enclosed space such as a basement. Large release of hydrogen sulphide into the atmosphere would also have a profound impact on the atmosphere's oxidation state and, through various chemical reactions, would end up increasing the abundance in air of methane (a greenhouse gas) while depleting the stratosphere of its ozone, which filters out biologically harmful ultraviolet radiation.

These arguments seem stronger than others that focus on bringing global primary productivity to its knees. For instance, it has been suggested that the demise of certain bacteria in the oceans might lead to the success of others such as the blue-green algae without a significant net decrease in biomass. Blue-green algae are simply not that good to eat – even for crustaceans and molluscs – and this change in the distribution of species could be enough to lead to catastrophe. Knoll and colleagues argue, rather, that a combination of factors – low oxygen and high carbon dioxide in the oceans, bacterial production of hydrogen sulphide and global warming – conspired in mass extinctions.

Rates and timescales of environmental change are also a key in understanding mass extinctions. If changes are slow, taking place over tens to hundreds of thousands of years, then genetic adaptations of organisms can keep pace. In the words of Knoll and his colleagues, 'effective kill mechanisms act fast'. Each pulse of a LIP would pump more sulphur dioxide and carbon dioxide into the atmosphere – the carbon dioxide would accumulate in the atmosphere leading to progressively stronger greenhouse conditions punctuated by intermittent cold snaps due to the short-lived effects of a sulphuric acid dust layer in the upper atmosphere. The environment and climate would make increasingly profound swings with each successive eruption, and the rates of change would be far too rapid for genetic mutations to enable adaptations that would make it possible for the organisms to cope.

Focusing on the Permian–Triassic boundary extinction (linked to the eruption of the Siberian Traps), Knoll and his colleagues argued that carbon dioxide did play a leading role as a kill mechanism. For many marine organisms, the physiology of building skeletons is critical. Bear in mind that one of the most common substances forming shells and coral structures is calcium carbonate, and that the carbonate must somehow be extracted from the seawater in which the organism lives. In modern oceans, the animals most susceptible to increasing dissolved carbon dioxide in the water are the ones that build extra-large skeletons compared to the quantity of body tissue they have to support them, and with limited capacity to buffer their body fluids against changing seawater chemistry. This class includes corals, certain sponges, most sea urchins and various kinds of shelly fauna. On the other hand, fauna that grow skeletons out of something other than calcium carbonate and with the potential to control their physiology fare better. Molluscs, crustaceans and assorted fish fall into this group.

Knoll and colleagues examined the fossil record either side of the Permian–Triassic boundary to see which critters made it through and which met their end. The results were startling: 85% of the genera that built outsized calcium carbonate skeletons died out (including species familiar to all first-year geology students: the rugose (horn) corals, most rhynchonellid brachiopods and crinoids (also known as sea-lilies)). On the other hand, a mere 5% of the genera with little or no calcium carbonate in their skeletons perished (including lingulid brachiopods, polychaete worms, sea cucumbers and conodonts (a toothy variety of now extinct fish)). Intermediate between these two groups were animals that built more modest skeletons out of carbonate, and had some potential for protecting their body tissue against changes in seawater chemistry (including sea urchins, ammonites, molluscs and arthropods). 54% of such genera went to the wall.

While the varieties of beasts of the ocean world selected for annihilation lend support to the hypothesis of carbon dioxide poisoning, more recent work suggests that physiological responses of calcifying organisms in the ocean are really quite complex and not so easy to predict [102]. Knoll's view that 'skeletal physiology really was destiny during the end-Permian catastrophe' does therefore require rigorous testing as better data become available. There are also a number of other factors that likely played a role in the Permian mass extinction. These include stagnant conditions in the ocean (identified from occurrences of very fine-grained, laminated seabed deposits indicative of minimal water disturbance) and extremely low oxygenation of the water, which would have drastically reduced the available marine habitat area. These stifled and suffocating seas have become known as 'Strangelove oceans' associated with limited primary productivity and the virtual shut-down of the biological pump that transfers organic carbon from dead plankton in the surface water to the deep ocean. Via complex biogeochemical feedbacks in the oceans involving nutrient supplies and bacterial blooms, these characteristics would have exacerbated hydrogen sulphide build-up in surface waters with the damaging consequences already discussed. This leads to the view that hydrogen sulphide and carbon dioxide likely worked together during this and other extinction events [103].

As sulphidic conditions took hold in the oceans, eventually the gas should have been released into the atmosphere. This offers one possibility for extinction on land, which leads to the next issue: any kill mechanism proposed for the oceans must be compatible with what was happening to the biota on land. While land plants can be adversely affected by high carbon dioxide due to acidification of soils, on the whole, high carbon dioxide in the atmosphere should be beneficial to vegetation since it would enhance photosynthesis. The more likely substantial impact of increased atmospheric carbon dioxide is on global temperatures, leading to what has been dubbed the 'End-Permian Inferno'. Temperature affects chemical reaction rates and thus influences not only photosynthesis but indeed the metabolism of most organisms. Variations in temperature around the world play a major role in the geographic distributions of flora and fauna. Profound and rapid environmental change leaves populations with only three options: cope, migrate to a more favourable environment (if one exists) or die. In one study cited by Knoll and colleagues, modelling of the known extinctions of land plants and animals over the past 30 years predicts that by 2050 between 15 and 37% of them will be extinct due to continued global climate change, with the losses greatest in temperate regions. A further recent study of 201 million-year-old rocks (laid down during the timespan of the flood basalt eruptions of the Central Atlantic magmatic province) provides further compelling evidence for the link between volcanism, a carbon dioxide 'super greenhouse' and mass extinction [104].

One further contributing factor to consider is that the stratospheric chemistry of volcanic clouds is very complex (Section 3.2). Accompanying the sulphur emissions will be large quantities of reactive gases including hydrogen chloride and hydrogen bromide. For the Deccan Traps, Self and colleagues estimated a release of around one megatonne of hydrogen chloride per cubic kilometre of magma. That could have resulted in emissions of up to one gigatonne of hydrogen chloride per year. Figures for hydrogen bromide are unavailable but both gases could be expected to act with the sulphate aerosol in the stratosphere to destroy ozone. The role of stratospheric ozone in limiting potentially damaging levels of the Sun's ultraviolet radiation, especially so-called UV-B wavelengths (0.29–0.31 thousandths of a millimetre), is well known. Increased ultraviolet penetration through the atmosphere can affect the growth, health and reproduction of many organisms living at or near the surface of the land or oceans. This has led some researchers to suggest the possibility of 'ultraviolet catastrophes' [105]. Even in the case of geologically puny eruptions such as that of Pinatubo in 1991, models suggest that, locally, dosages of UV-B radiation increased by up to 40% [106]. It is difficult to test whether this could have been an effective kill mechanism from the fossil record, but it remains likely that volcanically induced increases in ultraviolet radiation can only have added extra environmental stress to organisms that were already struggling to cope.

6.6 HOT LIPS AND COLD SLIPS

Although we have focused on dominantly basaltic LIPs, the moniker is also applied to describe some very large volcanic provinces composed mostly of silicic lavas and tephra. To distinguish them from their basaltic counterparts, they are referred to as SLIPs (for silicic LIPs) and sometimes as 'ignimbrite flare ups'. The most intensively researched SLIP formed episodically from 50 to 16 million years ago and stretches 3500 km from México to the southwestern USA. The total volume of ignimbrites is at least 0.4 million cubic kilometres, three quarters of which is represented by the Sierra Madre Occidental of México

(Figure 6.1). Three bursts of 20,000–50,000 cubic kilometres of ignimbrite each built the Sierra Madre Occidental 46–43, 38–27 and 24–18 million years ago. Individual ignimbrites have typical volumes of around 600 cubic kilometres. Considering that a comparable amount of tephra would have been associated with each (a volcanological rule of thumb is that for each kilo of ignimbrite there is a kilo of phoenix cloud fallout somewhere on the planet), it adds up to around 700 magnitude-eight super-eruptions during the period of activity; roughly one every 40,000 years. During the peak in volcanism, around 34–33 million years ago, the eruption rate would have been closer to one every 15,000 years.

Steven Cather, from New Mexico Institute of Mining and Technology, and his colleagues have suggested that, during the peak in activity, the plumes of ash associated with these gigantic explosive eruptions might, via an extraordinary biological route, have led to severe climate cooling 33.5 million years ago [107]. This global event lasted for around 300,000 years and heralded the glaciation of Antarctica. Cather and colleagues calculate that the prevailing easterly trade winds at the time would have deposited a time-averaged 75 million tonnes per year of ash across the equatorial Pacific Ocean. This would represent about half the total fallout of dust over the ocean. Volcanic ash typically carries iron salts on its surface that can be readily dissolved into seawater (Section 2.2.2). According to Cather and colleagues, Sierra Madre Occidental ash might have carried about 0.023% of water-soluble iron.

The biological significance of iron in the oceans is that it is a vital nutrient used in the synthesis of chlorophyll. Without sufficient dissolved iron, primary productivity in the oceans is low. Near coasts, there is typically plenty of iron washed in from the weathered land surface and from upwelling ocean currents. In the middle of the oceans, the main source of iron is wind-blown dust, sourced from the world's deserts. The iron distribution in the oceans is, therefore, highly variable and, today, around 30% of the open ocean can be classified as iron-deficient. Consequently, these parts of the ocean, including the Southern Ocean and the tropical and North Pacific, do not teem with life as do their counterparts elsewhere.

Sporadic volcanic eruptions thus have the potential to fertilise the oceans – bringing essential minerals to otherwise nutrient-limited waters. In theory, the fertilisation effect might be large enough to cause global cooling: if primary productivity in the oceans 'blooms' then the carbon dioxide dissolved in the sea is used in photosynthesis to make organic carbon. Much of this carbon ends up being recycled in the oceans but a fraction is buried as sediment on the ocean floor. To compensate for their loss of carbon dioxide, the oceans draw carbon dioxide from the atmosphere; the consequent reduction in a key greenhouse gas cools global climate.

Returning to the argument for the Sierra Madre Occidental volcanism, 0.023% of 75 million tonnes of ash per year equates to a time-averaged annual flux of iron of 17 kilotonnes. An oceanographic rule of thumb tells us that one gram of iron supports production of half a tonne of organic carbon. So, 17 kilotonnes of iron yields 8.7 gigatonnes of organic carbon per year. As we noted, only a fraction – around 0.1% – ends up being buried on the seafloor, equating to 8.7 megatonnes of carbon burial. Accumulated over 300,000 years, that would run to 2600 gigatonnes of carbon burial. This figure is already 2.6 times as much withdrawal of atmospheric carbon dioxide as is needed to account for the global cold snap.

Cather and colleagues go on to point out further coincidences in the geological record between glaciations and silicic LIPs, going back as far as 465–428 million years ago when ice sporadically advanced across the continents. This period overlaps with the peak in volcanism, around 457–448 million years ago, that produced very widespread ash beds known as bentonites. Unfortunately, it is challenging to test this theory given uncertainties in ages of the various geological formations. And while the ash fertilisation theory seems viable, it has to be balanced against the potential toxicity of ash in the environment. As well as iron, several other trace metals can be leached from ash grains, including copper, which is especially toxic. Indeed, the release of volcanic trace metals has even been suggested as one of the kill mechanisms involved in mass extinctions. Thus, predicting the precise biological effects of an eruption tens or hundreds of thousands of years ago on the flora and fauna of the oceans is fraught with difficulties.

6.6.1 LIPs, 'volcanic winters' and 'snowball Earth'

Even further back in time, there is geological evidence for glaciations even more extreme than anything we have discussed so far – when the ice advanced to the equator and the sea froze over. If the hypothesis is correct, the tropics might have looked like present-day Antarctica; the hydrological cycle would have effectively shut down and much of the Earth's flora and fauna would have perished. The evidence for Earth turning into a giant snowball is mainly drawn from the identification of widespread glacial deposits collectively referred to as tillites in many regions of the Earth that were, at the time, close to the equator (as deduced from palaeomagnetic measurements of the rocks). Not all scientists accept the arguments for a global 'icehouse' but we can be certain, at least, that widespread glaciations affected the planet around 754, 635 and 580 million years ago.

What brought Earth to such an icy state is also vigorously debated but factors are likely to include the geographic distribution of the continents at the time and strong chemical weathering, which would have drawn carbon dioxide from the atmosphere. This is an extremely important process in modulating Earth's atmospheric carbon dioxide abundance so it is worth briefly explaining how it works. Over geological timescales, the principal source of carbon dioxide to the atmosphere is volcanism. Cloud droplets and falling rain dissolve a small amount of carbon dioxide from air, which then dissolves to form a weak acid. Streams and rivers are thereby able to erode chemically the rocks over which they pass as well as the load of suspended mineral matter that they transport. Taking the average composition of the Earth's continental crust as a calcium silicate (a generic granite), the chemistry acts to remove carbon dioxide from the atmosphere and transfer it, in the form of bicarbonate ions, to the oceans, which will be used by marine organisms in building their skeletons:

$$CaSiO_3 + 2CO_2 + H_2O \rightarrow Ca^{2+} + SiO_2 + 2HCO_3^- \quad [6.1]$$

As we have seen, a fraction of this marine carbonate ends up being buried on the seabed, representing a net loss of atmospheric carbon dioxide. This reduces the atmosphere's greenhouse effect, reducing surface temperatures.

Robert Stern, from the University of Texas at Dallas, and his colleagues have suggested that what triggers the Earth to cool in the first place is the effect of sulphuric acid aerosols injected into the middle atmosphere by massive volcanic eruptions [108]. Although there is no decisive proof of the argument, they cite circumstantial evidence in the overlapping ages of the ancient rocks of East Africa and Arabia and the widely dispersed glacial deposits of the 'Marinoan' glaciation around 635 million years ago. This is interesting, not least since volcanism is thought to be responsible for eventually switching off the freezer and thawing out snowball Earth. Because the hydrological cycle essentially grinds to a halt during a global icehouse, the carbon dioxide emitted from volcanoes remains in the atmosphere. There is no continental weathering to extract it from the air and bury it on the seabed so, instead, the carbon dioxide levels build up. After around ten million years, the atmospheric carbon dioxide abundance should reach around 12% and the greenhouse effect become so strong that all the ice melts very rapidly – much faster than has happened during the terminations of the recent ice ages.

These hypotheses, taken together, suggest it is possible that volcanism has played a leading role in the evolution of life on Earth. Many scientists contend that life is likely to have originated in the hot fluid emissions from volcanic vents such as the black smokers of oceanic ridges (Section 1.1). These same hydrothermal environments, along with volcanically and geothermally heated areas, may have provided vital, ice-free refuges for ancient life-forms that enabled them to survive through global glaciation. Once the ice melted, the fossil evidence suggests that life went through extraordinary radiations and diversifications as the surviving multi-celled organisms exploited a warm planet, vast fluxes of nutrients carried into the oceans thanks to the resumed hydrological cycle and the abundance of mineral-loaded, glacially scoured rock powder, and the initially scarce competition.

Bringing this closer to home, one of the factors that led to the rise of mammals (and ultimately our own species) was the elimination of the dinosaurs. Had it not been for the eruption of the Deccan Traps, these rather successful predators might still be thriving today; in which case, this book would probably not have been written . . .

6.7 SUMMARY

If ever we could justify the use of the word catastrophe, it would surely be in describing the mass extinction events that punctuate the geological record. We have seen evidence that these resulted from processes occurring within the Earth's deep interior that act like a gigantic heat pump; every 20 million years or so, mantle plumes stir up the mantle leading to the eruption of unimaginably vast amounts of lava, accompanied by gases that pollute the atmosphere, land and oceans. In effect, the Earth periodically poisons the biosphere from the inside out. Although the evidence is only circumstantial, there are also reasonable hypotheses that link volcanism to both the initiation and termination of 'snowball Earth' glaciations. If volcanism did switch the Earth's climate between icehouse and greenhouse, then, by implication, it played a leading role in the evolution of life many hundreds of millions of years ago.

The detail surrounding the processes by which new mantle plumes end up, in some cases, almost wiping out all species is extremely complex but a key assassin is likely to be carbon dioxide. In the oceans, it poisons many species that need to secrete carbonate skeletons, while in the atmosphere it leads to global warming, with the attendant losses of biodiversity that are now readily apparent in our own age. We know that it is the rates of environmental change that are crucial in determining whether species can adapt and survive or else perish. The eruption of continental flood basalts involved hundreds or thousands of individual events recurring every few hundred years perhaps. Each would have resulted in a massive emission of sulphur into the atmosphere, likely to have been responsible for profound global cooling via the optical effects of the resulting stratospheric aerosol veil. But this cooling would have been relatively short-lived – lasting at most a few decades. Then a more prolonged period of warming would arise from the carbon dioxide accumulating in the atmosphere. The combined effects of rapid climate change, extreme fluctuations in seawater carbon dioxide and oxygen contents, and bursts of methane, hydrogen sulphide and toxic metals flushing the environment, cannot have conspired to make life easy.

It is important to recognise that while all significant extinction events coincide with LIPs, not all LIPs are associated with mass extinction events – some plumes are killers, others not. A notable benign example is the 133-million-year-old Paraná-Etendeka continental flood basalt province that was erupted across what are now parts of Brasil, Namibia and Angola prior to opening of the South Atlantic Ocean. Despite its great volume of lava, the contemporaneous extinction rate was very low. Another, more recent (30-million-year-old) LIP, the Ethiopian and Yemeni Traps, is implicated in 'bioclimatic' events but it is its subsequent development into the East African Rift system that we explore in the next chapter.

7

Human origins

'Roughness', generated by tectonic and volcanic movement characterises not only the African rift valley but probably the whole route of early hominid dispersal.

G. King and G. Bailey, *Antiquity* (2006) [109]

Over the last two million or so years, Earth has experienced tremendous climatic swings forced principally by subtle changes in the planet's axial rotation and orbit around the Sun. The last peak in glaciation was roughly 23,000 years ago, with a return to temperate conditions 11,500 years ago. The glacial periods were often associated with severe aridity in eastern Africa, drastically altering the composition of ecosystems and geographical ranges of human and other species. On the other hand, the warm intervening periods witnessed expansion of savannah, woodlands and lakes. Any attempt to understand human origins and evolution has to take account of this oscillating climate and its associated ecological and physical impacts (including waxing and waning of ice sheets and changing sea level).

An important question is: why does eastern Africa figure so prominently in the story of human evolution? And why, in particular, the East African Rift Valley? Some geologists and archaeologists argue that tectonic and volcanic activity might be the answer. They suggest that continuous rejuvenation of the rift by eruptions and earthquake faulting provided a unique geological driver for human origins and evolution.

There are many unresolved puzzles in palaeoanthropology in part because the human fossil record is so patchy. New hominin finds can abruptly overturn prior consensus. Discussion of the role of volcanism in fostering human adaptation and evolution, and in sporadically spurring migration, is therefore highly speculative. Nevertheless, the arguments are absorbing and cannot be simply dismissed, and it is the purpose of this chapter to review them.

7.1 THE EAST AFRICAN RIFT VALLEY

Around 30 million years ago, a new mantle plume (Section 1.1) impinged on the African Plate. The crust domed upwards and, eventually, partial melting of the uncommonly hot mantle yielded massive outpourings of basalt lava flows (Figure 6.2), punctuated by occasional super-eruptions of pumice and ash [110]. The uplifted region began stretching until it more or less split along a north–south line of weakness [111]. Continued extension caused fracturing along earthquake faults such that sections of the Earth's crust rotated like books slipping sideways on a bookshelf. In this way, the crust weakened and thinned.

All this action concentrated in such a way as to form the Red Sea, the Gulf of Aden, Danakil Depression and the East African Rift Valley. In the case of the Red Sea and Gulf of Aden, clearly the stretching process went all the way in rupturing the continental crust, filling the rent with new oceanic floor and culminating in creation of a new plate boundary. But across the East African Rift the stretching was slower paced and, though the crust has been greatly thinned beneath the valley floor, it remains composed of continental rocks rather than oceanic basalts. The Danakil Depression of Ethiopia, Eritrea and Djibuti, which geographically and tectonically joins the Red Sea and Gulf of Aden to the rift (forming a 'triple junction'), is right at the geological transition between continent and ocean. Indeed, much of the depression is 100 metres or more below sea level, and was beneath the waves 125,000 years ago.

The African Rift actually has two parts, the eastern branch, which runs through Ethiopia, Kenya, Tanzania and Malawi, and the western branch, which passes through eastern Congo and along its border with Tanzania. In the western rift, volcanic activity has been focused at Rungwe in southern Tanzania and the Virunga mountains that straddle the border between the Democratic Republic of Congo, Uganda and Rwanda. One of the Virunga volcanoes, Nyiragongo, hosts the most spectacular lava lake on Earth. In contrast, the eastern rift is peppered with volcanoes along much of its great length. Many lie within the valley but some massive volcanoes have grown on the flanks of the rift, including Mt Kilimanjaro (the highest mountain in Africa), Mt Meru, Mt Kenya and Mt Elgon. It is a geological curiosity that many of these volcanoes appeared in the last few million years. This burst of

Figure 7.1 Nabro volcano on the border between Ethiopia and Eritrea has an eight-kilometre-diameter crater. Its eruption chronology has yet to be ascertained.

magmatism continues today - seismic imaging techniques have revealed that much of the rift is underlain by magma bodies.

Between Eritrea, Ethiopia and Kenya, there are more than two dozen silicic caldera volcanoes of likely Quaternary age. Among them are the spectacular double caldera system of Mallalle and Nabro on the border between Ethiopia and Eritrea (Figure 7.1), Ma'alalta, Kone, Shala and Corbetti in Ethiopia, and Suswa and Menengai in Kenya. Caldera diameters range from 5 to 17 kilometres and certainly some of the volcanoes are associated with presently active magmatic systems [112]. As we shall see, this geologically recent resurgence of volcanism, much of it highly explosive as demonstrated by the size of the calderas and widespread distribution of silicic tephra layers, may be particularly relevant for understanding human evolution and dispersals. Another very useful result of the volcanism is that the tephra provide some of the most accurate means for dating fossil-bearing sites (Section 4.1.3).

7.2 THE FIRST HUMANS

Eastern Africa is widely regarded as the 'cradle of humanity'. Vegetarian apes are thought to have emerged from the forest as eastern Africa became increasingly arid. Pre-human forms such as *Australopithecus* (the genus to which Dinkenesh, also known as Lucy, belongs) first appeared more than four million years ago, starting the primate revolution of upright gait. In fact, the earliest direct evidence of walking on two feet is found in the hominin footprints at Laeotli in Tanzania - which are preserved in a 3.6-million-year-old tephra bed [113]! With increased brain size and physiological adaptations to heat stress, the first transitional humans (such as *Homo habilis*) appeared around 2.5 million years ago. By 1.8 million years ago, pre-modern humans in the form of *Homo erectus* were on the scene. They ranged widely within and beyond Africa, and meat contributed significantly to their diet.

During the period that the pre-modern forms of *Homo* were evolving, they became increasingly adept at knapping stone tools. These were employed to support a more carnivorous diet, sourced by hunting or scavenging herd animals of the savannah. However, this supposed behavioural transition from forest to savannah dwelling faces a contradiction: how could archaic or pre-modern humans living on the open plains of Africa have avoided the attention of faster and fiercer predators (lions, for instance)?

The pioneering geographer Carl Sauer exposed this paradox more than four decades ago, arguing that 'early man' 'was not specialised for predation; he was inept at flight or concealment; he was neither very strong nor fast. His daytime vision was good but he avoided moving about at night. He was not a forest creature; he lacked a protective pelt against cold, or thorns; he sweated most profusely on exertion and in dry air and so could not live far from water' [114]. Sauer supposed that instead of thriving on the plains, pre-modern humans exploited coastal areas. Others argued that patches of forest must have provided refuge.

More recently, the seismologist Geoffrey King (from the Institut du Physique du Globe in Paris) and archaeologist Geoff Bailey (from the University of York) have proposed that the tectonically and volcanically active environment of the East African Rift Valley played an essential role in early human ecology [109]. They emphasise the 'underlying geodynamic processes [that maintain] mosaic environments and the distinctive topographic features associated with them'. In particular, they speculate that the topographic roughness of terrain broken up by lava flows and earthquake faults, and its renewal and maintenance by repeated eruptions and earthquakes, stimulated human evolution. Such roughness would have enabled 'a ground dwelling bipedal omnivore [to] gain tactical advantage over faster moving quadrupeds, find protection from carnivore predators and outmanoeuvre herbivore prey, thereby gaining access to the vast meat reservoir of the savannah plains, while also benefiting from other resources'.

King and Bailey highlight four key aspects of the physical environment that promoted these developments: diversity of habitats and food resources; abundant and accessible water; security; and opportunities for trapping prey. They suggest that the tectonic and volcanic activity sustained in the African Rift for millions of years (throughout the period of human evolution) has constructed a unique geological setting meeting all the environmental needs of humans, while contributing 'powerful selective pressures favouring the human trajectory'.

The uplift of eastern Africa and the Horn of Africa by one or more mantle plumes led to stretching and cracking of the crust to form the familiar rift topography, and partial melting that fuelled volcanism. Repeated extensional movements on fault lines dropped the valley floor, constructing towering escarpments and deep lakes. In this way, tectonism maintained a large rain-capture system, funnelling fresh water to lakes and rivers, and allowing plant and animal life to flourish in an otherwise arid region. At the small scale, fault scarps bounded depressions that captured nutrient-rich sediments eroded and transported from the escarpment.

Frequent lava-flow eruptions, like the earthquake faults, built barriers and obstacles on the valley floor, adding sharpness to the terrain. This complex landscape offered protection from predators but at the same time provided natural traps to corral prey [115]: 'it is not difficult to see how an unspecialised ground-dwelling predator, dependent on powers of observation and intelligence rather than of speed and strength, might use local barriers and enclosures created by a topography of lava flows and fault scarps as a secure place for ground nesting, for feeding at leisure on food brought in from elsewhere, as a source of local food and water and, in time, as a means of actively manoeuvring and diverting live animals from the edge of more open terrain into natural enclosures and traps'.

Once solidified, the volcanic rocks provided suitable materials for putting together a tool kit. No other region of the world offers such a large-scale or long-lived environment of volcanic activity and extensional tectonics. If one part of the rift became inactive, elsewhere earthquakes and eruptions would resume, maintaining a landscape in flux. A measure of the dynamism of the rift is evident in the location of some key archaeological sites where early human fossils have been recovered. For instance, Hadar, in Ethiopia (where Dinkenesh was found), would once have been on the valley floor but is now part of the rift escarpment.

As they increasingly exploited the complex rift-valley environment, hominins would have become less dependent on forest cover as a refuge and more capable of venturing into unfamiliar terrain. King and Bailey consider that it is no coincidence that some of the earliest finds of early humans outside of eastern and southern Africa are in the Jordan Rift of the Levant. The East African Rift Valley makes for an obvious, long-distance migration corridor; the Red Sea, too, is a rift.

7.3 THE MIDDLE STONE AGE AND MODERN HUMANS

By 600,000 years ago, populations with larger brain size were evolving, leading towards the appearance of our direct ancestors, 'anatomically modern humans' (*Homo sapiens*). The earliest finds come from the Omo valley in Ethiopia and are 195,000 years old (dated using the argon–argon method on volcanic ash layers intercalated with the fossil-bearing sediments) [116]. Modern humans introduced the Middle Stone Age. This represented a technological revolution compared with the prevailing (Early Stone Age) 'Acheulean' industry, characterised by roughly hewn chunks of rock used as hand axes. (I have joined archaeologists on fieldwork and seen them stoop to collect these implements amongst other stones on the ground, and must admit that I wouldn't have spotted them. Some do bear a crude resemblance to a tapered blade but they are very clunky and many look little more than chipped cobbles to the untrained eye.) In contrast, the Middle Stone Age toolkit was remarkably diverse. Individual tools were far smaller, and their shapes and sizes reveal a medley of uses (suggesting exploitation of a greater variety of natural resources). They were in vogue in eastern Africa until around 40,000 years ago when they, in turn, became obsolete thanks to innovations of the Later Stone Age.

In eastern Africa there are around 60 sites where Middle Stone Age artifacts have been found *in situ* in sedimentary rock strata (and not merely scattered on the ground). These span an area of 6.5 million square kilometres. Of these, fewer than 40 are well dated, and nearly all are in the East African Rift Valley. This geographical distribution is widely regarded to indicate the preferred environment of our distant ancestors, not merely a consequence of better fossil preservation in rift environments (thanks to high sedimentation rates).

A major characteristic of Middle Stone Age implements is retouching: the deliberate, repeated fashioning of the tool yielding a much sharper edge than achieved by Acheulean industry. Careful study of these innovative tools reveals some subtle but important trends through the period they were in fashion. Firstly, after 125,000 years ago, heavy-duty choppers, cleavers and hand axes disappear from recovered assemblages. It is also evident that, over time, raw materials were increasingly being sourced from distant quarries rather than local outcrops. Heftier tools have been linked to more heavily built species of human but they are also associated with manual tasks such as wood-working. The latter association is significant since it implies the availability of wood to work, and hence the possibility that humans were living in forested or wooded habitats. Smaller tools are not only more portable – convenient for nomadic lifestyles – but they also open up further ecological niches for exploitation. In particular, smaller pieces can be hafted on to wood for use as arrows, spears and darts.

Human fossils have been found at a dozen or so of the Middle Stone Age sites in the East African Rift Valley. However, while bones and teeth older than 125,000 years vary greatly in anatomical proportions and have been attributed to various species and subspecies of the genus *Homo*, only remains of *Homo sapiens* have been identified in younger finds. There is also a pronounced change in the palaeoenvironmental setting of sites around this threshold date: before, they all indicate proximity to rivers or lakes but, subsequently, humans were evidently occupying caves, rock shelters, hillsides and even calderas (for example on Kone volcano in Ethiopia). The scarcity of coastal archaeological sites is perhaps surprising but very likely reflects sea-level changes associated with glacial cycles. During a glacial period, global sea level drops as more of Earth's water ends up in ice sheets and glaciers. When the ice melts, the sea rises again. Thanks to the present-day interglacial conditions, sea level is high, and thus a vast archive of artifacts and remains of our prehistoric ancestors is surely submerged.

Much has been written about the environments in which *Homo sapiens* lived. One significant line of evidence for their preferences, however, is that all the known Middle Stone Age sites are associated with woodland palaeoenvironments. There are numerous explanations, among them the availability of food and water, the respite from the Sun afforded by tree cover and the protection from lions and other big predators of the savannah.

Refuges are thought to have played a vital role in sustaining populations (not only of humans) through the millennia of environmental stress applied episodically by glacial cycles (Figure 7.2). As the ice sheets spread southwards from the Arctic, large regions of Africa presented a harsher environment than found today. At times, the Sahara desert doubled in size and much of eastern Africa transformed into barren desert, endless grassland or something in between. Woodlands, forests and 'green corridors' would have been in short supply. One likely refuge during periods of such environmental stress was the tropical rainforest of the Congo Basin and other parts of central and western Africa. However, while rainforests grow biomass faster than any other ecosystem on the planet, most of it is inedible. Humans cannot digest woody plant tissue, while fruit are mostly out of their

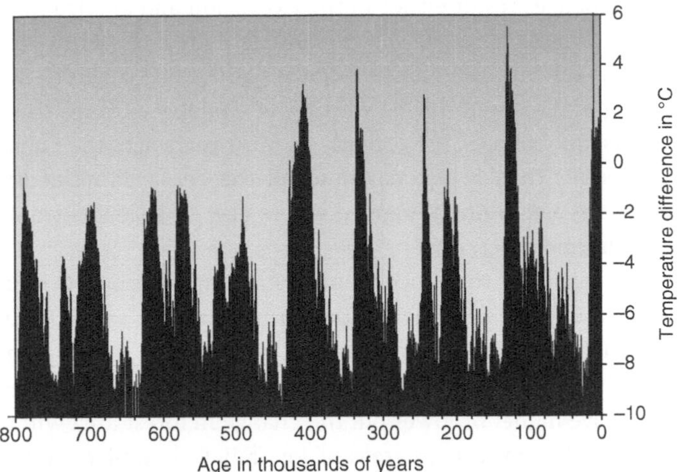

Figure 7.2 Ice-core derived temperature anomalies for Antarctica revealing glacial and interglacial periods for the past 800,000 years. The temperature scale indicates the difference between the measured temperature at a given time and the average temperature for the past millennium. Note the cold period beginning around 180,000 years ago. Source: EPICA Dome C ice core 800,000-year deuterium data and temperature estimates obtained from World Data Center for Paleoclimatology, Boulder, and NOAA Paleoclimatology Progam; contributor: V. Masson-Delmotte, LSCE/IPSL; IGBP PAGES/WDCA contribution series # 2007–091.

reach high in the canopy (where they are enjoyed instead by birds, insects and other primates). Furthermore, despite the astonishing animal diversity in rainforests, there is not so much in the way of available biomass compared with an equivalent area of a typical safari park. In other words, tropical rainforest is unlikely to have supported high densities of human foragers, hunters and gatherers. The generalised picture of primate residence in Africa during the Middle Stone Age is that the rainforests of central and western Africa were home to apes, while the woodlands and savannah of eastern and southern Africa were settled by humans.

In temperate parts of the Earth, when climate cools or warms up with the waxing and waning of the ice sheets, habitats and their associated ecosystems can migrate towards the poles or the equator so as to maintain the environmental conditions they thrive best in. Today, many northern Europeans behave in a similar way, at least on a temporary basis, in response to the changing seasons. In the tropics, climate changes result either in the expansion and amalgamation, or contraction and fragmentation, of habitats. Increased aridity and desertification, for instance, confine species in smaller regions, whereas greening of the desert during warm, wet climates expands the ranges of plants and animals. (As recently as 5000 years ago, the Sahara was much wetter than it is today, and all the beasts familiar to game reserves in eastern Africa roamed where now there is little more than sand dune and desert dust.)

Of particular relevance to understanding the impacts of climate change in eastern Africa is the role of topographic relief. In areas of high relief, changing climate results in relatively small changes in areas of different habitats – for instance, the highland forest line might move higher or lower but the change in forest area will tend to be limited. In contrast, in areas of low relief (on a plateau or in the lowland plains), a modest change, for example in annual average temperature, might result in a vast area of vegetation change. Many Middle Stone Age sites have been found at the boundary between habitat types ('ecotones'): specifically between upland forest and open canopy woodland, and between woodland and savannah. This pattern is especially evident in the Ethiopian Rift Valley, and it might just be a response to the relative environmental stability available at the interface between two ecosystems. Another advantage of living close to such an ecotone is that it increases the range of resources available.

7.3.1 Human migrations: push and pull

University of Southampton archaeologist, Laura Basell, has suggested that volcanism played a significant role in dispersals of *Homo sapiens* within and beyond Africa [117]. Around 180,000 years ago, the ice sheets advanced as Earth plunged into a glacial period. During its peak, much of Africa dried up. The Sahara expanded, the great lakes desiccated and grasslands replaced woodlands. With their preferred environments drastically diminished, our ancestors were almost extinguished – the population may have fallen to a few tens of thousands. The survivors took refuge in the highlands of Ethiopia and western Kenya where woodlands persisted; they tracked shifting ecotones in the rift valley, keeping close to oases near remaining lakes and rivers, which of course would also have attracted game animals. Wide tracts of desert and grassland now separated the habitable areas, isolating and fragmenting the communities of *Homo sapiens*. This isolation had the effect of increasing genetic diversity between communities.

That the deserts proved to be insurmountable barriers for our Middle Stone Age ancestors is widely acknowledged by palaeoanthropologists – as a species we still are poorly adapted to desert environments. Although tribes such as the Bedouin and Tuareg thrive in the deserts today, they do so thanks to the domestication of camels (enabling travel between oases). Without drinking water, the desert is tolerable only for a matter of days and migration would thus only have been possible via corridors of suitable climate, hydrology and topography.

Around 125,000 years ago, however, the Earth warmed again: glaciers and ice sheets retreated and the desert bloomed. With populations growing, *Homo sapiens* spread to the restored woodland and savannah habitats, following the water and food resources. Basell describes this as the 'pull' of habitat availability, which would have encouraged mobility and the manufacture of smaller, more portable and capable tools. These, in turn, provided increasing opportunities to exploit what the environment had to offer. The East African Rift Valley represented a 2000-km-long, lake-strewn, game-rich corridor of suitable habitats, as good as purpose-built for human settlement and migrations. It is uncertain whether other species and subspecies of *Homo* survived the glacial period. If so, their remains have yet to be found in Africa and it is plausible, therefore, that they lost out to *Homo sapiens* in competition for the dwindling resources. (In Asia, other *Homo* species such as *erectus* did persist through the glacial period, though, in that part of the world, they were yet to share territory with *Homo sapiens*.)

Although we have regrettably few firm age estimates, we know that numerous silicic volcanoes in the East African Rift Valley and in the Afar Triangle were intermittently growing, exploding and collapsing during the last few hundred thousand years. In the Ethiopian Rift Valley there are a dozen or so geologically young calderas. One of the largest, and among the very few that has been dated, is O'a caldera, formed around 240,000 years ago (approaching the time of first appearance of *Homo sapiens*) during 'a paroxysmal rifting episode' [118]. The collapse volume suggests an eruption of the order of 120 cubic kilometres of magma (roughly, a M_e 7.5 event). Numerous other silicic calderas have been sunk into the Ethiopian Rift Valley since then. Their magma compositions are also associated with very high sulphur contents so the eruptions were likely to have changed global climate (Section 3.3). It is very likely that the now flourishing Middle Stone Age populations of *Homo sapiens* experienced several episodes of cataclysmic caldera volcanism that devastated their territories.

Widespread fallout of tephra might have acted to isolate communities, and the larger eruptions would have led to prolonged ecological sterility and abandonment by bird and beast alike. Basell suggests that the environmental instability arising from such volcanism may have promoted adaptability in the affected *Homo sapiens* communities. Survivors would certainly have had some problem-solving to accomplish as they surveyed the sterile ashen terrain, stirred only by dust devils. She also argues that these disasters could have been enough to 'push' them in search of a new home, both within and beyond Africa.

The restoration of clement climate 125,000 years ago also coincides with the first known foray of modern humans beyond Africa. Their characteristic fossils have been found in deliberate burial sites in the Levant, at Skhūl and Qafzeh, where they are dated to *circa* 120,000 years ago [119]. It may have been the combination of environmental factors – the greening of the desert and the local to regional devastation caused by mighty silicic eruptions – that inspired this first exodus.

7.3.2 Out of Africa (again)

It seems that the *Homo sapiens* pioneers in the Levant died out without venturing into Eurasia. The fossil record of this offshoot simply dries up *in situ* leaving no descendants. So when did our direct ancestors leave Africa on their definitive world tour? Where did they exit Africa from, and how did they make their way across the Indian Ocean and/or Asia to reach Australia?

Thanks to modern DNA sequencing techniques, and since all that is required to collect a sample is to take a swab from the mouth with a cotton bud, by now countless analyses of human DNA have been made around the world: from the highlands of Papua New Guinea to the Kalahari desert, from Scandinavia to Tierra del Fuego. They tell a remarkable story of our origins and dispersal. The first general observation is that, given the size of the human population today, human genetic diversity is surprisingly low. It is much less than the diversity expressed in the genes of other great apes, for example, despite the lower population numbers of chimps, orang-utans, gorillas and so on. In addition, people from the same population can have more genetic diversity than people from different populations, and the highest diversity of all is found within and between indigenous African populations.

According to anthropologists such as Stephen Oppenheimer from the University of Oxford, the genetic evidence demonstrates the geographic spread and distribution of our ancestors, as well as an idea of timing [120]. There are many varieties of genetic clues but a particularly important source of information comes from 'non-recombining' DNA such as mitochondrial DNA (mtDNA), which is contained in structures called mitochondria found inside cells, and only passed down the maternal line. Mitochondria fuel cells via the energy obtained from food, and they contain just a fraction of the DNA found inside the nuclei of cells. Nevertheless, human mtDNA has around 16,500 'base pairs' that make up the structure of the double helix. Since it is non-recombining, in theory, an exact copy is passed from mother to child. However, rare spontaneous errors in copying the gene result in mutations that, over hundreds of generations and more, build up diversity in the mtDNA of a population. One practical application of mtDNA sequencing is the identification of human remains.

In studies of human palaeodemography, mtDNA samples are collected from individuals from populations around the world. By comparing the sequences, base pair by base pair along the DNA, for all the samples, it is possible to establish a statistical picture of within- and between-population genetic diversity, and then to draw up a family tree showing the relatedness of different populations. Once that is done, dates can be assigned to the tree based on estimates of the speed of the 'molecular clock' that controls the rate of those random mutations. The basic means by which the clock is calibrated is to divide the genetic difference between humans and chimpanzees by the time since they diverged on the primate evolutionary tree (approximately six million years ago). One of the triumphs of this technique was to show that the genetic lines of all living humans can be traced back to a 'Mitochondrial Eve' who lived in Africa as recently as 108,000 years ago according to one measure of the rate of genetic drift [121].

Further genetic evidence seems to confirm the extinction of those first emigrants whose remains were interred at Skhūl and Qafzeh approximately 120,000 years ago: there are no mtDNA lineages outside Africa that date anywhere near as old as this. Instead, the mtDNA found in non-African populations points to dispersal from Africa about 60,000–55,000 years ago. Again, however, based on an alternative calibration of the mitochondrial clock, it could be as early as 70,000 years ago [122]. As discussed in the next chapter, the date matters considerably in understanding the potential human impacts of one of the Earth's largest known 'super-eruptions': the eruption of the Younger Toba Tuff, roughly 73,000 years ago.

Whatever the exact timing (or date of multiple waves of migrations), it was the emigrants that went on to interbreed with

Neanderthals, as has been shown from analysis of the Neanderthal genome (using fossil material) [123]. Many non-Africans living today have inherited as much as 4% of their DNA from Neanderthals. Genetic data further suggest that our ancestors had colonised Papua New Guinea and Australia 50,000 years ago, implying a remarkably rapid advance across or around the landmasses and islands of the Indian Ocean.

Unfortunately, the challenge to reconstructing human migrations outside Africa is the lack of fossil evidence. With the data available, it is possible to argue for various routes, timings of migrations and motivations for dispersal. Oppenheimer points out that around 85,000 years ago, and then again at 73,000 years ago, sea level was as much as 100 metres lower than today. The Red Sea would have been far saltier owing to the evaporation associated with this drop in global sea level and its relative isolation from the Indian Ocean. Modern humans would have found their marine coastal resources depleted and been compelled to seek new areas. They could have crossed the Bab-el-Mandab straits by rafts to reach the Gulf of Aden, which would have been richer in nutrients and corresponding ocean productivity thanks to the winter monsoon. Coral reef 'stepping stones' may have emerged with the low sea level and, if so, might have facilitated the voyage.

This period also coincides with what has been called a 'troop-to-tribe' transition in human social organisation around 70,000 years ago (Section 8.4). It is marked by an abrupt increase in archaeological finds of sharp bladelets suitable for composite tools. These were manufactured from fine-grained rocks, notably obsidian, and represented a step-change in technology. Since obsidian outcrops are not found round every corner, the use of the material also indicates establishment of long-range exchange systems. In turn, this implies the emergence of a new degree of regional and reciprocal information sharing, resource exploitation and cooperation. As noted above, the evidence for Middle Stone Age occupation of the coastline is limited because much of the associated archaeology now lies beneath the waves. However, there are sites where former shorelines have been uplifted, exposing archaeological materials that attest to the harvesting of marine resources. One such site is at Abdur in Eritrea, which has been dated to 125,000 years ago. The ancient beach rock there contains abundant shell middens and obsidian tools that must be of corresponding age. Given this evidence that humans were enjoying the riches of the intertidal zone, it is easy to imagine that they would have followed the coast looking for more of the same, especially since shellfish supplies can be rapidly depleted by over-exploitation. Oppenheimer argues that this would have turned early humans into beachcombers, always on the move in search of untapped shellfish resources.

Where rocky promontories hindered their travel, they could have used their rafts to reach pristine shores beyond. He further suggests that this mode of migration essentially propelled modern humans all the way to Australia. Only somewhat later, perhaps as early as 45,000 years ago, they joined the Neanderthals in Europe for the first time. They were in Britain by 40,000 years ago. Meanwhile, they had populated China and, around 22,000–25,000 years ago, they crossed the Bering land bridge to begin peopling the Americas (reaching southern South America around 12,500 years ago).

7.4 SUMMARY

Like its subject, the study of human origins and behaviour has its own evolutionary path. Recent archaeological, genetic, palaeontological and palaeoenvironmental research has brought major surprises concerning our ancestry (including the discovery of the 'Hobbit' in south-eastern Indonesia and the genomic evidence for interbreeding of Neanderthals and modern humans). Genetic and archaeological dates and human evolutionary trees are subject to frequent revision.

While the human story is not an exclusively east African one, it is nevertheless hard to deny the pre-eminence of the region in understanding why and how our species evolved. One possible explanation is the evolutionary stimulus provided by the dynamic landscape of the East African Rift Valley. The topographic complexity afforded by faulted valleys and lava flow barriers built an environment where early humans could both hunt and hide. Meanwhile, the lakes on the rift-valley floor attracted game and provided water. It is even thought by some that the pathways of early human migration within and beyond Africa were shaped by the geometry and geography of the rift.

While for much of the time, the East African Rift Valley provided an attractive setting for human occupation, there must have been many colossal lava effusions and explosive eruptions during the time period of human evolution. Whole segments of the rift valley would have been smothered with searing ignimbrite, while a much wider area would have received substantial tephra fallout from phoenix clouds. One such event formed the O'a caldera in Ethiopia, around 240,000 years ago, and it could well have been significant in isolating populations and presenting new challenges to pre-modern humans at the cusp of modernity.

The argument that volcanism and associated tectonic activity in the East African Rift Valley played a role in human origins and migrations is compelling, though it remains as difficult to put definitive flesh on the argument as it is to reconstruct accurately the facial appearance of archaic human species from a handful of bones and teeth. This reflects the comparative paucity of volcanological investigation carried out in the region. Dozens of silicic calderas formed within the rift system during the past few millions of years but very few are well dated. Nor have their eruptions been characterised in the level of detail applied to caldera volcanism elsewhere in the world (such as the USA, central Andes, New Zealand and Italy). Establishing the tephrostratigraphy and tephrochronology, particularly in Ethiopia, is a crucial step needed to investigate further the potential impact of volcanism on human populations. In particular, it would make it possible to identify the source volcanoes of some of the thin tephra deposits used to date archaeological sequences, as well as tephra layers identified in drill cores from lake beds and from the bottom of the Red Sea and Indian Ocean. Perhaps more footprints will be revealed beneath the ash!

8

The ash giant/sulphur dwarf

When the modern African human diaspora passed through the prism of Toba's volcanic winter, a rainbow of differences appeared.

S. H. Ambrose, *Journal of Human Evolution* (1998) [124]

The eruption of the Younger Toba Tuff in northern Sumatra 73,000 years ago expelled seven thousand billion tonnes of silicic magma (M_e 8.8) and, by some estimates, as much as three gigatonnes of sulphur (more than 300 times as much as Pinatubo's 1991 eruption). It made a sizeable contribution to a 100-kilometre-long by 30-kilometre-wide hole in the ground that is today occupied by Lake Toba. This enormous magnitude ranks Toba as one of the largest single eruptions ever documented, along with the 27.8-million-year-old Fish Canyon Tuff associated with La Garita caldera in the San Juan Mountains of Colorado (M_e 9.1–9.2) and the two-million-year-old Huckleberry Ridge Tuff of Yellowstone (M_e 8.7–8.9). The much more recent Younger Toba Tuff eruption took place close to the time Middle Stone Age *Homo sapiens* left Africa prompting far-reaching debate concerning the impacts and consequences for modern human adaptation, cognition and migration. Using geological and modelling evidence, this chapter reconstructs the cataclysmic events, and scrutinises the evidence for Toba's impact on the environment, climate and our human ancestors.

8.1 THE ERUPTION

Today, the Toba region is a place of rugged peaks, lush forests and bamboo groves, and emerald rice terraces enclosing the traditional villages of the Batak people. And then there are the inviting blue waters of Lake Toba itself. Unless you are looking down from the International Space Station, only the cliffs of brilliant white pumice hint at what lies

Figure 8.1 The 100-kilometre × 30-kilometre Toba caldera is situated in north-central Sumatra around 200 kilometres north of the equator. It is comprised of four overlapping calderas aligned with the Sumatran volcanic chain. Repeated volcanic cataclysms culminated in the stupendous expulsion of the Younger Toba Tuff around 73,000 years ago. The lake area is approximately 1100 square kilometres. Samosir Island formed as a result of subsequent uplift above the evacuated magma reservoir. Such features are called resurgent domes and are typically seen as the culminating phase of a large eruption. Landsat image from NASA's Earth Science Enterprise Scientific Data Purchase Program.

below the surface – the crater is really too big to appreciate from the ground (Figure 8.1).

8.1.1 When did it happen?

Age estimates for the Younger Toba Tuff are remarkably consistent given the diverse dating techniques that have been applied. The best current estimate for the age of the eruption is 73,000 years with an uncertainty of a few thousand years, which places it in the Late Pleistocene Epoch of geological time and, in paleoanthropological contexts, the Middle Stone Age (Middle Palaeolithic). Craig Chesner from Eastern Illinois University, and Bill Rose from Michigan Technological

University were amongst the first to estimate the size of the eruption [125]. Based on thicknesses of tephra on Sumatra and in the Indian Ocean, their rough estimate of 2500–3000 cubic kilometres of dense magma (M_e 8.8) has yet to be refined.

The Younger Toba Tuff was not the first monumental eruption from Toba. It was preceded by the Oldest Toba Tuff (0.84 million years ago) and Middle Toba Tuff (half a million years ago). Toba deserves the sobriquet 'super-volcano'! The Younger Toba Tuff ignimbrites (Section 2.3) form a plateau that slopes gently away from the caldera and covers an area of at least 30,000 square kilometres of northern Sumatra. Most of the Younger Toba Tuff is not welded as some ignimbrites are, and contains pumice blocks up to a metre across. Exposures of the tuff in caldera walls are up to 400 metres thick, and the central part of the caldera is occupied by Samosir Island, a product of magmatically promoted uplift since the eruption (Figure 8.1).

Toba has not erupted historically but earthquakes occur around the southern rim of the caldera associated with movement of the Great Sumatra (Semangko) Fault. This fracture zone runs up the backbone of Sumatra. It is associated with subduction of the Indian Plate beneath the Sunda Plate and the 'megathrust' that produced the giant 2004 Sumatra–Andaman earthquake and tsunami (there is even a hint that the shock triggered eruptions of two volcanoes in the region). Between 1920 and 1922, earthquakes caused damage along the southwest shore of Lake Toba, and a magnitude 6.5 temblor on 25 April 1987, centred at the southern end of Lake Toba, was linked to increases in steam emissions along the south and west shores of the lake. Geophysical studies of the caldera have identified a vast magma chamber still residing beneath the tranquil lake and its verdant shores.

8.1.2 What was it like?

The trick to producing a super-eruption of a few thousand cubic kilometres of silicic magma is to keep the lid on its magma chamber for a very long time. This allows such a large quantity of magma to accumulate and brew in the crust. In the case of Toba, the magma chamber, some ten kilometres below the surface, stewed for at least 100,000 years before its catastrophic leak [126]. The eruption is thought to have started as the chamber roof foundered. This opened up circular fractures through which pyroclastic currents gushed, spreading to the coastlines northeast and southwest and covering an area of up to 30,000 square kilometres. If correct, this implies that the caldera formed progressively during the eruption and not catastrophically afterwards. Evidence in favour of this interpretation includes the generally symmetrical distribution of Younger Toba Tuff around Lake Toba, and the absence of tephra-fall deposits from a Plinian eruption (although it is possible that any fallout was eroded by the massive pyroclastic currents).

The aftermath of the eruption has been likened to an 'enormous fire' as a vast area of vegetation was set ablaze [127]. It is judged that the Younger Toba Tuff ignimbrite would have been as hot as 550 °C when it formed, a couple of hundred degrees cooler than the actual magma temperature. Considering the aftermath of Pinatubo's 1991 eruption it is certain the ignimbrite remained hot for years given its thickness (up to several hundred metres). Its lack of welding would have resulted in frequent disturbance, spawning smaller pyroclastic currents. There must have also been many months of explosions due to the interaction between the hot deposits and rain or surface water. The scene of destruction across such a vast scale is hard to imagine – a far cry from the languid lakeside views enjoyed by residents of Samosir Island today.

The immense phoenix clouds rising from the pyroclastic currents produced thick fallout deposits. Toba ash has been identified in many deep-sea sediment cores (Figure 8.2), some as far distant as the Arabian Sea, 4000 kilometres away. It is also found on the Indian mainland where it represents an exceptionally valuable time marker for archaeological studies. At least 1% of the surface of the Earth was buried to a depth of ten centimetres or more. According to one estimate, the equivalent dense magma volume of this deposit amounts to 800 cubic kilometres [128]. The volume of ignimbrite has been estimated from its characteristic thickness of 50 metres outside the caldera (1000 cubic kilometres bulk volume) and the 400-metre-thick deposit within the caldera (another 1000 cubic kilometres). The crude estimate of the total dense volume of tephra thus amounts to 2800 cubic kilometres, of which the phoenix-cloud fallout accounts for about 30%.

Such wide tephra dispersal points to unusual dynamics of the phoenix cloud. However, the physics of such massive ash plumes are poorly understood and conflicting descriptions prevail. One explanation for the widespread fallout is that the eruption took place during the monsoon, with southwesterly winds blowing tropospheric ash towards the South China Sea, while stratospheric ash was transported to the west. However, monsoon circulation cannot explain the transport of ash grains of the size found so far: they would have settled out much sooner. An alternative idea considers that super-eruption

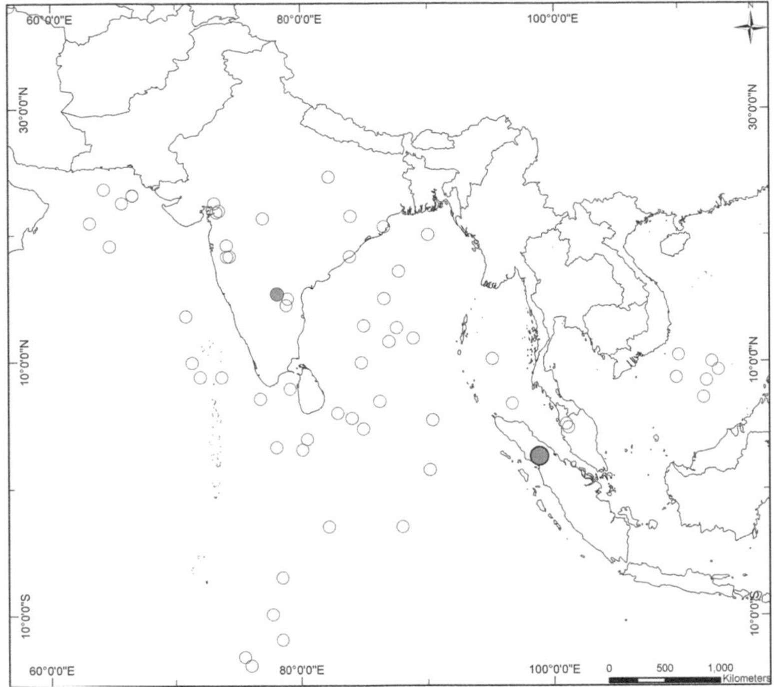

Figure 8.2 Distribution of recovered Younger Toba Tuff phoenix-cloud fallout deposits (open circles). Note how the distribution spans the Arabian Sea, the central Indian Ocean basin and the South China Sea. Filled circles indicate Jwalapuram in India (Section 8.5) and Toba itself. Courtesy of Emma Gatti.

umbrella clouds spread according to the Coriolis effect enabling them to attain diameters of 6000 kilometres centred above the volcano, and barely deviated by stratospheric winds [129]. (For comparison, Pinatubo's 1991 eruption cloud was about 1000 kilometres across; Figure 3.2.) However, this model too has been questioned because it only represents a single plume whereas a phoenix cloud should be fed by innumerable sources rising off pyroclastic currents [130]. Multiple updrafts of air and ash should generate a very different dynamical regime. Three-dimensional modelling of such a cloud suggests it would intrude into the atmosphere close to tropopause height. This is substantially lower than predicted heights from simple one-dimensional models, which cannot differentiate between the maximum height reached by a plume and the neutral buoyancy height where most of the fine ash and gases become concentrated. Since it affects the distribution of emitted sulphur gases, this difference has important implications for understanding the climatic effects of super-eruptions.

One way to gauge the intensity of the Younger Toba Tuff eruption would be to estimate its duration. This has been attempted in an ingenious fashion based on the sizes of feldspar crystals in ash layers retrieved from the seabed of the Indian Ocean [131]. The ash particle size distribution in any tephra-fall deposits depends on distance from the volcano, intensity of the eruption and wind patterns experienced in transit. Generally, marine tephra layers consist of particles of a few thousandths up to a few tenths of a millimetre across. From theory, the settling rates of such particles in water should vary between a few centimetres per second to less than one millimetre per second. Thus, for typical ocean depths of a few kilometres, it would take between a day and a year for ash particles to hit the seabed. Based on the Toba crystals' calculated settling velocities, and the known water depth, the eruption was estimated to have lasted between 9 and 14 days. Combined with the total magnitude estimate, this implies a staggering average eruption intensity of 12.8 (equivalent to a discharge of seven million tonnes of magma per second). Peak rates would have been even greater. Sadly, the determination could be flawed because there is good evidence that real settling rates of ash grains in water are much faster than theoretical ones.

This discrepancy arises because experimental and theoretical research on settling rates has tended to focus on dilute concentrations of particles. However, these probably fail to represent conditions in dense particle concentrations raining out of a real volcanic ash cloud. To explore this further, the volcanologist and oceanographer Steven Carey (University of Rhode Island) set up a water-tank experiment in which tephra fallout was simulated by sprinkling Pinatubo ash on to the water surface via a chute [132]. Particles dropping through the water column were then imaged with an ultrasound device. This revealed sedimentation rates up to ten times faster than those predicted for single particles, explained by rapid sinking of dense mixtures of water and ash particles under gravity.

Notably, these results explain rapid sedimentation rates that were actually observed for Pinatubo ash fallout in the South China Sea [133]. By chance, sediment traps were operating on the seabed at the time, nearly 600 kilometres west of the volcano. The collection was fully automated using pre-programmed traps moored at water depths of 1200 and 3700 metres. By recording the time and height at which the Pinatubo plume passed above the traps' location using satellite data, and modelling the atmospheric fallout of tephra, the delivery time of ash to the sea surface above the traps could be calculated. This revealed three important findings. Firstly, fine ash fell to the sea surface much earlier than predicted. This was due to its aggregation by electrostatic and other forces, which produces ash clumps with the same aerodynamic properties as much larger single particles. As they fall through the water, the aggregates break apart so that the final deposit includes two ash grain size populations. This might explain the distributions of particle sizes observed in some of the deep-sea Younger Toba Tuff samples.

The second result was that the timings of trap closure indicated sinking rates of particles greater than two centimetres per second (nearly two kilometres per day). These are hundreds of times faster than theoretical descent rates for particles only a hundredth to a tenth of a millimetre across. Third, the grain size distributions in the two traps were almost identical, demonstrating that any vertical sorting by size or density of particles was completed within a kilometre or two of descent in the water column. The good news from these findings is that sedimentation of tephra on the seabed from such volcanic clouds is effectively instantaneous, demonstrating the reliability of deep-sea ash layers as widespread time-markers. The bad news is that the estimate of the Younger Toba Tuff eruption duration based on theoretical settling rates for single particles is unreliable.

8.2 SULPHUR YIELD OF THE ERUPTION

Much of the literature on the Younger Toba Tuff eruption argues or assumes that it must have had colossal impacts on the atmosphere and climate – far more extreme than the effects of Pinatubo in 1991. However, while there is no doubt from the size of the Toba caldera, and the extent of ignimbrite and phoenix-cloud deposits, that a monumentally large eruption took place some 73,000 years ago, it is far less certain how much it perturbed the Earth system.

As ever, critical for evaluation of Toba's climatic impacts is estimation of its sulphur yield to the atmosphere. This has proved far from straightforward and most of the claims for extreme climate impacts are predicated on some very casual estimates of the sulphur output. These include an initial 'guess' of the sulphur yield based on Toba pumice chemistry, from which it was estimated that for each tonne of magma, 0.5 kilogrammes of hydrogen sulphide gas were released into the atmosphere. (The reducing character of the Toba magma favoured this sulphur species over sulphur dioxide.) Scaling these values up by the estimated total eruption magnitude (M_e 8.8) suggested an eruptive emission of at least 3.5 gigatonnes of hydrogen sulphide (along with 540 gigatonnes of water vapour). In sulphur terms, this would make the Younger Toba Tuff nearly 400 times larger than the 1991 Pinatubo eruption. With hindsight, there is no petrological basis for this estimate.

A second estimate for the Toba sulphur output was based on fallout in the Greenland Ice Sheet Project 2 (GISP2) ice core [134]. A large sulphur anomaly was identified in the GISP2 record at a depth corresponding to an age of 71,000 (give or take 5000) years (Figure 8.3). However, while its magnitude seems befitting of one of the largest known eruptions, no tephra were recovered that might have re-inforced the match. Using the atomic bomb calibration

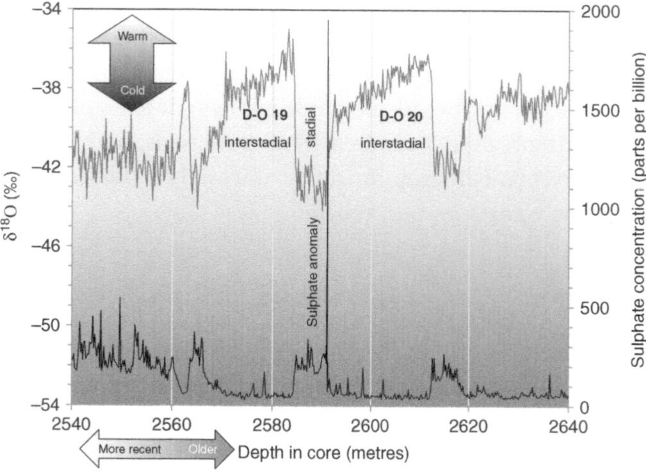

Figure 8.3 GISP2 record of sulphate (lower curve) and oxygen isotopic variation (upper curve) spanning the so-called Dansgaard–Oeschger (D–O) cycles 19 and 20. Core age increases with depth so increasing age is to the right of the graph. More negative oxygen isotope values correspond to lower temperatures. The sulphate measurements are uncorrected for non-volcanic contributions but the putative Toba spike is very clear. Note the cold 1000-year-long 'stadial' and the sulphate spike (the bounding warm periods are called 'interstadials'): cooling was well underway before the eruption. Oxygen isotope data from the Quaternary Isotope Laboratory, University of Washington. Sulphate data from NCDC/NOAA.

(Section 4.2.2), the fallout in the ice suggested an aerosol cloud of between 1.7 and 3 gigatonnes of sulphuric acid.

However, an emerging view suggests that the Younger Toba Tuff eruption, despite its vast magnitude, might have delivered a comparatively puny pulse of sulphur into the atmosphere. This reflects the silicic (rhyolitic) composition of the Younger Toba Tuff magma. High silica magmas tend to contain little sulphur in the first place, and cool, reducing magmas like Toba's should hold even less. What made both the El Chichón (1985) and Pinatubo (1991) eruptions uncommonly sulphur-rich was highly oxidising conditions in their magmas. These considerations led experimental petrologist Bruno Scaillet (University of Orléans – also known as 'the man who recreates volcanoes') and his colleagues to suggest that Toba's sulphur yield was almost trivial [135]. They estimated that as little as 35 megatonnes of sulphur were liberated by the eruption, 1000 times lower than the previous estimates! Indeed, this amount is only three or four times more sulphur than released by Pinatubo in 1991, even though the Younger Toba Tuff eruption was 600 times larger in magnitude. This much lower estimate also agrees with new evidence based on analyses of melt inclusions (Section 4.1.5) trapped in quartz crystals in Younger Toba Tuff pumice [136]. These reveal low quantities of sulphur. It looks increasingly likely that while the Younger Toba was an ash giant, it was a sulphur dwarf!

This begs the question how to reconcile the discrepancy with the ice-core-based estimate of sulphur output. One possibility that cannot be discounted is that the Younger Toba Tuff has been mis-identified in the GISP2 record. One of the main criteria for its designation was that the largest sulphuric anomaly in the 110,000-year-long record seemed 'appropriate for the great size of the eruption and the estimated large amount of atmospheric loading associated with it'. This runs into circular reasoning. On the other hand, if the sulphur fallout doesn't belong to Toba, where did it come from? Could it be a large, sulphur-rich eruption in Alaska or Kamchatka? But, if so, why hasn't any ash been identified in the layer? One modelling study suggests that high-latitude ash clouds can be especially strongly affected by heating and cooling (by day and night), and rotation caused by the Coriolis force [137]. This could result in a situation where an ash cloud would sit above the volcano for days, warming and spinning and consequently depositing its ash in a more focused area on the ground. Meanwhile, the much finer sulphuric acid particles would separate and be dispersed further afield by stratospheric winds. In such a way, a large, high-boreal-latitude eruption might produce a major sulphate anomaly in the Greenland ice without an accompanying band of fine ash.

A further curiosity is that no one has yet correlated the large sulphur spike in the GISP2 core with comparably aged layers in other polar ice cores. Even though this important work was published in 1996, no further glaciochemical evidence has emerged in the literature to corroborate the Toba identification in the ice cores. My guess is that, before long, ice-core data will shed new light on the matter. Best of all would be the smoking gun of Younger Toba Tuff ash grains in the ice.

8.3 CLIMATE CHANGE

In the 1970s, there was much discussion of the prospects of 'mutually assured destruction' in the event of a nuclear-weapons exchange between the USA and Soviet Union. In fact, the atmospheric impacts of nuclear exchanges continue to be studied by some of the leading atmospheric scientists of the day. Among the old scenarios is the idea of a 'nuclear winter', in which the pall of soot and dust, which is carried up into the stratosphere by multiple weapons strikes, intercepts sunlight and chills the planet.

Michael Rampino and Stephen Self extended this idea to super-eruptions. Toba was an obvious candidate, especially since the deep-sea sediment cores revealed that Younger Toba Tuff ash fallout coincided with the start of the last glacial cycle about 67,500 years ago. They argued in a 1992 paper [138] that Toba triggered a 'volcanic winter', which accelerated glaciation that was already underway thanks to cyclic changes in the Earth's rotation and its orbit around the Sun. However, since then, much higher-resolution sediment and palaeoclimate records have emerged that show the Toba eruption preceded the last glacial period by several thousand years. This is very clear, too, if we accept that Toba's sulphate fallout has been correctly identified in the GISP2 core. The layer precedes a warm period, referred to as an interstadial, that comes before the prolonged glaciation ('D-O 19' in Figure 8.3). Proponents of Toba as the harbinger of extreme global change switched attention to the millennium of cold (stadial) climate that appears to follow the sulphuric acid spike in the GISP2 core [139]. Here again, however, close inspection of the juxtaposed sulphate and temperature records shows that cooling, which took a century or two, was well underway by the time of the eruption (Figure 8.3).

Rampino and Self also suggested that not only did the eruption accelerate the onset of the last glaciation, but that the deteriorating climate beforehand may have triggered the eruption in the first place (Section 14.3.2)! As water shifted from the oceans to the ice sheets, global sea level dropped. In principle, such an effect could have influenced stress distributions in the Earth's crust and reduced the forces that kept the Toba magma reservoir sealed. However, the depth of the reservoir (more than ten kilometres), its distance from the sea (over 100 kilometres) and the relative weakness of the Sumatran crust argue against this mechanism. But another potential link between sea-level change and the Toba eruption could be via earthquake activity along the Great Sumatran Fault that runs the length of the island. If stress redistributions accompanying falling sea level resulted in very large earthquakes on the fault, the movements could feasibly have triggered the eruption.

8.3.1 Climate models

With this preamble behind, it should be clear why scepticism of any numerical simulations purporting to be Toba scenarios is warranted. That is not to say the models don't work (though they do have to be extraordinarily complex to represent coupled responses of the atmosphere, oceans and cryosphere) but that their simulated effects are only meaningful if the input parameters are correct. In the case of the Younger Toba Tuff, we really don't know the sulphur output, eruption intensity, season of the eruption or prevailing climate state. Another set of issues to bear in mind is that, since we have not observed a super-eruption in modern times, we cannot be certain of the physical and chemical processes that occur in such a mega-plume. It was recognised a long time ago that stratospheric volcanic clouds may act in a 'self-limiting' way. Even simplified aerosol photochemical and microphysical models show that for a 100-megatonne injection of sulphur dioxide (roughly five times more than Pinatubo in 1991), condensation and coagulation produce larger sulphate particles. These are less efficient at scattering incoming sunlight, and fall back to Earth faster. Also, because of the exponential relationship between optical depth and light scattering (Equation 5.1), there are diminishing radiative forcing returns for larger and larger stratospheric aerosol loads. Combined, these effects mean that larger sulphur emissions do not result in proportionally larger climate effects. With these caveats in mind, we can proceed to review attempts to understand the climatic and environmental impacts of the Younger Toba Tuff eruption. Note their increasing sophistication over time.

The original simplistic estimates of climate response to the Toba eruption suggested a dramatic 'volcanic winter' with immediate and severe 'hard freezes' at mid latitudes followed by several years of global cooling averaging about 3 °C, with up to 12 °C cooling focused on high northern latitudes [140]. It was argued that the cold climate endured for decades following the eruption due to increased snow and ice cover and perturbed ocean temperatures.

The first model capable of linking ocean and atmosphere responses to a super-eruption at the global scale was reported by Gareth Jones and his colleagues at the UK Met Office and University of Reading [141]. They set up the Hadley Centre General Circulation Model with present-day atmospheric conditions (except for carbon dioxide abundance which was set to pre-industrial levels) and then let it run for 50 years (of simulated time) with a 100× Pinatubo scenario. The results suggested five to seven years of strong effects. Earth's reflectivity increased from around 30% to 70%, mostly due to the scattering of light by stratospheric aerosols. The corresponding optical depth of the atmosphere reached a massive 15 (Section 5.3.1), eight months after the eruption. This would imply a very bizarre sight: the Sun's disc would be invisible in such a turbid atmosphere, while the rest of the sky would appear brighter than usual due to forward scattering by the aerosol. The amount of sunlight reaching the surface dropped 58% in the first year of the simulation, corresponding to loss of 60 watts per square metre of Earth's heat budget (averaged over the globe for a year). This is a huge deficit, equivalent to a global monthly average cooling near the surface of 10.7 °C. After a decade, the average global temperature remained 2 °C below normal and, even 50 years after the event, the Earth was 0.28 °C cooler than usual owing to ocean cooling. The regional temperature anomalies were yet more extreme, with annual average cooling of up to 17 °C for Africa and 9 °C for Europe (which was only spared an even colder climate by an enhancement of the Gulf Stream).

Because of the cooling, sea-surface temperatures decreased by around 6 °C globally in the first model year, while the capacity of the air to hold moisture fell. This resulted in huge decreases in rainfall – globally by about 50% in the first year after the eruption. The effect was strongest in some of the normally wettest parts of the world – in the Amazon, central Africa and southeast Asia rainfall dropped 90%, recovering after a couple of years. Meanwhile, snow fall increased 30%

in Europe and South America, taking around five years to return to normal levels. In North America and Eurasia, snow covered the ground north of 50° N throughout the first two summers after the eruption, and sea ice doubled in extent, encasing the coastlines of Japan and the Pacific northwest of the USA. This all sounds dramatic but the one important phenomenon that did not arise in the simulation was glaciation. Despite extended snow and ice cover, once the aerosol settled from the stratosphere, it could not sustain itself. Another significant result is that although the model was only run for 50 years, there was no indication that such an eruption would result in a millennium of climate change as some had inferred from the GISP2 record.

An important drawback of this model is that it took a very simplistic view of aerosol evolution, excluding chemical and microphysical processes that play major roles in a volcanic cloud's evolution. Alan Robock of Rutgers University and his colleagues sought to redress these shortcomings with an improved climate simulation [142]. As well as enhanced treatment of aerosol evolution, they included the effects of vegetation in the model, a significant step forward. They ran the model for several sulphur dioxide emissions. Their 100× Pinatubo scenario yielded very comparable results to those of Jones and his colleagues – a maximum global average surface cooling of around 10 °C, dropping to 2 °C after a decade, and a 45% drop in global precipitation, recovering after a few years. In addition, they identified massive impacts on forests – mid-latitude tree cover shrank considerably while broadleaf evergreen trees and tropical deciduous trees virtually disappeared.

For a 300× Pinatubo scenario (their preferred representation of the Younger Toba Tuff emission), the maximum globally and annually averaged cooling reached 15 °C, and global temperature remained 5 °C below normal up to five years after the eruption. Then, they ran the model with 900× Pinatubo's sulphur dioxide emission but even that failed to initiate glaciation!

The latest modelling effort suggests much less dramatic consequences. Claudia Timmreck, from the Max Planck Institute for Meteorology in Hamburg, and her colleagues ran a 100× Pinatubo scenario for 50 years for three initial climate states: El Niño, La Niña and 'neutral' (neither El Niño nor La Niña) [143]. The key to their approach was to include a more realistic global aerosol model capable of representing the evolution of the aerosol size and its optical depth (Section 5.3.1). This showed only slightly delayed aerosol formation compared with the Pinatubo case (so no significant hydroxyl radical limitation) but, much more importantly, that the aerosol particles were large – with an effective radius of at least one thousandth of a millimetre for a couple of years. This is double the value observed after the Pinatubo eruption and emphasises the point that denser sulphur clouds grow larger particles. As sulphuric acid vapour is formed (through oxidation of sulphur dioxide), it preferentially condenses onto existing particles, growing them, rather than forming miniscule new particles. These larger aerosols are less effective at reflecting sunlight back to space, more effective at absorbing infrared emission from the Earth and radiating it back (the greenhouse effect) and gravitate to the lower atmosphere and surface much more rapidly.

The consequences of this model were a shorter (only four to six years) and weaker period of global cooling. Aerosol optical depth peaked at 0.35 during the first half of year two of the simulation and the peak forcing at the top of the atmosphere reached −18.5 watts per square metre. This is a fraction of the peak calculated in the model of Robock and colleagues (−140 watts per square metre). Consequently, the maximum globally averaged cooling amounted to just 2.5 °C (compared with 10 °C cooling from previous models), though significant regional effects were still observed, for instance in mid-latitude continental areas (which cooled by up to 10 °C in summer). Global rainfall dropped significantly for the first two years and the Indian monsoon failed. This sounds dramatic but the monsoon rains vary as much according to the phase of the El Niño–La Niña cycle.

To sum up all these results, it is clear that modelling the climate effects of a super-eruption is challenging. Non-linear aerosol effects may be especially important. A further factor that has been little explored is the climate response to ash on the ground. Fresh ash is as reflective as snow and the fallout from a super-eruption could blanket an area of continental scale (ten million square kilometres). This will reflect sunlight adding to surface cooling due to stratospheric aerosol. One clear point to emerge, however, is that none of the scenarios, not even the incredible 900× Pinatubo simulation, initiated glaciation. And none indicated anything like a millennium or even century of climate change.

8.3.2 Palaeoenvironmental evidence

The next question is whether there is evidence to support any of these model outputs. The problem in answering this question is that while climate models can simulate month-by-month evolution of the stratospheric aerosol and its atmospheric and climate effects, the geological record provides scant opportunity to get anywhere close to such time resolution. An exception is the ice-core record (though, as noted above, some doubt remains as to whether Toba fallout has been correctly identified). If we do accept the Toba tag in the Greenland core as valid, the sulphuric fallout spans a period of six or seven years. That is coherent with the lifetime of stratospheric aerosol in the 'big impact' models [141,142] but diffusion over time of an originally finer band of sulphuric acid cannot be ruled out.

After ice cores, the next set of records to consider are the cores and profiles collected both from sediment sequences on land and beneath the seabed. In theory, by examining the characteristics of the sediments below and above the ash it is possible to infer environmental changes arising from the eruption. For instance, if the ash bed coincided with an abrupt change in fossil pollen or plankton species or abundance, it might reflect the environmental and ecosystem impacts of either the ash fall itself or the immediate or longer-term climatic effects of the volcanic cloud. However, detailed examination of a climate transition spanning a century or two demands very high time resolution. The rates at which sediments accumulate are low, such that a few millimetres thickness of deposit can represent a century of time - any critters that burrow into the sediment, or vegetation that takes root in it, can rapidly mess up a thousand years of environmental history. So, although numerous studies do reveal a warm-to-cold transition in close association with Toba ash, along with evidence for dropping sea-surface temperatures in the South China Sea and eastern Indian Ocean, soil erosion in Indochina and aridity in China, with resolution, at best, of around a century, it is challenging to attribute such changes directly to Toba. More problematically, these are the kinds of signals we would expect to see anyway during a global warm-to-cold transition. Since that was underway well before Toba, the eruption, big though it was, looks incidental to the global change.

In one recent study focused on the Toba question, the Quaternary scientist Martin Williams, from the University of Adelaide, and his colleagues examined sediment sequences containing Younger Toba Tuff ash at sites in north-central India and in sediment cores from the Bengal Fan [144]. For the terrestrial sites, they measured carbon and oxygen isotopic ratios in calcified nodules and root casts found below, within and above the ash. The results suggested replacement of forest, which had thrived prior to the eruption, by grasslands or wooded grasslands. Pollen analysis of the offshore sediments also pointed to reduced tree cover, as well as cooling and at least a millennium of drought. They concluded that the Toba eruption led to these changes but again, one could turn the interpretation around and suggest that the findings substantiate the known weakening of the Asian monsoon associated with warm-to-cold global transitions.

Furthermore, it may be difficult to disentangle immediate consequences of inferred climate change from the direct ecological effects of a major ash fall. The depth of Toba ash fallout across peninsular India was around ten centimetres. This was likely sufficient to decimate existing vegetation, inviting subsequent colonisation led by grasses as discussed in Section 2.9. Recovery depends on many factors including climate, depth and mineralogy of ash, and so on, but can potentially take a couple of centuries or so. Thus, rapid shifts in isotopic composition following ash fallout are open to different interpretations.

8.4 THE HUMAN STORY

In the last chapter, we saw how critical the period between around 80,000 and 60,000 years ago was in terms of migrations of Middle Stone Age *Homo sapiens* beyond Africa towards south and southeast Asia. Furthermore, the routes must have involved travel around or across India, which, as we have already seen, received a good deal of ash fallout 73,000 years ago. Were modern humans already in India by that time? It is not surprising then that attention has focused on the possible influences of the Toba eruption and its associated climate and environmental effects on the ancient demography of our species.

Mike Rampino and Middle Stone Age expert Stanley Ambrose considered that 'given the magnitude of the effects of volcanic winter on global climate and primary productivity it would be remarkable if this cataclysmic event did not affect tropical humans and other species' [139]. They argued that soot from the burnt vegetation, as well as very fine ash and sulphuric acid particles, would have drastically reduced sunlight falling at the Earth's surface in the weeks after the eruption. This would have reduced photosynthesis, which, combined with the cold conditions, would have decimated vegetation. Tropical forests would have been most vulnerable to chilling with potentially severe and lasting damage resulting from temperature drops of up to 10 °C. Temperate forests may have fared little better, suffering especially during cold springs and summers when new shoots are less resistant to chilling.

Ambrose had already been thinking through the implications of the Toba eruption for human populations and had published an influential paper in the *Journal of Human Evolution* in 1998, titled 'Late Pleistocene human population bottlenecks, volcanic winter, and differentiation of modern humans' [124]. His claim was so profound – essentially that Toba's eruption almost wiped out our ancestors – that it spurred much of the recent research on super-eruptions (as well as a number of television documentaries). The argument runs as follows. Studies of mitochondrial DNA (mtDNA, see Section 7.3.2) and other genetic characteristics point to a bottleneck in our ancestral population between 50,000 and 100,000 years ago, when the population would have been as low as a few thousand. (With such low numbers, our ancestors would have merited mention on the list of endangered species.) Ambrose suggests that 'the extraordinary magnitude of the bottleneck effect recorded in our genes' was a direct consequence of the environmental catastrophe precipitated by the Toba eruption:

> Famine caused by the Toba supereruption and volcanic winter provides a plausible hypothesis for Late Pleistocene human population bottlenecks. Six years of volcanic winter, followed by 1000 years of the coldest, driest climate of the Late Quaternary, may have caused low primary productivity and famine, and thus may have had a substantial impact on human populations ... Many local human populations at higher latitudes and in the path of the ash fallout may have been eliminated.

He suggested that Neanderthals might have fared better in these conditions due to physiological adaptation to cold, while the equatorial region of Africa offered the most extensive refuges for humans to bunker down and see out the bottleneck. Release from the bottleneck, that is a population explosion, did not begin until about 50,000 years ago, coinciding with tremendous leaps in stone tool technology. The Later Stone Age opened up new possibilities to exploit food supplies, enabling populations to grow and expand their territory. In re-stating his ideas in a later article, Ambrose added [145]:

> Capacities for modern human behavior were undoubtedly present during the last interglacial, but the stable environments of this period did not foster widespread adoption of the strategic cooperative skills necessary for survival in the last glacial era. Modern humans may have eventually developed such strategies during the last ice age, but they were crucial for survival when volcanic winter arrived. We are the descendants of the few small groups of tropical Africans who united in the face of adversity.

This idea that severe environmental downturn in subsistence societies represents a strong selective driver for cooperative social behaviour has many adherents. Most important, arguably, is reciprocity between rather than within groups. Visits and exchanges enable pooling of resources across a wider territory, and provide opportunities for gathering and sharing news and innovations, as well as support during times of hardship. Matt Rossano, a psychologist at Southwestern Louisiana University, has supported Ambrose's case, drawing on a wide range of evidence from experimental psychology, archaeology, anthropology and primatology [146]. The essence of his argument is that the 'unprecedented resource and social stress' resulting from the Toba eruption contributed to the development of ritual practices, which, he believes, played a crucial role in human cognitive development; what Ambrose refers to as the 'troop-to-tribe' transition. In particular, he believes that the social rituals of our ancestors helped to build 'working memory', leading to more complex social groups that, in turn, enabled new cooperative behaviours, such as trading. Anatomically modern humans could remember better and adapt their behaviour to new situations; ultimately this capacity led to the use of symbols and the ability to share meanings - the basis for communication, language and culture.

Behind the argument is the belief that ritual behaviour - especially where the ritual involves inhibition of a more innate pattern of behaviour - enhances the capacity of the human working memory. When our brains are engaged in tasks that require conscious, focused attention, it is regions of the frontal lobe, known to be associated with our working memory, that are particularly active. As an example, Rossano cites experiments in which subjects viewed erotic films while their brain function was being monitored. As expected, the experience elicited the response of parts of the brain known to be linked to sexual arousal (the amygdala and hypothalamus). But when they were asked to suppress any sexual response, the frontal lobe fired up, indicating its capacity for top-down, inhibitory control of more natural responses. Ritual behaviour is an example of such wilful control.

In another fascinating experiment, the brain imaging procedure was applied while expert flint knappers were engaged in making Acheulean tools (Section 7.3). This did not elicit signals from the 'working memory' regions of the brain. Thus, social factors, more than technological ones, may have led to the working memory development that fostered our unique human cognition.

In traditional societies today, there are three kinds of ritual practice that build social cohesion: (i) ceremonies related to trust-building, conflict resolution and peace-making; (ii) adolescent rites of passage that establish group commitment; and (iii) shamanism that contributes to community or individual healing. It has been argued, for instance, that initiation ceremonies are essential in times of adversity: a broad view of the practice worldwide today suggests that the harsher the environment, the more eye-wateringly severe are the tests of endurance demanded. In terms of cognition, their significance is that getting through such ghastly challenges requires mental control over natural responses. In Rossano's view, this inhibitory control is a 'critical hallmark of the enhanced working memory capacity necessary for symbolic thinking'. More inhibition requires more attention focus and more control.

He cites recent discoveries at a cave site in Botswana, dated to around 70,000 years ago, as possible evidence for the development of human ritual practices in response to the trauma of the Toba eruption. At the entrance to the cave is a six-metre-long rock that appears to have been deliberately fashioned so that, with flickering firelight, it would have cast the terrifying shadow of a serpentine phantasm. The context of the site, he argues, is suggestive of rituals still practised in traditional societies, in which altered states of awareness act to heal, impart sacred knowledge or terrify.

How did we develop the tendency for ritual behaviour in the first place? Rossano reminds us that ritual practices are widespread in apes and act to reinforce trust, group harmony, and social relations:

> The wealth of social rituals present among our primate cousins indicates that our hominin ancestors were pre-adapted for using ritualized behaviour as a means of social bonding and could call upon a rich repertoire of them in their everyday social life.

Rossano cites one curious practice engaged in by male baboons as a good example of a behaviour that helps to bond a community. Referred to as 'scrotum-grasping' it involves two baboons striding up to each other and taking it in turns to present their hind-quarters, allowing the other to hold his testicles. This is a procedure that requires a high degree of trust. It tends to be practised by older baboons and builds strategic alliances against younger, stronger males in order to secure mating opportunities. Younger males usually fail to execute the greeting and are less successful in forming social alliances than old timers. In Rossano's words 'no scrotum-grasping, no social alliances'.

8.4.1 Counter-arguments

There is great difference of opinion over the interpretation of the various data pertaining to human palaeodemography. Two duelling factions include the palaeoanthropologists who claim that fossils are the only direct evidence of evolution, and the geneticists who argue that all the living have ancestors, while fossils may have no descendants. Although genetic studies of living humans point to population bottlenecks, interpretations of different kinds of DNA found within and outside the cell nucleus are contradictory. A possible explanation may be that present-day human genetic variation is the product of an extended period of low, but reasonably stable, population size – a 'long-necked bottle'.

Another point of contention in this debate is that if environmental catastrophe following the Younger Toba Tuff eruption was responsible for a human bottleneck, then there ought to be comparable genetic evidence in other species, such as the great apes. Population bottlenecks at various times are turning up in the lineages of all kinds of creatures including elephant seals, pin-worms, koalas, fruit flies and anchovies. However, a detailed genetic survey of African apes (chimpanzees, bonobos and gorillas) highlighted striking differences in patterns of mtDNA variation between these species and humans [147]. In spite of their much smaller populations and limited ranges, the African apes retain far more genetic variation than we do. Only Tanzanian chimpanzees show anything like the narrow range of mtDNA diversity seen in humans. These data make it clear that human genetic history is dramatically different from the histories of our closest relatives, and that the great apes did not all experience a coincident bottleneck. The ecological ranges, and habitat preferences and tolerances of humans and apes, are not the same, so we should not expect all species to respond in the same way to a global environmental catastrophe. But if we accept that humans are smarter and more adaptable than chimps and gorillas, we might argue that our ancestors would have fared better during a volcanically induced cold spell – contradicting the apparent trends in genetic diversity.

Turning to fossil evidence, the palaeontological record of various land mammals (other than humans) living in southeast Asia before and since the Toba eruption is equivocal [148]. Unfortunately, the sites from which the various fossils were found are not well dated. Nevertheless, it appears that in Sumatra itself, the orang-utan survived Toba, along with macaques, gibbons and the Asian tapir. In fact, there were no apparent extinctions on Toba's home territory. Nor were any extinctions evident in another study of the fauna of the Mentawi islands, located off the west coast of Sumatra and just 350 kilometres from Toba. Significantly, all these animals live in rainforest habitats suggesting that whatever impacts the Toba eruption had on the ecosystems of Sumatra, mammal populations found refuges in which to survive. Thus, whatever the adversities presented by the Toba eruption, humans did not have a monopoly on coping and adaptation strategies. The few species for which there is some evidence of extinction lived in open-forest habitats – in southern China, a species of elephant, and one of rhinoceros disappeared from the record, as did a pig and an elephant from Vietnam, and another pig, a bear, an ape and a couple of monkeys from Java. The extinguished species may have been in long-term decline before the eruption, which merely provided the *coup de grace*.

One other piece of minor evidence relates to the most extraordinary palaeoanthropological revelation of the last decade – the unearthing from a cave on Flores Island (Indonesia) of a tiny human known as *Homo floresiensis*. Hailed as 'the most extreme hominin ever discovered', the 'Hobbit', as it is more popularly known, stood only a metre tall and had a brain the size of an orange. Interpreted as either an archaic pre-erectus human or the result of stunting of *Homo erectus* through isolation, the Hobbit appears to have reached Flores as early as a million years ago [149]. How it got there remains a mystery but the species survived on the island until around 17,000 years ago. It thus withstood any adverse impacts of the Younger Toba Tuff eruption. Nevertheless, its eventual demise may have been at the hands of volcanism, since the youngest fossils were found sealed beneath a dark layer of tephra!

8.5 FOCUS ON INDIA

One key to moving forward on Toba's environmental and human impact (as well as its human consequences) is to find continuous sedimentary sequences of high accumulation rate spanning the few thousand years immediately before (and after) Toba ash fallout. One candidate could be the Jwalapuram site in the south-central Indian state of Andrha Pradesh. I have visited it on two occasions with an international team of archaeologists and Quaternary scientists. The site is truly spectacular from a scientific perspective, and will surely yield penetrating insights into the Toba story once all the data have been sifted through, analysed and interpreted.

The site is very close to the Jurreru River, which is now barely more than a trickle, and in a valley enclosed by modestly sized hills of red-brown sandstone. Not only does Jwalapuram preserve an exceptional sequence of pre- and post-eruption sediments but workers from the nearby village continuously quarry the site to a depth of a few metres in order to mine the ash (which is sold as an abrasive ingredient for detergents). Thus, new excavations exposing new perspectives into the architecture of the deposits are guaranteed on each visit. The first shock on reaching the site is the great thickness of ash, which reaches up to 2.5 metres. Bearing in mind that Jwalapuram is over 2600 kilometres from Toba this cannot possibly all represent the tephra fallout and signifies a prolific redistribution of ash by wind and/or water once it had settled.

My contribution to research there involved use of a set of chisels and paintbrushes to exhume the original ground surface immediately below the ash fallout, as well as internal layers within the ash beds. Temperatures approaching 40 °C in the shade (except that there was no shade!), the pall of dust from the brushwork, and persistent, biting flies, made it uncomfortable work. But I can attest that peeling back an ancient ground level last seen at the precise time it was encapsulated in fallout from one of the largest known volcanic eruptions has been one of the most engrossing and exciting pieces of fieldwork I have ever been involved in (and there wasn't even a volcano in sight).

Palaeoenvironmental analyses of hundreds of samples collected at Jwalapuram go a long way to reconstructing what that ancient environment was like [150,151]. Prior to the eruption, the area was already cooling and drying, consistent with the apparent timing in the GISP2 ice core. Immediately before Toba exploded Jwalapuram was nevertheless wetter than it is today, with small lakes edged by swampy areas. In between lay wooded areas with trees growing up to several metres in height. In 2009, I found impressions of their leaves at the upper surface of what appeared to be the initial fallout of ash from the eruption cloud (Figure 8.4). This 'primary' ash layer attains a thickness of just under ten centimetres, still an astonishing depth considering the great distance from Toba. The position of the leaves on the surface of this ash veneer, rather than beneath it, indicated that they were not pre-Toba leaf litter but that the heavy ash fall had killed off the foliage. Probably within a matter of days of the eruption, the stricken trees shed their leaves, which were then entombed by thick layers of very nearly pure, grey-brown ash carried in from the surrounding hills. Meanwhile, numerous buried bugs struggled to reach the surface.

Figure 8.4 Imprint of a leaf at the top of the Toba ash fallout layer at Jwalapuram. Its position shows that it was not part of the leaf litter beneath the tree but rather that the ash fall resulted in defoliation.

These 'reworked' ash layers tell their own story. Closer inspection shows that they form five individual packages that are discriminated by intervening dark-grey bands of very hard rock (Figure 8.5). Whereas the ash may be scraped easily with the fingers, the cemented layers need a hammer to break them up. Each package is probably the result of heavy monsoon rains washing the loose ash covering the landscape down into the valley. Towards their tops, the size of ash particles diminishes and there are short, vertical cracks propagating down from the hard band, which have been mineralised with calcite. These small fissures are tell-tale signs of aridity – they are fossilised examples of the polygonal cracks that form in mud when it desiccates. The repetition of these units suggests that in the first years following the eruption, the monsoon was active and the re-deposited ash quickly choked up the river valley. Alternatively, the hard-bands indicate sporadic wetting of the surface by rainfall. Even if the initial fallout of ash didn't kill off the trees, they couldn't have survived the first influx of reworked sediment.

What happens above these five units is equally interesting. The next unit of sediment is around a metre thick and its appearance and texture differ dramatically from the underlying ashes. It still contains Toba ash but this is mixed in with more sandy material eroded from the surrounding hills. Its colour is pinkish and there are two starkly contrasting interpretations of it. One interpretation sees it as the result of a resumption of wet conditions but it could equally be a silty sediment deposited by the wind and indicative of aridity (a 'loess' deposit).

The timing of all this is more or less instantaneous as far as can be discerned from the various dates obtained for the sediments. The next deposit above the orangey-brown layer has been dated to 74,000 years

Figure 8.5 Exposed section through the Toba ash at Jwalapuram. Only the ten centimetres or so at the base (at the level of the foot of the ladder) is fallout from the ash cloud. The rest has been washed (and perhaps blown) in from the blanket that must have covered the entire Jurreru valley. The prominent horizontal bands in the light-grey reworked ash are much harder than the surrounding sediments and represent periods of aridity and desiccation. The sequence therefore appears to show five or six monsoon cycles (the thickness of the ash packages decreases as less and less is left on the hill slopes for erosion). At the ankle level of the man on the ladders, there is a transition to a darker (orangey-brown) deposit. This unit contains ash mixed in with other sediment, and may represent a period of aridity. Above it is an even darker band of sediments composed of sands and silts that indicate a return to more humid conditions. The provisional dating of the section suggests that the whole sequence spans less than a few thousand years.

with an uncertainty of 7000 years up or down, thus we cannot say whether the pinkish layer started to accumulate one year after the influx of almost pure ash, or a thousand years later. But for what it is worth, my best guess is that what happened goes something like this: in the time before the eruption, the environment of the valley floor was one of small lakes, marshy areas and open woodland. The fallout buried the land surface beneath about ten centimetres of ash. The coatings on foliage quickly led to all the leaves being shed and probably the trees died.

Any early climatic consequences – within the first years – were not such as to terminate the monsoon. Rainfall appears to have vigorously washed ash from the low sandstone hills into the valley where, over the period of the next five years, it progressively smothered the withered tree trunks and branches. There are at least two ways to explain what happened next. One is that the valley simply became so choked with reworked ash that the local river drainage reorganised itself, cutting off that part of the valley floor and resulting in a long hiatus in sedimentation. Alternatively, it is possible that five years of climate forcing due to Toba's stratospheric aerosol, combined with the effects of reflective ash on the ground, hit the regional heat budget hard enough to weaken or even shut down the Indian summer monsoon. If this were the case, maybe it did profoundly change the environment and ecosystems on the subcontinent.

As well as a remarkable sedimentary sequence amenable to palaeoenvironmental analysis, the Jwalapuram site is rich in archaeological materials, notably flakes, points, scrapers and cores prepared from chert, chalcedony, quartz and limestone. These are found both above and below the Toba ash horizon. Statistical analysis of their type and dimensions shows closer affinities to the contemporary Middle Stone Age lithic industry of sub-Saharan Africa than other regional Neanderthal or Late Acheulean stone assemblages. This has been taken to imply that modern humans were in India prior to the eruption. Furthermore, the tools found beneath and above the ash are very similar in their characteristics suggesting that later generations of indigenous survivors of the eruption eventually resettled the area once natural recovery of the habitat was complete. However, no human fossils have been found in association with the sites to confirm the attribution of the artifacts and, if modern humans were indeed in India this early, they did not contribute to today's genetic diversity (which points to a single 'out-of-Africa' migration between 60,000 and 55,000 years ago, i.e. well after the Toba eruption). Thus, either the tools were made by modern humans who exited Africa prior to Toba but left no descendants or they were the handiwork of Neanderthals or perhaps an archaic Asian population.

8.6 SUMMARY

What makes Toba so interesting right now is that we really don't have all the answers to what happened and what the consequences were. In fact, we don't even have that many of the pieces of the puzzle to reconstruct events. In particular, human fossils demonstrably belonging to the Toba period have thus far eluded discovery (indeed, specimens of any human fossils are few and far between in India). What we do have is plenty of intriguing ideas and some exciting new sites and modelling capabilities. However, while the eruption does raise immensely significant questions for understanding human origins, we do need, in my view, to recognise that the case for Toba wreaking global climate havoc of a magnitude greatly removed from that of Pinatubo in 1991 is far from convincing. Rather, the evidence seems to be mounting that the claims for huge sulphur output and extreme global climate change have been overblown.

Furthermore, it is obvious in deciphering the human story of the period that if one is looking for climate-determined scenarios of human demography and behaviour there was plenty of global change afoot between 80,000 and 60,000 years ago that had nothing to do with Toba. Amidst the unambiguous evidence for severe climate variability independent of the eruption – first the millennium of cold conditions around 73,000 years ago, and then the long period of glacial conditions – pinning any blame on Toba for super-imposed effects requires data of exceptional temporal resolution.

The results obtained from climate models for various Toba eruption scenarios are highly dependent not only on the prescribed sulphur yield, but also on the unknown dynamics of super-eruption plumes, and the precise state of the atmosphere and climate prevailing at the time of the event. Significantly, even models for massive sulphur clouds do not predict anything like a thousand years of atmospheric change.

The narrative of the Younger Toba Tuff eruption and its potential impacts highlights two human propensities. Firstly, we imagine that a bigger eruption would have a bigger impact on the atmosphere, climate, environment, ecology and so on. Secondly, we are quick to spot coincidences and attribute cause and effect. The problem with the first expectation is that it is the sulphur output that really determines the extent of any climatic effects, and not the quantity of ash ejected. And with the second, we have to accept that the estimated dates of genetic bottlenecks, archaeological sites, fossils, even the eruption itself are no more than estimates, each with its limited certainty. It is worth noting that another large event (M_e 7.7), which took place around 84,000 years in Guatemala – the Los Chocoyos event that formed Atitlán caldera – may have released far more sulphur than

Toba owing to its Pinatubo-like magma composition. Maybe this is one we should be looking at.

Nevertheless, these uncertainties and doubts do not rule out the possibility that the Younger Toba Tuff eruption did have substantial consequences for our human ancestors (who perhaps - allowing for uncertainties in estimated rates of genetic drift - had already colonised western and south Asia beforehand). And perhaps the experiences of our ancestors, the physical and psychological ways they found to cope with the trauma, really did shape the cognitive skills that we enjoy today. When I worked at the Jwalapuram excavations, exhuming the ancient land surface beneath the ash, it certainly made me wonder about the community that dwelt there, and what people must have thought when the Sun failed to appear one morning, and the land had turned to lifeless, grey powder.

9

European volcanism in prehistory

> ... we cannot sideline catastrophic environmental change in our reconstructions of prehistoric culture history.
>
> F. Riede, *Journal of Archaeological Science* (2008) [152]

Most of us tend not to think of Germany as a country prone to acts of volcanic violence. And yet, about 12,900 years ago, it witnessed a powerful pyroclastic eruption on the scale of Pinatubo's in 1991, right in the heart of the Rhineland.

On the other hand, we are well accustomed to volcanic activity in the Campanian province around Naples. There have, nevertheless, been some events very much greater than the infamous 79 CE convulsion of Vesuvius that sealed the fates of Herculaneum and Pompeii. In particular, approximately 39,300 years ago, the Campi Flegrei (Phlegrean Fields) volcanic system ruptured, disgorging up to 80 times more magma than its neighbour in the year 79.

Lastly, the eastern Mediterranean, too, is home to numerous dormant volcanoes. The best-known is Santorini, whose Bronze Age eruption has been linked to the demise of one of Europe's first great civilisations - the Minoans - and even to the Atlantis legend. This chapter focuses on these three exceptional European volcanic eruptions and considers the evidence for the nature and extent of their human impacts.

9.1 THE CAMPANIAN ERUPTION AND THE HUMAN REVOLUTION IN PALAEOLITHIC EUROPE

The Campanian province of southern Italy is today considered one of the regions of the world at highest risk from volcanic activity. Its volcanoes include the Campi Flegrei, Vesuvius and the island of Ischia (Figure 9.1). Around 39,300 years ago, the Campi Flegrei was the source of a M_e 7.4–7.7 eruption (equivalent to between 100 and 200 cubic kilometres of dense magma). Events began with a Plinian eruption whose column reached a maximum height of 40–44 kilometres above sea level. Later, the eruption became dominated by emission of turbulent pyroclastic currents with sufficient vertical extent and energy to surmount rugged topography and travel up to 80 kilometres from source, covering an area of around 30,000 square kilometres. This phase of the eruption was accompanied by foundering of the crust above the magma chamber, leaving a caldera 30–35 kilometres in diameter. Meanwhile, the phoenix clouds flooded into the middle atmosphere. Tephra layers from the eruption span an area of between two and four million square kilometres, stretching across the central and eastern Mediterranean Sea, Greece and Bulgaria, and northeast across the Black Sea to Russia (Figure 9.2). This massive volcanic blast – the largest in Europe for more than 100,000 years – is known as the Campanian Ignimbrite eruption.

Even before its identity was known, the tephra (of intermediate-silicic 'trachytic' composition) had been widely used as a marker horizon across the central and eastern Mediterranean Sea, and had become

Figure 9.1 The Campanian Plain and Bay of Naples. The cluster of craters to the left of the photograph belongs to the Campi Flegrei – site of the Campanian Ignimbrite eruption 39,300 years ago. Vesuvius, on the right, is much more notorious than the Campi Flegrei but the size of the Campanian Ignimbrite dwarfs all eruptions in the province. Downtown Naples is approximately halfway between the two volcanoes. Image from ASTER Volcano Archive, NASA/JPL.

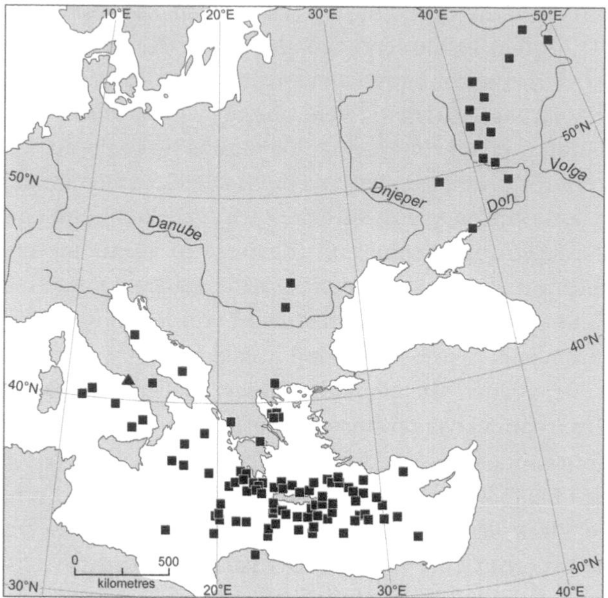

Figure 9.2 Map showing sites (square symbols) where fallout from the Campanian Ignimbrite eruption has been identified. The location of the Campi Flegrei is shown by the triangle symbol. Modified from reference 157 and used with permission of Elsevier.

known as the 'Y5' layer. Because of its distinctive composition, widespread occurrence and its date, it is increasingly seen as providing a critical timeline in archaeological and palaeoenvironmental contexts.

9.1.1 Climate impact

We lack firm evidence for the climatic effects of the Campanian Ignimbrite eruption. However, the eruption's sulphur yield has been estimated from petrological arguments as 1.2 gigatonnes - more than 120 times Pinatubo's 1991 sulphur output. Given the explosiveness of the eruption we can be certain that a significant sulphuric acid aerosol layer was generated (though its fallout has yet to be convincingly identified in any ice-core records). However, with the source at latitude 42° N, the climate forcing would likely have depended rather strongly on the season of eruption. This expectation is borne out by a pertinent climate model for a 100-times-Pinatubo sulphur injection representing a hypothetical eruption of Yellowstone in Wyoming (whose latitude is similar to that of Campania; Section 3.3.4). The other complicating factor is that very shortly before the eruption, Earth had plummeted into a cold phase due to a surge at the front of the North American ice sheet. This disgorged a flotilla of sediment-laden icebergs into the Atlantic Ocean, which rapidly melted releasing an estimated 2.3 million cubic kilometres of ice-cold freshwater. This was enough to redirect the Gulf Stream southwards, causing a rapid and severe cooling of climate referred to as 'Heinrich Event 4'. Thus, at the time of the Campanian Ignimbrite eruption, the Earth's climate system may have been in an acutely sensitive phase, which could have accentuated the effects of the volcanic aerosol.

9.1.2 Human impact

The period of the Campanian Ignimbrite eruption is one of immense interest in archaeology and palaeoanthropology. Around 40,000 years ago, two major trends in human prehistory were underway – one biological, the other cultural. They define the 'European Palaeolithic Shift', an important chapter in the 'human revolution' that saw modern humans extend their range across the planet. Anatomically modern humans had arrived via the Caucasus from western Asia and it was not long before the Neanderthals had disappeared from Europe (the last vestiges of their populations are known from Gibraltar and dated to around 28,000 years ago), possibly outcompeted for resources by the interlopers [153] (but see below). The other change was tremendous innovation in stone-tool technology – the Middle Palaeolithic toolbox was replaced over time by Upper Palaeolithic industry [154] (note that for historical reasons only, the terms 'Middle Stone Age' and 'Late Stone Age' are used to describe African archaeological contexts while the equivalent terms 'Middle Palaeolithic' and 'Upper Palaeolithic', respectively, are used in Europe).

In Europe, the Middle Palaeolithic period is strongly associated with 'Mousterian' technology, characterised by knapping of stone flakes and their subsequent retouching to produce a range of side-scrapers and points. The only human skeletal remains found in association with Middle Palaeolithic tools belong to Neanderthals. The Upper Palaeolithic period emerged in Europe by around 40,000 years ago. Its showpiece technology, the 'Aurignacian', focused on manufacture of small blades (used for making composite tools), end-scrapers and points made from bone and antler. The culture was also associated with an explosion in use of personal ornaments made from shell and other materials. Tools representing the transition from Middle to

Upper Palaeolithic include types of both industries, and hybrids between the two. Upper Palaeolithic sites are associated with both Neanderthals and modern humans, as are the 'combined' Upper/ Middle Palaeolithic toolkits. However, the Aurignacian appears to represent an exclusively modern human industry.

Among the revolutionary innovations of the Upper Palaeolithic were the use of fibres and bone needles, enabling manufacture of clothing and nets for fishing. Being able to clothe infants, for instance, would have significantly improved survival rates through harsh winters. Furthermore, sewing and net-making are activities that don't require the kind of athleticism, strength and stamina required for hunting wild animals, thus enabling younger and older people to contribute to obtaining resources for their communities. These were huge advantages for modern humans over Neanderthals, for example.

The archaeological record either side of the Campanian Ignimbrite fallout provides us with two important pieces of information: firstly, at the time of the eruption, lithic industries present a spectrum of Middle to Upper Palaeolithic technology, firmly placing the period within the 'European Palaeolithic Transition'. We already know that as much as 10,000 years before the eruption, experimentation and innovation were transforming the ways in which humans could hunt and forage, enabling them to exploit previously untapped food resources, and new landscapes and habitats, thereby extending their geographical range. The second indication is that immediately above the ash layer in southern Italy and possibly parts of the Balkans, Greece and Bulgaria, the sediments are archaeologically sterile, pointing to abandonment that must have signalled wider disturbances in population density and geographical distribution.

One exceptional archaeological site where the Campanian Ignimbrite tephra was only recently identified is at Kostenki' in southwest Russia, where excavations are directed by Andrey Sinitsyn of the Institute for the History of Material Culture in St Petersburg [155,156] (Figure 9.3). Here the ash layer is typically five to ten centimetres thick and contained within sequences of ancient soil horizons and river sediments rich in archaeological materials. Tools exhumed from beneath the ash horizon, in layers dated 45,000–42,000 years ago, show earliest traces of Aurignacian technology, in the form of chisels, large retouched blades, end-scrapers and micro-blades, as well as bone points, awls and ornaments, and carved ivory. But they also include assemblages belonging to the 'combined' Middle and Upper Palaeolithic style. A few teeth, attributed to *Homo sapiens*, have also

Figure 9.3 Excavation 14 at Kostenki' in Russia revealing Palaeolithic cultural layers and, shown between the pairs of arrows, the Campanian Ignimbrite ash fallout. The ash layer is typically 5–10 centimetres thick but was re-worked after the initial fallout and reaches up to 25 centimetres at this site.

been sifted out. Pollen samples indicate that, prior to Heinrich Event 4, the environment was already oscillating between warm conditions, associated with broad-leafed vegetation, and much colder, drier periods. At two of the Kostenki' excavations, a number of artifacts, including earliest Aurignacian bladelets, retouched blades and scrapers, as well as elongated bone beads (decorated with spiral patterns), perforated shells and a few fox teeth, have been found within the ash (Figure 9.4). It is conceivable these were discarded by people who actually witnessed the fallout.

In the layers of alluvium and soils above the ash horizon, and dated to 40,000–30,000 years ago, are definitive Aurignacian scrapers and bladelets, along with further 'combined' toolkits and even some typical Middle Palaeolithic assemblages of scrapers and points. Additionally, bone shovels and some of the oldest-known eyed needles (also made from bone) were recovered. Some fragments of tibia and pelvis and a tooth, and the partial skeleton of a child found in a burial pit, have been tentatively identified as belonging to *Homo sapiens*. However, the very mixed assemblages of artifacts found in deposits laid down after the Campanian Ignimbrite eruption indicate a long period of cultural and, perhaps, biological transformation.

For anthropologist Francesco Fedele, from the University of Naples, and his colleagues, 'the occurrence of a significant environmental accident at 40,000 [years ago], during a glacial period and within an on-going or incipient epoch of cultural change (and gene flow?), necessarily raises questions' [157]. They believe that the combination of the

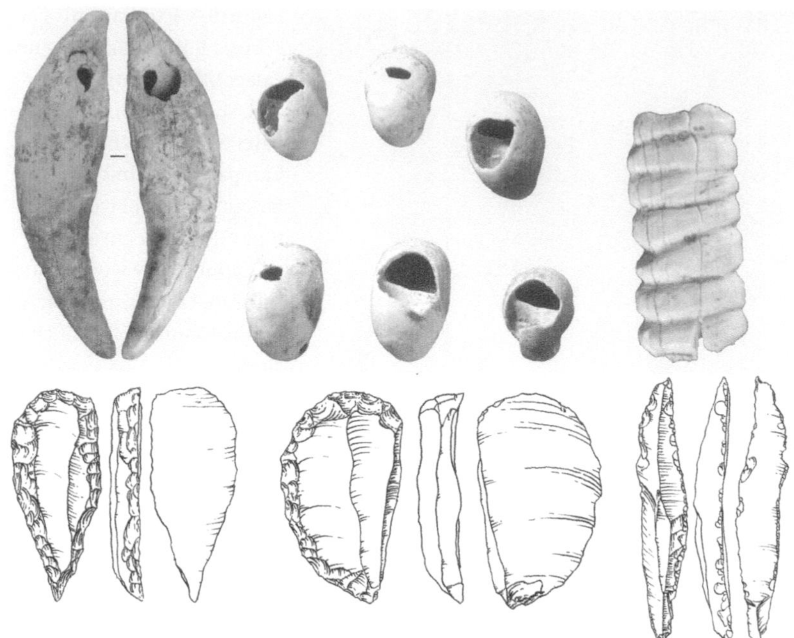

Figure 9.4 Cultural materials found within the Campanian Ignimbrite ash layer at Kostenki' attributed to Early Upper Palaeolithic. These include ornaments made from shell, tooth and bone, and lithic assemblages attributed to the Aurignacian industry with 'Dufour' microliths. Courtesy of Andrey Sinitsyn.

Campanian Ignimbrite eruption and Heinrich Event 4 played an important role in the European Palaeolithic Shift; while either event in isolation would have been dramatic, their coincidence greatly amplified the human disruption. Shrinking forests and invasions of arid and cold steppe arising from the severe effects of Heinrich Event 4 are evident from pollen records [158]. These ecosystem disturbances already had human populations on the move in much of Europe. They had to adapt to transformed habitats, and cope with their repercussions for the nature and availability of food resources. Where ash fallout from the eruption was more than a few centimetres thick, it would have smothered wild foods and grasslands, possibly wiping out herds of deer and other animals that were hunted by contemporary humans. Fedele and colleagues speculate too that access to water resources and sources of stone prized for tool manufacture would have been disrupted. They suggest that northern Europe may have been 'depopulated to the point of desertion' because of the climatic and ecological perturbations.

The demographic displacements – especially migrations beyond the zone affected most strongly by the eruption – would have put the survivors in unfamiliar territories. Some areas became abandoned, others over-crowded. While co-existence has probably always represented a challenge to humans, it also provides a strong incentive to experiment in new social practices and belief systems. While times of hardship and enforced stress may have disrupted traditional means of interaction within and between groups, they may, too, have stimulated sharing and transmission of information. These permutations in learning, understanding and ideology would have selectively affected the survivors according to their ability to adapt. In this respect, Fedele and colleagues see the Campanian Ignimbrite as acting as a 'filter and catalyst' for human cognitive evolution – favouring innovation and cooperation.

One manifestation of this behavioural change was standardisation in the manufacture of miniaturised tools called 'micro-blades' and the discovery of new techniques for hafting them onto wood. This enabled construction of sophisticated composite tools suited to specific tasks. There was also an increased use of personal ornament in the form of pendants and beads, and 'image-making' in the form of figurines and symbolic carving. These artistic expressions represent a developing sense of identity among human groups and are seen by Fedele and colleagues as the outcome of the more crowded social environments imposed by the post-eruption landscape and climate. Since art is 'critically related to both cultural transmission and collective emotion', they argue that these developments may even have inspired the emergence of ethnicity and personal identity.

More recently, Liubov Golovanova, from the ANO Laboratory of Prehistory in St Petersburg, and colleagues have argued that the Campanian Ignimbrite contributed to the extinction of Neanderthals [159]. Their conclusions are based on archaeological studies of deposits found at the important Palaeolithic site of Mezmaiskaya cave in the northern Caucasus. Although they did not find any signs of fallout from the Campanian Ignimbrite eruption, they did identify a basaltic tephra layer of a similar age, which they suggested might have been erupted by Mt Kazbek in the central Caucasus. Despite abundant Neanderthal remains below the tephra, none were found above. This led them to hypothesise that the combined impacts of ash fallout and abrupt climatic deterioration 'drastically destroyed the ecological niches of Neanderthal populations' leading to 'the mass death of animals and hominins'. They further argued that modern humans were essentially waiting in the wings in the Middle East and north Africa, thereby avoiding a 'nuclear winter' in Europe. Along with social and technological innovations that made modern humans more resilient in severe conditions, their subsequent conquest of northern Eurasia was, they suggested, promoted by the 'Neanderthal population vacuum in Europe'.

To report that this hypothesis is controversial would be an understatement. In a commentary on what he called a 'flawed manifesto', Fedele wrote: 'This facile, pre-Darwinian catastrophism shuns the challenge of exploring displacement instead of replacement, ecosystemic stress, selection for innovation - a whole gamut of possibilities'. Certainly, aspects of the argument are confusing and, in view of the dating uncertainties for the site, it is stretching the point to connect local effects of an eruption in the Caucasus with regional impacts of the Campanian Ignimbrite eruption.

9.2 'CULTURAL DEVOLUTION' AND THE LAACHER SEE ERUPTION

Laacher See is a volcanic crater lake in the Eifel district of Germany, some 40 kilometres south of Bonn. The reasons for volcanism in the region are complex but related to the collision between the African and Eurasian plates. In addition to uplift of the Alps, the sustained tectonic forces opened a roughly north–south trending rift system that reaches as far north as the North Sea. Although volcanism in Germany can be traced back for millions of years, it has been concentrated in bursts that were associated with the loading and unloading of the continent by ice sheets during glacial advances and retreats. Changing the distribution of ice cover can influence the physical stresses on magma chambers, acting as a potential eruption trigger for already 'ripe' magma reservoirs (Section 14.3.2).

The ancient land surface that lies beneath the Laacher See tephra has been picked over by palaeontologists, palaeoecologists and archaeologists. These investigations have provided a detailed picture of the environment and its human occupants immediately before the area was devastated. Impressions of leaves, carbonised plants and trees and pollen indicate there were lush, open woodlands of birch and aspen with a dense undergrowth of ferns, herbs and mosses [160]. Towards the river banks, willows and bird cherry flourished, while at higher elevations the drier conditions favoured pine forests. The forests abounded with deer, elk, wolves, horses, bears and boars; pike, perch and chub shimmered in the Rhine; and voles, moles, shrews, marmots and beavers scurried around the river banks and meadows.

The prevailing climate was similar to present-day conditions in the region – temperate and humid – though it was somewhat cooler in winter. The human population – known archaeologically as the Federmesser Group – was sparsely distributed and subsisted from foraging and hunting. They enjoyed a range of the woodland fauna, which they shot with both spears and bows and arrows. Quite possibly, they got wind that something was going to happen from the onset of seismic shaking and the appearance of gas vents. Then, sometime in late spring or early summer (judged from impressions of leaves and fruit in the ash, and from matching the tree rings of stumps preserved in the ash to the established dendrochronological sequence), Laacher See's subterranean magma met the ground water table generating tumultuous hydrovolcanic blasts. The best estimate for the date of the eruption (from a plethora of argon–argon and radiocarbon measurements) is a few decades after 11,000 BCE [161].

The initial hot blasts flattened trees up to four kilometres from the vent. They plastered the defoliated debris with steaming, greenish mud and assaulted anything left standing with a hail of lava blocks as the magma teased out a route to the surface. Once the conduit was established it fed a sustained Plinian eruption of intermediate (phonolitic) composition magma for around ten hours. The estimated eruption intensity of 11.5 suggests that the plume would have easily reached the stratosphere, perhaps reaching peak heights of 35 kilometres above sea level before subsiding towards the tropopause. It then dispersed to the northeast, and later to the south, blackening the skies above central Europe. Activity continued for several weeks or even months with alternating Plinian and hydrovolcanic phases, and the discharge of ground-hugging pyroclastic currents that plastered valleys with sticky tephra up to ten kilometres from the crater. One valley was choked with a 60-metre-thick dump of pyroclastic current deposits (Figure 9.5). The total magnitude of the events, M_e 6.2, is similar to that of Pinatubo's eruption in 1991. The petrological method for determining gas outputs is poorly suited to Laacher See's alkaline magma composition (a type known as phonolite) but a minimum of 1.9 megatonnes of sulphur were released along with 6.6 megatonnes of chlorine. In the GISP2 ice core in Greenland, there are two very concentrated sulphuric acid fallout horizons whose timing is compatible with the age of Laacher See but, as yet, there is no strong evidence to identify either of them.

Figure 9.5 Fifty-metre-thick sequence of pyroclastic rocks from the Laacher See eruption at Wingertsbergwand. Courtesy of Felix Riede.

Ninety per cent of the total volume of the deposits is in the form of tephra fallout. Near the two-kilometre-diameter crater formed by the eruption, the deposits reach a thickness of over 50 metres. Five kilometres away, the deposits are still ten metres thick. For a distance of around 60 kilometres to the northeast and 40 kilometres to the southeast, all plants and animals must have been exterminated. The fallout has been identified across an area exceeding 300,000 square kilometres, stretching from central France to northern Italy, from southern Sweden to Poland (Figure 9.6). This makes the tephra an invaluable chronological tool for many archaeological and palaeoenvironmental studies of this period in central and northern Europe. Meanwhile, so much tephra accumulated in the river basin that the Rhine was dammed, forming a lake 80 square kilometres in area. Eventually the dam failed catastrophically – perhaps triggered by an earthquake or the shockwave of another volcanic explosion – and mudflows surged at least 50 kilometres downriver.

9.2.1 Climate impact

As with the Campanian Ignimbrite, we know little about the climate impact of the eruption but the relatively high latitude (around 50° N) would have restricted the dispersal of the volcanic aerosol in the stratosphere to the northern hemisphere, as suggested by atmospheric modelling [162]. Nevertheless, the models suggest a year or two of the northern-hemisphere summer-cooling, winter-warming response

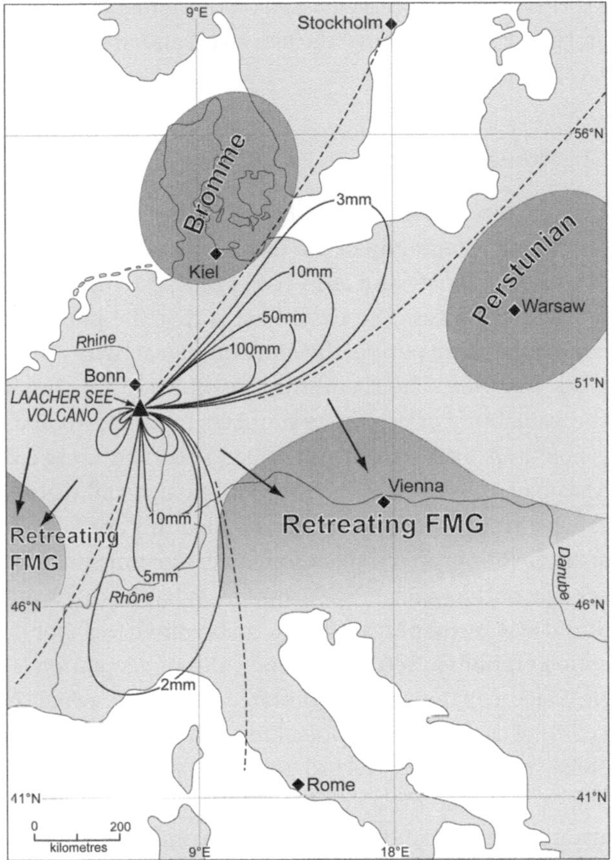

Figure 9.6 Isopach map showing tephra fallout distribution for the Laacher See eruption. Territories of cultural groups discussed in the text and their migration directions are indicated. FMG indicates people of the Federmesser Group. Adapted from reference 160 and used with permission of Elsevier.

that is typical of tropical explosive eruptions. Anomalies identified in tree-ring data from Switzerland are consistent with several years of cold summers, though the anomalies could also reflect stunted growth resulting from acidification of the soils associated with Laacher See's tephra fallout. Various lines of sedimentological and palaeoecological evidence also suggest increased precipitation in Germany in the years following the eruption, and up to two decades of environmental disruption due to a combination of changed surface reflectivity (that affects local climate), acidification, soil erosion, forest fires (whose incidence seems to have increased based on the abundance of charcoal particles in lake sediments above the ash layer) and protracted ecological recovery [163].

9.2.2 Human impact

One of the ways that archaeologists try to unravel demographic and cultural history and geography in the distant past is to plot up a timeline of recognised artifacts and sites based on the temporal distributions of radiocarbon dates. The basic idea is that the more recorded dates there are for a given period, the more human activity prevailed in that region at the time. The theoretical justification for the method requires the availability of very large numbers of radiocarbon dates to overcome potential biases from particular sites and archaeological contexts that have been researched in particular detail. Felix Riede, an archaeologist from Århus University, applied this approach to published northern European radiocarbon dates spanning about 3000 years either side of the Laacher See eruption. What he discovered was a stark cultural and geographical pattern that pointed to major impacts of the eruption on human settlement and cultural adaptation [152].

In all, Riede collated and recalibrated 139 dates, which he subdivided according to eight geographical regions and/or cultural categories. He then compared the record for each with the well-defined eruption date to provide evidence for change in the population and cultural history for each category. For the Thuringian Basin in central Germany, just a few hundred kilometres from Laacher See and beneath the main northeasterly axis of the tephra fallout, the data suggested depopulation or even abandonment after the eruption. In contrast, in southwest Germany and France, it appeared that populations increased, suggesting that the Federmesser people affected by the eruption migrated there. But the most interesting pattern revealed by the analysis was that two cultures – known as the Bromme and the Perstunian – emerged after the eruption. In terms of their toolkits, the Bromme of southern Scandinavia and Perstunian of northeast Europe are rather similar, even equivalent. But both differ markedly from the, until then, predominant technology of the Federmesser people (Figure 9.7). Riede sees the Bromme and Perstunian toolkits as evidence of a loss of skills in tool manufacture – a 'cultural devolution' or degeneration.

This conclusion was reached on the basis of careful measurements of the dimensions of nearly 500 tools relating to the different

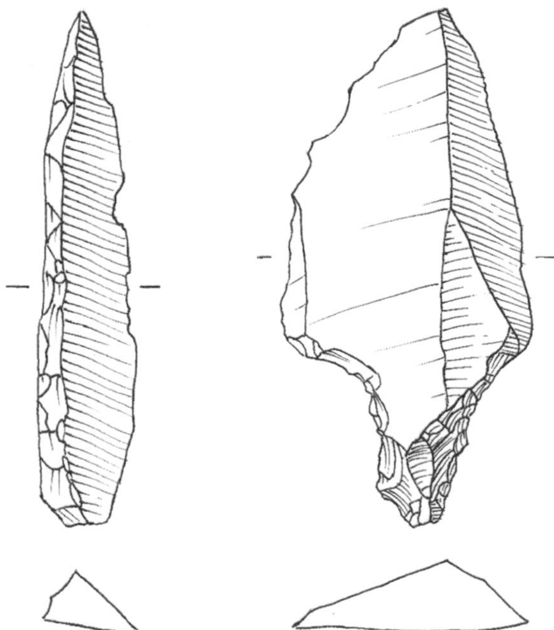

Figure 9.7 (Left) Federmesser or arch-backed point. These tools were used as arrowheads prior to the Laacher See eruption. (Right) Bromme point or large tanged point used as dart-tips before the Laacher See eruption. These became the exclusive weapon type following the eruption. Courtesy of Felix Riede.

groups of the period. Unfortunately, despite some rather ingenious (and presumably messy) experiments involving the firing of Bromme points into a variety of meat (including boar's head, pork with bones, sheep and fish, as well as reeds, bushes and trees) from a range of distances, it has not been proven conclusively how the Bromme people hunted. Nevertheless, in general terms it is thought that narrower points were used as tips to arrows whereas broader points would have been hafted onto sticks and poles to make darts or spears. Maximum width proved to be the key parameter that discriminated between the Federmesser and Bromme tools. Whereas the width distribution of Federmesser points peaked at about 10 centimetres, with plenty of wider points and subsidiary peaks at around 20–30 centimetres, the Bromme points showed a broad distribution of widths with a peak at around 20 centimetres, and very few pieces narrower than 15 centimetres. Riede concludes that the Federmesser people used different tools for different jobs – including bows and arrows – while their descendants in southern Scandinavia, the Bromme, relied on a single weapon system, probably darts that were used for hunting elk. Of particular note, the lack of narrow blades suggested the Bromme people had lost bow-and-arrow technology despite its considerable advantages over dart throwing as a hunting technique.

How did this simplification of the Federmesser toolkit come about? Riede suggests that the impact of the tephra falls on natural resources (especially prey animals such as elk and deer) compelled people to migrate from afflicted areas. This splintered and isolated social groups, as suggested by the localised cultures of the Bromme and Perstunian. Even before the eruption, the human population in northern Europe was thin on the ground. Its further disintegration would have resulted in smaller, more dispersed populations, curtailing social interactions within and between groups. This, it is argued, reduced the effectiveness of social learning within groups and led to a loss of skills over time. To explore this idea further, the extraordinary cultural trajectory recognised for another prehistoric forager-gatherer society, the aboriginal Tasmanians, provides a valuable comparison.

Tasmania

The anthropologist Joseph Henrich, from the University of British Columbia, examined the prehistory of the Tasmanians and developed a compelling model to explain some remarkable evolutionary aspects of their material culture [164]. Starting at the beginning, it seems that prehistoric Australians first occupied Tasmania around 34,000 years ago, when sea level was low enough that they could simply walk there from southeastern Australia – a distance of 200 kilometres. With sea-level rise at the end of the glaciation, the Bass Strait reformed, and the descendants of the first settlers became isolated in Tasmania for 10,000 years. When the Europeans arrived, they encountered a cluster of societies numbering around 4000 people in total. They also found that their toolkit represented 'the simplest technology of any known contemporary human group'. Archaeologists and ethnologists have since argued that not only was it simple compared with the wider world, it was also very limited compared with their contemporaries 200 kilometres away on the mainland, and with their own ancestors who had crossed the land bridge tens of thousands of years earlier.

At the time of European contact, the Tasmanians used only spears, rocks and throwing clubs for hunting and fighting. Their complete toolkit comprised of just 24 items. Meanwhile, their contemporaries on Australia had everything the Tasmanians possessed plus hundreds of other specialised tools, including nets, canoes, drinking bowls and mounted adzes. Bone tools are found in archaeological contexts back to 18,000 years in Tasmania but they began to disappear from the technological repertoire 8000–7000 years ago. By 3500 years ago, bone tools were not just out of fashion, they no longer existed on Tasmania. Meanwhile, on the mainland, their distant cousins made use of an immense array of bone artifacts. For the Tasmanians, this had numerous consequences, one of which was that they stopped making cold-weather garments to cope with the seasonally harsh maritime climate, probably because they no longer had bone needles to sew with. Instead, they wore wallaby skins attached with scraps of skin, and daubed exposed skin with grease. Their ancestors knew how to make warm clothes, and to the north, on the mainland, people continued to wear snug outfits made from possum skin.

The Tasmanians changed their diet over time, too. Eight thousand years ago, they fed on fish, which were second only to seals in terms of overall calorific intake. But by 5000 years ago, fish bones decline in archaeologically excavated middens (ancient rubbish piles) and, by 3800 years ago, fish was dropped from the menu altogether. And yet, across the Bass Strait, consumption of fish persisted, while fishing technology shifted from spearing to the use of hooks (made from seashell), nets and traps. In fact, it is said that when the Europeans arrived in Tasmania, the native people were disgusted by the very sight of the foreigners eating fish.

Henrich explains the Tasmanians' loss of material culture as the result of their small population size and long isolation from the rest of Australia once the land bridge was severed. Out of contact from the rest of their world, they lost valuable skills and technologies – even the boomerang – because, Henrich argues, they no longer had 'critical mass' in terms of the opportunity for social learning and, hence, cultural transmission. Humans acquire skills by imitating successful, influential and skilful members of their group. However, they make mistakes, perhaps through failing to recall something quite right; or perhaps because they didn't spot a particularly deft manoeuvre at the time they were being shown an unfamiliar technique. (I can readily identify with this since I can never remember, no matter how many times I am shown, how to tie fancy knots while erecting tents on fieldwork.) Sometimes the mistakes are lucky ones that lead to improvements but more often the result is that the imitation fails to come up to the standard of the expert's original (the tent collapses in the first stiff wind). This is especially true of hard-to-figure-out skills.

Furthermore, humans experiment. A lot of experimentation ends up proving a waste of time and effort but occasionally it leads to the Eureka! moment of innovation and opportunity. In a large population or at least one connected to the wider world through social, political and economic networks, the accidents and successful experiments, over time, result in sustained innovation in technology, skills, beliefs and practices. Henrich argues that, cut off from the mainland, the Tasmanians simply were too few in number to sustain or evolve their culture. Rather, they steadily lost skills through the propagation of errors in cultural transmission played out over many generations. Across the Bass Strait, the large pool of interacting 'social learners' generated rapid cultural change promoting high levels of skills and technology.

Not everyone accepts this model or Henrich's broader aim of building a unified evolutionary science of culture. In the case of the 'cultural degeneration' hypothesis for the Tasmanians there are several aspects that could be picked away at. For instance, perhaps cold-weather gear was dispensed with simply because it proved unnecessary as climate ameliorated after the end of the last glaciation. (Wearing limited clothing and application of grease to exposed skin sounds not unlike present-day human culture at Bondi beach, for instance.) Possibly, the Tasmanians abandoned fishing because they were more partial to seals and mutton birds, or maybe because an influential clan leader once survived eating contaminated fish and thereafter instigated a taboo on their consumption. Maybe they simply found out that they could do without all the fancy components of their original Swiss Army Knife (who needs the toothpick and tweezers?) and focused their efforts on improving the basic and most essential tools. But even if all this were true, it still leaves unanswered the question of why the Tasmanians' cultural trajectory was so different from that of their contemporaries just 200 kilometres away on the mainland. It seems, thus, that Henrich's model has real substance. And if it is applicable to the prehistory of the Tasmanians then maybe it also helps to explain why the Bromme and Perstunian cultures lost the bow and arrow following the fragmentation of their ancestral communities beneath the ash expelled from Laacher See.

On a larger scale, demography is seen by some as the key to understanding behavioural changes such as those represented by the Middle to Upper Palaeolithic Transition (Section 9.1). Rather than environmental stresses forcing cultural and technological innovation (Section 7.3.1), it is the periods of stability that allow populations to grow and expand. This increases the networking of communities and the opportunities for social learning and skill accumulation [165]. In some ways the argument is subtle since it still implies that environmental stress shrinks and fragments populations.

9.3 ERUPTION OF SANTORINI AND DECLINE OF THE MINOAN CIVILISATION

The shape of Santorini (also known as Thera) immediately gives a clue to the origin of this most romantic of Aegean islands. It is a partially submerged caldera with spectacular walls of alternating yellow, white, grey and red volcanic strata, open to the west and set in the glittering Aegean Sea (Figure 9.8) between Mainland Greece and Turkey. In volcanological circles, it is best known for its major outburst, the 'Minoan' eruption, named after the famed Bronze Age society that thrived at the time of the paroxysm some 3600 years ago. More specifically, the eruption coincides with the very latest stage of the cultural period known as the 'Late Minoan IA'.

The human impacts of this caldera-forming eruption were first recognised in the 1930s, when the archaeologist Spyridon Marinatos began exhuming a Minoan port town at Akrotiri, on the southern flank of Santorini. Minoan 'Akrotiri' (its ancient name has never been identified) had been a prosperous town, perhaps until it was damaged in an earthquake some years prior to the climactic eruption. The townsfolk appear to have suspected impending doom – at least no victims have been found, suggesting that Akrotiri's residents abandoned the town

Figure 9.8 Santorini caldera open to the sea either side of Therasia Island (part of which is visible in the upper right of the photograph). The dark islet behind the ship is Nea Kameni Island.

before it was buried by thick tephra fall and pyroclastic-current deposits. On the other hand, so much tephra remain unexcavated that it is entirely possible that victims will be located eventually.

Archaeological evidence indicates that within a few generations of the eruption, during the 'Late Minoan IB' period, the Minoan society, which at its height had spread across the Aegean, with its palaces on Crete, became increasingly unstable, culminating in widespread destruction of its elite buildings by fire. Eventually, the rulers succumbed to invasion from the Greek mainland. Mycenean culture transformed the Minoan palaces and the earlier script, 'Linear A', was replaced by Linear B, an early form of Greek. As far as is known, writing was used for administrative purposes, for instance, in recording the storage and redistribution of goods. However, to date, only Linear B has been deciphered.

Many scholars have ventured arguments to explain how the eruption might have contributed to this major socio-cultural upheaval. These include the physical impacts of earthquakes, ash falls and tsunami (triggered by pyroclastic currents entering the sea and by caldera formation), food shortages and famine, and the lasting psychological stresses of coping with disaster. Santorini has repeatedly been linked to the mythology of Atlantis and various other Greek legends, but while the circumstantial evidence might seem to add up, it remains very hard to prove the point.

9.3.1 The Minoan world

The Minoan civilisation was centred on the island of Crete, and reached its apogee in the Bronze Age when its cultural influence, established through trading, spanned many ports of the eastern Mediterranean. It has been described as a very structured, hierarchical and rather complex society ruled from the main palaces on Crete, such as Knossos, Malia and Phaistos.

Three crops represented the basis of the Minoan food economy: cereals, olives and grapes, with a minor contribution from legumes. This basic diet was supplemented by fruit such as pears, figs and almonds; the domestication of sheep, goats and pigs; harvesting of intertidal seafood and fishing for sea bream and grouper; and hunting for wild deer and hare. The Minoans also collected honey, dried their fruit, salted olives, manufactured cheese from sheep and goat's milk, drank wine, and fashioned clothing from linen, wool and leather. They are even credited with inventing fitted garments. Cattle were kept though their main purpose seems to have been to work the fields rather than provide meat or milk.

Minoan art and architecture are highly distinctive and artisans seem to have been encouraged through the extensive trading networks that distributed raw materials, finished goods and foodstuffs across the Aegean. Exquisite ceramics, frescoes and figurines are known from the period and are a highlight of many museum galleries around the world. Further accomplishments of the Minoans include the earliest-known implementation of the flush toilet (at the royal palace at Knossos) and associated feats of hydraulic engineering, including sophisticated water-mains networks connected to underground water-storage tanks, and storm drains. Houses were typically built from mud bricks or dressed stone and had two or sometimes three floors (Figure 9.9). The roofs were flat and tiled.

All these achievements owed much to the Minoans' mastery of the sea: they sailed in a wide variety of boats and ships, which they used for fishing and transport of cargo and, presumably, when necessary, waging war, though there is very little clear evidence for military campaigns. Rather, they seem to have let off steam through sporting festivals featuring acts such as bull-leaping. While their power base was on Crete, the outposts of Minoan culture in the Cycladic islands (including Santorini) must have been vital both as trading hubs linked to Anatolia, Cyprus, Mesopotamia and Syria, and also as sources of important raw materials (including, in the earlier period, obsidian). The port town of Akrotiri on Santorini was likely to have held particular

Figure 9.9 View of the West House in Akrotiri with large upper-floor windows. The house had three floors and faces on to 'Triangle Square' (as it is known). Many loom weights were found here suggesting that weaving was carried out in the house.

significance given its population of several thousand people and its location in the south of the Cylades group (and just 120 kilometres from Crete). Elsewhere on Santorini, many further archaeological sites have been found, indicating that rural settlements and farms were spread across the island.

9.3.2 The eruption

Prior to the Minoan eruption, the southern part of the present-day caldera was a shallow bay that faced a group of islands. The bay connected to the north with a flooded caldera formed from a previous eruption, 21,000 years ago. A few years before the Bronze Age denouement, Akrotiri was struck by a damaging earthquake (Figure 2.11). The clearing away of debris and reconstruction were unfinished when the first hydrovolcanic blasts excavated a new pathway for magma to reach the surface, probably through a vent on one of the islands and towards the eastern wall of the present-day caldera. Once the conduit was established, a sustained Plinian eruption ensued, gaining in intensity through time (evident from increasing size of pumice chunks upwards through the associated deposits). The eruption column reached an estimated maximum altitude of 36 kilometres, from which it would have descended to its level of neutral buoyancy in the lower stratosphere. The plume was then carried towards the east and southeast by prevailing winds.

Fallout covered parts of the island in up to six metres of white, silicic (rhyodacite) pumice. Seawater seems to have intermittently sloshed into the vent, as atop the Plinian pumice fallout are alternating layers of hydrovolcanically generated pyroclastic-current deposits and tephra fallout. The former display spectacular sedimentary structures known as mega-ripples, and extend radially in all directions from the inferred vent location. They are also very fine grained reflecting the enhanced fragmentation of magma that arises when it mixes suddenly with water (as became widely known to the air-travelling public in 2010 during the subglacial phase of the eruption of Eyjafjallajökull in Iceland). The resulting deposits, which accumulated to a depth of 12 metres, are punctured by desk-sized lava bombs that must have traced ballistic trajectories from the vent to thwack into the soft and sticky pyroclastic beds. These characteristics indicate formation by successive, shattering blasts and associated 'base surges' (similar to the ground-hugging currents apparent in photographs of atmospheric nuclear-weapons tests) that would have readily scaled the complex topography of the island.

The vent was now a funnel filled with a turbulent slurry of ash, pumice, steam and water and the blasts progressively widened it, cranking up the eruption intensity. Repeated violent explosions resulted in further thick and unstructured deposits that today are easy to spot in the caldera walls. These white rocks are thought to have formed by a combination of cool pyroclastic currents (below 300 °C) and debris flows. They are full of blocks of lava ripped from the mouth of the volcano. Tephra continuously piled up around the vent, which eventually isolated it from the sea. This is apparent from the overlying ignimbrite, which reaches up to 40 metres thick on land and forms a gently sloping apron across the whole Santorini island group. Geophysical profiling of the seabed from a research ship showed that this layer extends out radially for 20–30 kilometres [166]. Towards the vent region it is 80 metres thick. This climactic phase of the eruption generated a soaring phoenix cloud and at some point a new caldera formed as the crust foundered above the evacuated magma chamber. When the dust had settled, there was a period of reworking of the deposits through the action of debris flows and mudflows.

The total size of the eruption, which probably lasted no more than a few days, is difficult to estimate since so much of the material is beneath the waves but is thought to have been around M_e 7.2 (or 60 cubic kilometres of dense magma). The entire island was entombed and Minoan tephra have been found as far away as the Black Sea indicating a minimum area of fallout exceeding two million square kilometres (equivalent to the size of México) (Figure 9.10).

9.3.3 Dating the eruption

Arguably, the most crucial issue to understand the human impacts of the Minoan eruption is precisely when it happened. To quote the archaeologist Manfred Bietak [167], 'any scholar who is able to present cogent evidence of this eruption date will deserve an archaeological Nobel Prize'! The essence of the controversy is that dates based on archaeological chronologies for the eastern Mediterranean are intimately linked to the long-established timeline for Egyptian Pharaonic dynasties (which is based on reign lengths from written and archaeological evidence tied to a handful of astronomical observations). The conventional archaeological view implies that the Minoan eruption

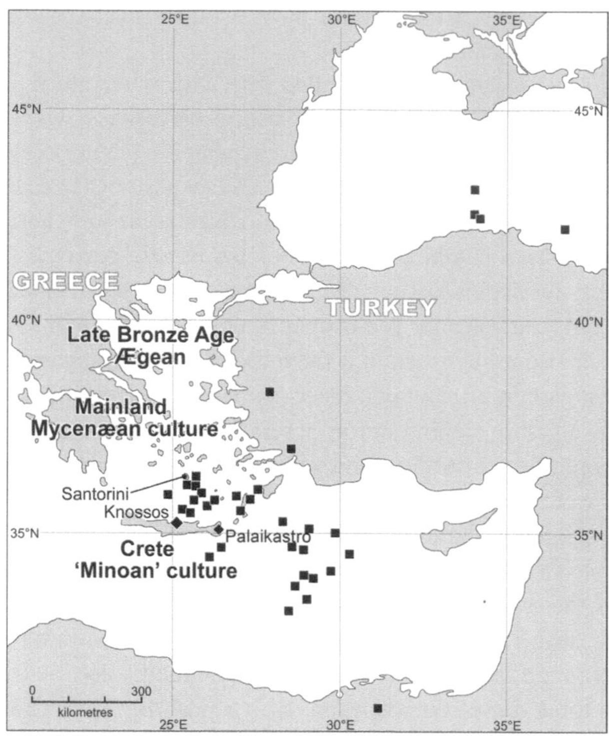

Figure 9.10 Map showing sites (square symbols) where fallout from the Minoan eruption of Santorini has been identified. Adapted from reference 173.

took place around 1520–1500 BCE (during the Eighteenth Dynasty of the New Kingdom of Egypt). This is a century or more later than the best radiocarbon evidence would suggest, which dates the event to 1663–1599 BCE [168] or 1627–1600 BCE [169]. While a recent radiocarbon-based chronology for Egyptian archaeological sites has pushed the start of the New Kingdom back to about 1560 BCE, earlier than historical estimates [170], a number of discrepancies remain. Regardless of the archaeological upheaval that having to adjust interlocking timeframes for prehistoric cultures across the eastern Mediterranean, Egypt and Near East would cause, the reason for casting scientific doubt on the radiocarbon dates is that uncertainties in the atmospheric carbon-14 production between 3500 and 3700 years ago make it particularly difficult to obtain accurate dates for this period. A further critical dating issue is how much time elapsed between the eruption and the destruction of the palaces at the end of the Late

Minoan IB cultural period. Again, there are high and low dates of 1490–1470 versus 1425–1420 BCE.

Dendrochronologists have also tried to date the eruption assuming it would have had a marked global climate impact. Bristlecone pines from the western USA reveal prominent frost-damaged rings dated to 1628–1626 BCE (Section 4.3), while ring widths in Irish bog oaks point to a decade of retarded growth beginning in 1630 BCE. Both these indications have been interpreted as signals of the Minoan eruption's climatic effects. Another set of ring anomalies dated around 1645 BCE has been identified in North American, Siberian and Finnish trees.

A recent study with fascinating implications has been reported by dendrochronologist Charlotte Pearson, from Cornell University, and her colleagues [171]. They analysed sections of partly carbonised juniper trees that had been excavated from the foundations of a Hittite postern gate in Porsuk in southeast Turkey – around 840 kilometres from Santorini and on the dispersal axis of the Minoan ash cloud. The ring widths of the trees showed a dramatic increase of as much as seven times normal widths, representing a terrific surge in growth beginning within a few years of 1650 BCE, and consistent with the 1645 BCE cluster of tree-ring anomalies from high northern latitudes. The effect lasted three years, and was followed by three to four years of return to normal growth rates. Considering that the trees concerned ranged in age from 19 to 23 years at the time of the growth spurt, whatever the cause was, it affected a mature forest ecosystem. The increased ring widths suggest improved growing conditions – how might this be consistent with volcanic forcing of climate? Trees growing in Turkey are likely to respond in a very different way to trees in the sub-Arctic or at altitude in the western USA. For Turkish trees, heat and aridity are more likely to curtail growth; thus, a cooler summer and milder winter with increased cloudiness and precipitation – as might be expected of the climatic impact of the Minoan eruption – could favour growth. But that wasn't the end of the case.

Pearson and colleagues wondered if the tree rings might still preserve any geochemical anomaly associated with volcanic fallout. To analyse the very low concentrations of various elements – down to abundances as low as a few parts per trillion – they used highly sophisticated microscopic and spectroscopic techniques. Firstly, they carefully prepared sections of several of the specimens, isolating the period of around three decades centred on the ring-width anomaly. Several elements were found to increase sharply in the year of the first growth anomaly: calcium increased by around 10–20% and remained high for a few years, and sulphur and manganese also showed similar trends. Other elements, including zinc and hafnium, spiked in the first year only and then returned to baseline levels. To explain the coincident chemical and ring-width anomalies requires an event that increases growth and alters the environmental chemistry. Fire could explain chemical changes but not the ring patterns. Volcanic-ash fertilisation, on the other hand, is a very plausible candidate. Further, at the range of 840 kilometres from source, the deposition from the Minoan fallout would have been relatively light – millimetres rather than centimetres thick – and thus more likely to have facilitated growth through nutrient supply than retarded it by brute force of smothering of foliage.

Chemical inputs taken up in the woody tissue of the trees could come from two sources: either new chemicals brought into the environment (for instance trace metals leached from ash) or the mobilisation of elements already contained in the soil, as a result of acidification. The coincident growth and sulphur, calcium and manganese uptake suggest release of nutrients from the soil due to increased acidity, favouring the latter possibility. However, experiments have shown that a range of trace metals can also be rapidly leached from ash grains, and hafnium is generally much more abundant in igneous rocks than in the kinds of limestone and metamorphic rocks of the local area where the trees were found. Either way, a combination of ash fertilisation and soil acidification could account for both the chemical anomalies and the growth spurt recorded in the tree rings. Combined with summer cooling, winter warming and enhanced precipitation, and increased soil moisture retention thanks to the covering of ash (mulching), a compelling case can be made to link the tree-ring anomalies to tephra fallout.

Unfortunately, there are no tephra deposits, nor sufficient quantities of rare earth elements in the tree rings, to provide a geochemical fingerprint that might be matched with known Minoan ash. And it is also possible the effects were not due to Santorini ash! There are plenty of volcanoes in Turkey whose prehistoric pyroclastic eruptions have been little studied to date. This may explain why Pearson and colleagues conspicuously avoid ascribing the date of 1650 BCE to the Minoan eruption. Nevertheless, they suggest that a volcanic origin of the anomaly is very likely and that 'given the approximate growth location of the trees and the current date of the anomaly, the most logical source would be the Minoan eruption ... which [would date it to] the mid–late seventeenth century BCE'.

Further efforts to date the Minoan eruption have been based on ice-core sulphate layers. Several sulphuric acid peaks in the Greenland ice cores have been suggested as possible Santorini fallout, including anomalies dated to 1623, 1642 and 1669 BCE. For a while, it seemed that the 1642 BCE anomoly in the GRIP core was associated with Minoan ash, though subsequent geochemical analysis fingered Aniakchak volcano in Alaska as responsible. The trouble with attempts to link signals across multiple records is that one dated event can suck in other similarly dated events via psychological propensity. For instance, when searching for the sulphuric acid signal associated with what is known to have been a very large eruption, the eye may well be drawn to the largest spike roughly around the time of interest. To illustrate the problem further consider that, over the last two centuries (for which the record of major volcanic events is reasonably good), eruptions of magnitude M_e 4 and above have occurred at the rate of about three per decade in the northern hemisphere. Assuming this rate is typical for the last few thousands of years, we may expect that there were around 45 northern-hemisphere eruptions exceeding this magnitude during the 150-year span of uncertainty for the Minoan eruption date. There would have been around five eruptions north of the Tropic of Capricorn greater than M_e 5. Without the 'smoking gun' geochemical fingerprint it is risky to put too much faith in an apparent correlation of ice-core sulphur with a particular radiocarbon date or tree ring.

Consideration must also be given to the sulphur content of erupted magmas since these are as important as column heights and eruption magnitudes in dictating climate change (Section 3.3). The low-iron silicic composition of the Santorini magma and its rather reducing conditions associate it with low sulphur content, so despite the great size of the eruption, it may have released comparatively little sulphur to the atmosphere and may, therefore, not be the source of a prominent ice-core acid layer.

9.3.4 Tsunami and human impact

No written records of the eruption have been uncovered to date. Perhaps if Linear A could be decoded some eyewitness accounts might emerge. However, none of the available texts scripted in Linear B refers to the eruption. Numerous scholars have argued that certain other literary sources might bear witness to the Minoan eruption, preeminent among them Plato's description of Atlantis, the story of

Deucalion and the flood, episodes in the Argonauts' tale, and passages concerning the Phaeacians' fate in the Odyssey as well as assorted Egyptian and biblical laments. Thus reads the key passage in Plato's *Timaeus*, translated here by the nineteenth century classicist Benjamin Jowett (who did much to make Plato an English classic, though he found that of all Plato's writings 'the *Timaeus* is the most obscure and repulsive to the modern reader, and has nevertheless had the greatest influence over the ancient and mediaeval world'):

> [On] ... this island of Atlantis there was a great and wonderful empire which had rule over the whole island and several others, and over parts of the continent, and, furthermore, the men of Atlantis had subjected the parts of Libya within the columns of Heracles as far as Egypt, and of Europe as far as Tyrrhenia. But ... there occurred violent earthquakes and floods; and in a single day and night of misfortune all your warlike men in a body sank into the earth, and the island of Atlantis in like manner disappeared in the depths of the sea. For which reason the sea in those parts is impassable and impenetrable, because there is a shoal of mud in the way; and this was caused by the subsidence of the island.

The fact that *Timaeus* was written 1300 years after the Minoan eruption does not in itself preclude the Atlantis story having some basis in events that befell ancient Santorini (as we saw in Section 5.2.3, tales depicting the much earlier eruption of Mount Mazama in Oregon are still retold in Native American communities). However, alternative interpretations of these stories are just as compelling, and many scholars would see Plato's Atlantis as no more than a philosophical allegory for a lost world.

The best chances of finding contemporary historical accounts of the eruption are from Egypt and Mesopotamia. In Egypt, depending on which eruption chronology you adhere to, the time of the eruption coincided with the end of the Second Intermediate Period, and the rise of the brothers Kamose and Ahmose, who founded the Eighteenth Dynasty of the New Kingdom. Meanwhile, in Mesopotamia, the Old Babylonian period was nearing its terminus, with the Hittites' sack of Babylon dated *circa* 1595 BCE. Unfortunately, there are virtually no surviving historical texts from the period. It has been suggested that hieroglyphs on a stela erected by Ahmose in the Karnak Temple bear witness to the Minoan eruption's climatic consequences in the guise of a great storm accompanied by flooding and destruction [172] but it seems more likely the events recorded refer to severe monsoonal flooding of the Nile, as still occurs from time to time.

In any case, our key concern is the role played by the eruption in the demise of the Minoan palaces. Given that ash fall on Crete was relatively light, and bearing in mind that the Minoan economy was very much focused on coastal and marine resources, the most plausible means by which the eruption might have disrupted the Minoan society in a major way is via tsunami triggered by caldera formation or intense pyroclastic currents striking the sea [173]. Accordingly, there have been many attempts to model Minoan tsunami scenarios, despite very limited evidence to specify the initial water displacements. In addition, searches for tsunami deposits have been carried out along the shores of Santorini, northern Crete and southwest Turkey though these have tended to yield contradictory findings.

The difficulty lies in discriminating ancient tsunami deposits from the sediments left by storm surges or even flash floods. Furthermore, coasts are unfavourable environments in which to preserve any kinds of surface deposit because of constant attack by wind and waves. However, meticulous field and laboratory studies undertaken by Hendrick Bruins, from the University of the Negev, and colleagues provide a strong case that tsunami did strike the major Late Minoan town of Palaikastro on the east coast of Crete [174]. Recognition of ancient tsunami in the geological record has been advanced significantly in recent years following studies of the deposits left by the 2004 Indian Ocean tsunami. Because tsunami are both destructive and constructive, they leave particularly diagnostic deposits, especially so when they impact settlements. Tsunami violently erode surface soils and unconsolidated geological formations as the waves crash onshore and race inland. The backwash then carries anything loose back towards the sea but, since most of the wave energy has now dissipated, much debris is dumped on the newly eroded surface. All kinds of wreckage accumulate in the resulting deposit - building rubble, litter, seashells, clay, silt, sand and boulders - mashed together in one chaotic mass.

Like many Minoan towns, ancient Palaikastro was built on the coast - most of it lay just a few metres above sea level. It was nestled at the foot of limestone hills and stretched part way round a wide natural bay. Bruins and colleagues closely examined several sedimentary sections in both the archaeological site, some 300 metres inland, and in the low-lying coastal cliffs. The coastal sequences were especially diagnostic, revealing a discontinuous layer, a few metres thick, of distinctly exotic sediments. In places, these were found wedged in-between ruins of Minoan walls or above stone floor slabs. The deposits contained jumbles of potsherds, wall plaster, building stones, gravel, sand, silt, cattle bones and a few broken seashells. The manner in which the larger fragments were stacked rather like roof tiles – a sedimentary fabric known as imbrication – demonstrates that the deposit had been laid down by a very energetic flow of water.

However, an origin by river flooding could be ruled out because the coastal strip is actually higher than the hinterland, so streams bypass the site. Furthermore, the deposit contained seashells and other marine microfossils, as well as rounded beach pebbles. Storm waves could also be ruled out because they are not known to produce such poorly sorted deposits (comprised of everything from boulders through gravels, sands and silt to clay-sized particles) or imbrication. Additionally, some of the deposits were found up to nine metres above the ancient sea level, three times the typical height of storm waves. In any case, the northeast-facing bay is protected from the predominant westerly direction of storms in the region.

One important feature of the deposit provided a strong temporal link with the Minoan eruption – fist-sized clumps of almost pure ash, geochemically fingerprinted to Santorini. These would have been generated by the scouring action of tsunami waves as they rushed inland. The fact that they were incorporated into the deposit indicated that ash was already lying on the ground after the Plinian phase of the eruption. The bones and seashells yielded dates precisely in the ballpark of other radiocarbon ages for the Minoan eruption. Swirling decorations on the ceramics were also diagnostic of Late Minoan style.

Three-hundred metres inland, archaeological excavations revealed further light-grey Minoan ash, mixed in with reddish-brown sediment and wedged between the ruined dwellings. This material was probably washed in as the tsunami swept ashore, consistent with the evident destruction of buildings at the coast. It is likely that most of the town was inundated – like other major settlements of the period on Crete, Palaikastro lacked a defensive wall. The cattle bones seemed to be remnants of food from the destroyed coastal houses, 'perhaps the last meal before the tsunami struck' [174]. It seems that similar events befell the town at Malia (not far from Knossos on the north coast of Crete), where the palace ceased to function in the Late Minoan period.

Recollecting the 2004 Indian Ocean tsunami gives a guide to the damage caused. Probably a large part of the Minoan sailing fleet – from fishing boats to merchant vessels and war ships – was lost or damaged. This would have disrupted both trade and fishing, and severely undermined the naval strength of the Minoans. Although good pollen records from the period are lacking for the region, timber was likely to have been in short supply, delaying reconstruction of both the built environment and fleet, leaving the palaces more vulnerable to invasion. Tsunami also contaminate agricultural land with sea salt, potentially rendering it useless for decades. A tsunami that struck in 1956 ruined vineyards close to modern Palaikastro for 20 years. Such impacts on the Minoans could have reduced the ability of the complex palace society to feed its population adequately.

In addition, although the tephra fall on Crete was relatively light – perhaps amounting to 15 centimetres depth on the east of the island, much less elsewhere – it is likely to have carried an abundance of chlorine and fluorine scavenged from volcanic gases at the eruption vent. This would have further contaminated soils and water supplies and, speculatively, might have killed domesticated animals by fluorosis.

Losing Akrotiri must have been a terrible blow to the Minoans on Crete – the city had played a major role in trade and redistribution of goods throughout the Cyclades and beyond. This alone would have contributed to destabilisation of palace authority on Crete. Slow recovery and increasing internal conflicts subsequently appear to have sapped the vigour of Minoan culture. Through the Minoan IB period on Crete, archaeological evidence reveals contraction and abandonment of settlements. The rooms of dwellings were subdivided so as to house more families. For archaeologist Jan Driessen from the University of Louvain these phenomena signal an increasingly insecure society, in decline from a combination of famine, plague, war and environmental catastrophe [175]. As food production and distribution slumped, the social and political scene fragmented. People erected walls to protect their livestock, or squirreled away hoards of bronze and other valuable goods, symptomatic of economic stress and contributing to a deepening of the crisis by disrupting exchange networks.

Religion, too, underwent remarkable transformations. 'Crisis cults' sprang up, community shrines were erected, certain caves were sanctified and new sacred symbols circulated. There is even some evidence for human sacrifice and ritual cannibalism. At Knossos, a mass of children's bones was discovered in a house in the western part of the town. There were 299 bones in all, with cut-marks characteristic of deliberate removal of flesh. Meanwhile, ceramics were decorated increasingly with marine motifs, including dolphins, octopi and shells, seen by some as indicative of the society's psychological trauma following the tsunami and inferred decimation of seafood resources [176]. (There is an interesting echo here of the introduction of Lapita pottery style in New Britain following the mid-second-millennium BCE eruption of Witori in Papua New Guinea discussed in Section 5.1.3.) Eventually, many of the Cretan palaces were destroyed by fire in what some archaeologists interpret as an anarchic holocaust aimed at the ruling elite.

We do not know the precise timing but within, at most, a few generations of Santorini's great eruption, the Mycenean Greeks appear to have taken advantage of the situation and extended their dominion to Crete. Their navy may have escaped the tsunami thanks to safe harbour in the Gulf of Corinth. It is possible to see this regional power shift as the culmination of a chain reaction of events starting with Santorini's eruption. But some read the evidence very differently, arguing that the Late Minoan IB period saw prosperity and cultural vigour and that the interval between the eruption and the destruction of palaces – up to a century – is too long to 'blame' Santorini for the Minoan decline. Longer-term climate change may also have played a role. There are even debates about the extent to which the Mycenean influence in Crete resulted from a military conquest or rather by assimilation of mainland Greek customs by conniving rulers at Knossos. Whatever the truth, Linear A script was transformed into Linear B to suit mainland language and the Cretan population came under an increasingly militaristic rule.

We know very little of the fate of the former residents of Santorini or the other Cyclades islands including Rhodes (where ash accumulated to an alarming depth of a metre). Kos and parts of the Anatolian coast must also have been affected by both heavy ash fall and tsunami. As noted above, the lack of victims found in the excavations at Akrotiri suggests abandonment before the eruption. Perhaps damaging earthquakes or the initial hydrovolcanic blasts prompted evacuation. If so, where did people flee to – to Crete, perhaps, or to neighbouring islands? Perhaps some even travelled to Turkey where they had trading partners. Did the Minoan eruption represent one of the first exercises in emergency mass evacuation?

9.4 SUMMARY

This chapter has jumped through the prehistory of Europe, focusing on three exceptional eruptions: the Campanian Ignimbrite, Laacher See and the Minoan eruption of Santorini. In each case, archaeological, geological and palaeoenvironmental evidence points to severe disruption of terrestrial and/or marine habitats and human lifeways.

The massive eruption of the Campanian Ignimbrite coincided with heightened climate sensitivity during Heinrich Event 4, and it has been suggested it played a role in the major cultural transition from Middle to Upper Palaeolithic in Europe. More controversially, it has been implicated in the extinction of the Neanderthals. These hypotheses remain to be rigorously evaluated but they have certainly stimulated further research on the human impacts of the Campanian Ignimbrite eruption. Thanks to recent successes in detecting and analysing cryptotephra (Section 4.1.4), there are tremendous opportunities for locating Campanian Ignimbrite tephra in natural shelters, caves and open-air archaeological sites around the Mediterranean and in central and eastern Europe. In addition, there is further scope to identify Campanian Ignimbrite fallout in long environmental records on land and in the oceans. In particular, the eruption has yet to be identified in the Greenland ice-core record but it is being actively searched for. Significant advances in our understanding of the European and north-African dimensions to the human revolution are likely to be made in the next years.

The Campanian Ignimbrite was not the last caldera-forming eruption of Campi Flegrei. Sixteen thousand years ago, the volcano erupted again to cover the surrounding region in the Neapolitan Yellow Tuff (widely used as a building stone in Naples and neighbouring towns and villages). The last eruption was a miniscule affair by comparison (though it claimed 24 lives) and took place in 1538 when the Monte Nuovo cone was born. Today, almost 1.5 million people live *inside* the caldera, presenting a problem that we return to in the final chapter.

The Laacher See eruption in Germany was a more localised affair but also offers a valuable opportunity to use radiocarbon 'dates as data' in reconstructing the associated 'cultural founder effects'. Of course, this means the interpretations rely on the quality and quantity of dates available and factors that bias preservation of archaeological materials.

The destruction of Akrotiri on Santorini provides the earliest evidence of the direct impacts of a volcanic eruption on the built environment. Bearing in mind that the site is nearly twice the age of Pompeii and Herculaneum, its archaeological significance is outstanding. The Minoan eruption reminds us, too, of the threat to the Mediterranean coastline of volcano-induced tsunami.

10

The rise of Teotihuacán

> Plinian ash falls, as opposed to lava flows or surges, provide the most interesting prospects for archaeology: When the column of tephra rises high into the atmosphere from the crater and then collapses over the landscape, the pumitic material blankets the agricultural fields, buildings, and activity areas in a way that effectively preserves them for future study. The local inhabitants flee, some with more luck than others, and in their haste they often leave behind many of their household goods, providing archaeologists with an impressive array of primary contexts that are particularly relevant to . . . understanding past lifeways.
>
> P. Plunket and G. Uruñuela, *Latin American Antiquity*, 1998 [177]

Teotihuacán - now located some 45 kilometres from México City - was one of the greatest cities of the ancient world, rising in the first or second centuries BCE and lasting into the seventh or eighth century CE. At its apogee in the fourth century CE it was home to around 125,000 people and covered an area of 30 square kilometres. By the third century CE, a construction boom had established an astonishing complex of pyramids - including the Sun and Moon pyramids, and the Pyramid of the Feathered Serpent. Despite this grandeur, we know rather little about the society that built and sustained this extraordinary civilisation, in part because, unlike the contemporary Mayan glyphs, the inscriptions found at Teotihuacán have yet to be deciphered. The city was partially destroyed around the mid-sixth century and the site became a place of mystical ruins for the much later Aztecs.

Situated in the northeastern part of the fertile Basin of México and surrounded by volcanoes and mountain ranges, Teotihuacán came to exert direct authority over an area of around 25,000 square kilometres around the city. But it also held immense prestige over a much wider region. It regulated or, at the very least, subordinated key trading posts and the routes between them, across much of Mesoamerica. The spread of Teotihuacán culture has been recognised in archaeological contexts across México and as distant as lowland Maya sites such as Tikal (in the Petén Basin of Guatemala) and highland sites such as Kaminaljuyú (Guatemala) and Copán (Honduras, 1200 kilometres away). An important sector of the city's economy concerned the procurement of obsidian and its manufacture into an array of tools, mirrors and sacrificial knives. Indeed, the location of Teotihuacán may have arisen from its proximity to a spectacular outcrop of green obsidian at Pachuca.

Among the other explanations for Teotihuacán's rise to power are the impacts of two volcanic eruptions: one relatively small but close by (Popocatépetl), the other much larger but more than 1200 kilometres away (Ilopango, in present day El Salvador). These events are the subject of this chapter.

10.1 POPOCATÉPETL

Yet another 'Pompeii' of the New World that has provided extraordinary insights into ancient lives, thanks to the preserving effects of tephra fallout, is the site known as Tetimpa, in México. It is located at an altitude of 2350 metres, and sits just 15 kilometres downhill from the summit crater of the magnificent 5400-metre-high Popocatépetl volcano (Figure 10.1). Popocatépetl is one of the many volcanoes stretching 1000 kilometres from the Pacific coast to the Gulf of México (and linked to the subduction of the Cocos Plate beneath the North

Figure 10.1 View of Popocatépetl from the Paso de Cortés. Photograph courtesy of Marie Boichu.

American Plate). It has a long record of influence on its surrounding population, richly reflected in local folklore. Tetimpa (whose name comes from the local Nahuatl language meaning a 'place filled up with rock') was first discovered when a farmer, lamenting the low productivity of his land, decided to strip off the feeble topsoil and underlying layers of tephra to reach an ancient and much more fertile soil layer beneath. Into the bargain, he sold the extracted tephra to a construction company. Archaeologists soon got wind of unearthed dwellings and artifacts but rather little was done to rescue them, and many other farmers followed the trend.

Alarmed at the rate at which the sites were being destroyed, and curious about the light they might shed on ancient human experiences and the impacts of volcanic eruptions on prehispanic society, the archaeologists Patricia Plunket and Gabriela Uruñuela, from the Universidad de las Américas in nearby Cholula, began careful excavations of the site in 1993. They have described Tetimpa as 'more like a ghost-town than a typical highland archaeological site', adding 'we don't excavate the remains as much as we sweep and dust. It's like travelling back in time to look at an ancient Mesoamerican village without the villagers' [178]. Now, with the fruits of two decades of research behind them, they have a much clearer picture of the ancient communities that lived under the threat of eruptions while exploiting the resources supported by the volcano [179,180].

Before disaster struck, sometime in the first half of the first century CE (during the final stage of the Preclassic period of Mesoamerican archaeology), Tetimpa had a population of 3000–4000 and covered an area of about four square kilometres. Its 600 or so dwellings, constructed with wattle-and-daub walls, were erected on waist-high stone platforms built in *talud-tablero* style with a sloping wall (*talud*) topped by a horizontal panel (*tablero*) (Figure 10.2). Households consisted of two or three buildings set at right angles to each other and surrounding a patio, a number of which featured domestic shrines in the form of miniature volcano effigies (Figure 10.3)! These were only about knee-high but they actually smoked via a chimney linked to a small furnace [181] (charcoal remains were found inside providing valuable radiocarbon dates, though the best date came from a carbonised corn-on-the-cob found inside a sealed jar). According to Plunket and Uruñuela, the miniature effigies were probably erected to appease Popocatépetl, though the volcano seems to have taken the gesture the wrong way, since Tetimpa ended up buried beneath a metre to a metre-and-a-half of yellowish intermediate (andesitic) composition pumice from a M_e 5.3 Plinian eruption [182] (Figure 10.4). So complete was the

Figure 10.2 Excavation of a Tetimpa house showing artifacts *in situ* in the patio. Note the tephra still covering the floors of the collapsed buildings that once stood on the stone-faced platforms. Photograph courtesy of Patricia Plunket and Gabriela Uruñuela.

Figure 10.3 Volcano shrine at a Tetimpa house. Photograph courtesy of Patricia Plunket and Gabriela Uruñuela.

destruction that, combined with the high elevation and low temperatures, and thus slow rates of weathering and soil formation, it took up to five centuries for the area to be re-occupied. The events might even be referred to in the Aztec 'Legend of the Suns', which describes how

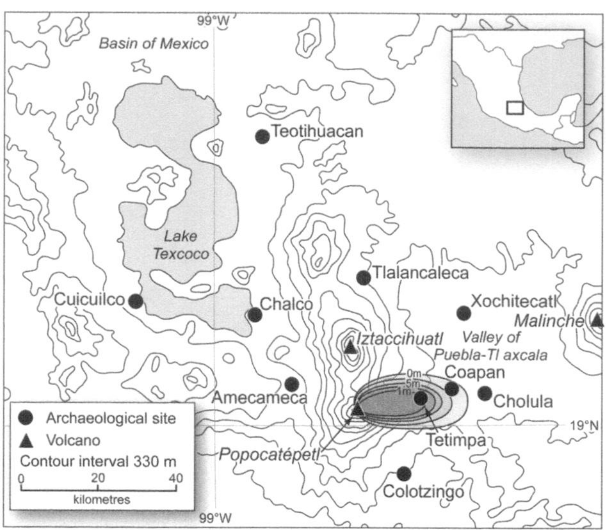

Figure 10.4 Map of central Mexico showing archaeological sites and distribution of fallout from the eruption of Popocatépetl that buried Tetimpa around 2000 years ago. Updated from reference 180 with permission of Elsevier.

the third epoch of creation was destroyed by a volcanic eruption 'when it rained fire, the rocks boiled, and houses burned', but 'with its destruction came the creation of a new world'.

At Tetimpa, at least the population seems to have foreseen the impending calamity since no casualties have been found. But there is more to the story than a simple abandonment of the village in the face of volcanic disaster. Plunket and Uruñuela came to realise that the archaeology suggested a degree of community preparedness for the eruption: '. . . analysis and reflection suggest that . . . we are viewing a very extraordinary period in the village's history, a time of high drama and intense anxiety caused perhaps by an increase in Popocatépetl's volcanic activity'. The arrangements of objects, for instance the face-up or face-down positions of *metates* (grinding stones) and ceramic bowls, showed that some families were in residence when the eruption occurred while others were away having stored them. A bowl left face up was usually filled with food whereas one face down was clean and either had been left to dry or was being stored.

Furthermore, several of the dwellings were in a poor state – the daub was falling off, patios were uneven and steps were collapsing. Despite evidence for some makeshift repairs it seemed that Tetimpa's householders were only sporadically in the village – they had perceived the risk, found somewhere safer to live and returned only to tend their cornfields and gardens. Further evidence for abandonment of the village prior to the Plinian eruption includes the lack of certain objects typical of other archaeological sites of the period. For instance, obsidian cores and tools, and figurines (which may have been toys) were few and far between at Tetimpa. These would have been useful or portable items that the refugees would have taken with them. Large grinding stones, however, had been left behind because they were awkward to move and perhaps, in any case, their owners expected to return.

More recently, this pattern of partial abandonment has been practised on the island of Montserrat in the West Indies. Soufrière Hills volcano burst into life in 1995 and, with occasional reposes, continues alternately to grow and destroy its lava dome. In the first few years of the eruption, the dome burgeoned, threatening an ever greater zone on the volcano's flanks with pyroclastic currents. Many homes fell within the exclusion zone declared by the authorities and their occupants moved to shelters or stayed with family elsewhere on the island. But the small plots of land, on which they grew sweet potatoes and mangoes or grazed a few goats or cattle, remained the basis of their livelihoods. They continued to go by day into the evacuated area to tend to their animals and gardens, perhaps in much the same way, and with many of the same fears, as the inhabitants of Tetimpa some 2000 years before them.

At Tetimpa, archaeological evidence also suggests the time of year that the eruption took place. Surrounding the dwellings were corn fields, evident from linear ridges in the buried soil that resembled the furrows of modern ploughed fields, and effective in limiting soil erosion. However, the fields were not planted, suggesting an eruption during the dry season, which lasts from October to May.

Other villages on the flanks of the volcano were also abandoned, as were several large towns at its base. Many tens of thousands of refugees were on the move, seeking respite. To the east, they resettled in the Puebla-Tlaxcala valley. To the west, though, in the Basin of México, something truly remarkable transpired. Up until this time, these people had been settled in scattered villages and towns, living off agriculturally productive soils and forests rich with foods for hunters and gatherers. They were about to trade their rural lifestyles and become city slickers in Teotihuacán. To Plunket and Uruñuela 'the intentional displacement of 50,000 people and the consequent abandonment of some of the best agricultural lands in the Basin of México by an emergent expansionist archaic state, no matter how aggressive and powerful its armies, no matter how persuasive its ideology [represented] an awesome and puzzling undertaking'. Could the eruption of, and disruption caused by, Popocatépetl have played a role in this massive demographic transformation? Plunket and Uruñuela think so:

> If they are of a sufficiently catastrophic scale, [disasters] can motivate social action, provide contexts for innovative agendas, promote new power relations, and allow the emergence of new leadership. . . . the Terminal Formative (100 BCE–CE 100) [was a time] of rapid social transformation in central Mexico; in this context, a major volcanic eruption would have served as a catalyst to accelerate those changes even more and highlight the inability of local authorities to deal not only with the forces of nature, but also with unforeseen social and political consequences derived from decision-making during the recovery process.

Teotihuacán grew to become the largest and most cosmopolitan city in the Prehispanic Americas, covering some 30 square kilometres. Some estimates suggest as many as a quarter of a million people once lived there. It consequently exerted huge influence in Mesoamerica during the period 150–450 CE, including on the Maya. It is possible even that some of the Mayan regions such as Tikal and Petén were conquered by Teotihuacán, though more likely the control was exerted politically. In 379 CE, within a year of the first recorded meeting between the authorities of Teotihuacán and the Maya at El Perú in the Petén, Yax Nuun Ayiin I (better known as Curl Snout) became Mayan ruler at Tikal. His ancestry was part Teotihuacán elite so perhaps Teotihuacán's diplomats had been planning this dynastic move for many years.

As well as constructing some exceptional pyramids in the finest *talud-tablero* style (Figure 10.5), Teotihuacán's artisans produced many outstanding objects from obsidian, among which are exceptionally ornate knives that may have been used in ritual human sacrifices. (The victims were probably warriors captured in battle and met horrific ends such as beheading, extraction of the heart and burial alive.) Teotihuacán eventually declined during the sixth century and the temples and civic-ceremonial buildings were sacked and burned around the middle of the sixth century.

Returning to the first century CE in the Valley of México, the short-term impacts of Popocatépetl's eruption would have been

Figure 10.5 The apogee of *talud-tablero* (slope-and-panel) architectonics is seen today in Teotihuacán's pyramids.

dreadful: towns and villages were buried in tephra, food would have been in short supply and seeds for planting destroyed. But the damage would also have had longer-lasting effects due to the burial of agricultural land (even if they'd still had seeds they wouldn't have been of much use); fires, ignited by pyroclastic currents, would have set forests ablaze; and mudflows, triggered by melting of ice and snow atop the volcano by hot tephra, would have choked river ecosystems and contaminated water supplies. The natural resources for agriculture and foraging would have been severely disrupted for decades and centuries until topsoil and forests became re-established. Clearly, this picture would have encouraged displaced people to find a new home. But why do so many appear to have resettled in Teotihuacán?

One theory is that the city recognised an opportunity to attract a large labour pool. Although the massive influx of people must have greatly challenged the capabilities of the authorities to cope, it seems they put the migrants to work, draining swamps to provide new agricultural land. In the view of Plunket and Uruñuela, 'the sudden abundance of labor, viewed against a dramatic and horrifying visual backdrop associated with traumatic dislocation, may help explain the surge in the building of civic–ceremonial architecture at Teotihuacán during the first and second centuries AD'. This boom period witnessed the erection of some of the largest buildings in the New World. Plunket and Uruñuela believe that these monumental building projects represented 'not only the creation of a new kind of settlement, but also the development of a new accommodation between humans and their gods'.

10.2 THE ILOPANGO ERUPTION

In El Salvador, archaeologists often have to dig through a prominent white volcanic ash layer, locally called the *tierra blanca*. The eruption responsible is referred to as the Tierra Blanca Joven. It is not yet securely dated but there is no question that it originated at the site of present-day Lake Ilopango in El Salvador (Figures 10.6; 4.2). Archaeologists working as early as the 1920s located numerous artefacts and structures belonging to the Late Preclassic and Early Classic periods buried by this horizon. They concluded that cataclysmic eruptions had rendered central Salvador uninhabitable for many decades, resulting in migration of the Maya people to the lowlands of the Petén and the Usumacinta valley in Guatemala.

The Ilopango eruption deposited around 84 cubic kilometres of (uncompacted) silicic tephra [183] (M_e 6.9). Radiocarbon dates fall into two groups: one 150–370 CE, the other 408–536 CE. The evidence has recently tipped strongly in favour of recent versus older ages as new radiocarbon dates have emerged, and it has even been suggested the eruption was responsible for a major sulphuric acid aerosol veil in the stratosphere reportedly observed in 536 CE (Section 11.1). Resolving the timing conclusively could have tremendous archaeological

Figure 10.6 Lake Ilopango in El Salvador – site of one of the largest volcanic eruptions of the past two millennia – viewed from space. The capital San Salvador is sandwiched by Lake Ilopango to the east and El Boquerón volcano to the west. (El Boquerón is part of San Salvador volcano and has grown up inside a 6-kilometre caldera formed about 40,000 years ago.) Lake Ilopango is approximately 12 kilometres across. Satellite image obtained from ASTER Volcano Archive, NASA JPL.

significance given the dispersal of tephra and the limited existing constraints on archaeological dating in the region. Pyroclastic currents travelled at least 45 kilometres, judging from preserved deposits, and must have devastated low-lying agricultural areas and settlements. But more widespread and longer-lasting damage was caused by the subsequent ash-cloud fallout, which blanketed an area of at least 10,000 square kilometres waist-deep in pumice and ash. Far less than this is sufficient to shut down traditional agricultural practices for decades.

Immediately prior to the eruption, western Salvador was densely populated and was experiencing a 'cultural florescence' [184]. The people lived in villages, farmed maize and had strong cultural and economic links with the Maya of the highlands and the Pacific coast of Guatemala. Many thousands of people surely perished when Ilopango erupted, and the landscape of western, central and eastern Salvador must have been transformed into a denuded and sterile mosaic of dustbowl and mud bath. During the rainy seasons, tephra would have choked major rivers killing off fish and turtles while mudflows would have spread across the lowlands. Recovery was slow, following the ecological succession of colonisation by algae and mosses, then ferns, grasses and sedges and, finally, herbs, shrubs and trees. The area was not resettled until two centuries had passed, by which time the ecological recovery had re-established soils suitable for agriculture.

Clearly, across El Salvador, the cultural trajectory was abruptly and profoundly disrupted. At Chalchuapa, 77 kilometres from Ilopango (and close to the present Guatemala border), tephra deposits are knee-deep. Until the eruption, Chalchuapa had been one of the most important ceremonial and regional trade centres in the southeast Maya highlands. It boasted several 15-metre-high pyramids arranged around plazas and surrounded by residential areas. Great quantities of obsidian were imported from Ixtepeque in southern Guatemala, and fashioned by local specialists into knives, scrapers and ceremonial objects. Jade was conveyed from the Montagua valley and redistributed, along with cacao, cotton and salt. Elaborate, painted ceramic vessels were also manufactured in Chalchuapa and exported throughout the highlands and as far as the Pacific lowlands and Petén of Guatemala, and Cuello in Belize. Following Ilopango's convulsion, Chalchuapa and its impressive pyramid at Tazumal were abandoned. Pollen extracted from sediment cores taken from a lake about 15 kilometres from Chalchuapa reveal that maize vanished from the assemblage immediately, and did not reappear until much later, indicating complete abandonment of agriculture in the area [185].

Ripples from this profound economic collapse were felt to the west, in the Jutiapa region of Guatemala, where the population declined sharply, while many sites on the Pacific slope and southeast (Pacific) coast of Guatemala were abandoned. The major site of Kaminaljuyú, much of which still lies buried (and tantalisingly out of the reach of archaeologists) beneath modern-day Guatemala City, also experienced huge demographic disturbances as a third of the urban population abandoned the city, perhaps due to food shortages and starvation. According to the Mesoamerican specialist Robert Dull, from the University of Texas at Austin, and his colleagues 'the more potent blow to the entire southern Maya realm probably came in the form of a collapsing regional economy and, as a result, the political destabilization and decentralization of a number of fledgling mercantile polities' (Figure 10.7).

Copán seems to have taken advantage of Chalchuapa's demise by accepting many refugees from the abandoned site but also seems to have thrived by strengthening its ties to Teotihuacán, by now Mesoamerica's greatest expansionist trading civilisation. If so, this would have been largely indirect, via the Petén and especially Tikal. It appears that K'inich Yax K'uk' Mo' was crowned king of Copán where subsequently he established the royal dynasty [186]. Copán became one of the most important Maya cities and experienced its own phase of monumental architecture. Further evidence for the migration of highland Maya following the eruption includes the arrival of foreign artefacts into the lowland settlements of Belize and northern Guatemala. Agriculture intensified and socio-economic and political structures were strengthened to cope with the influx of refugees, accelerating a sophistication of culture and society that may have already been underway. It seems the lowland Maya capitalised on the social disruption in the highlands. This is seen in changes in the contemporary trade routes in Mesoamerica [187]. Prior to the Ilopango eruption, these ran through southern Guatemala and into El Salvador. But later, lowland sites such as Tikal appear to have dominated trade, re-routing it across the Petén.

It is at this time that Teotihuacán, at the Mexican end of the trade route, also increased its influence in Tikal. It may even have conquered Tikal, home to some of the most exceptional Mayan temples of the Late Classic period. Teotihuacán's interests may have been to enhance its access to two major obsidian sources in the Mayan highlands (El Chayal and Ixtepeque) since by this time there were several hundred obsidian workshops in the city. In this way, we can view Teotihuacán's ever

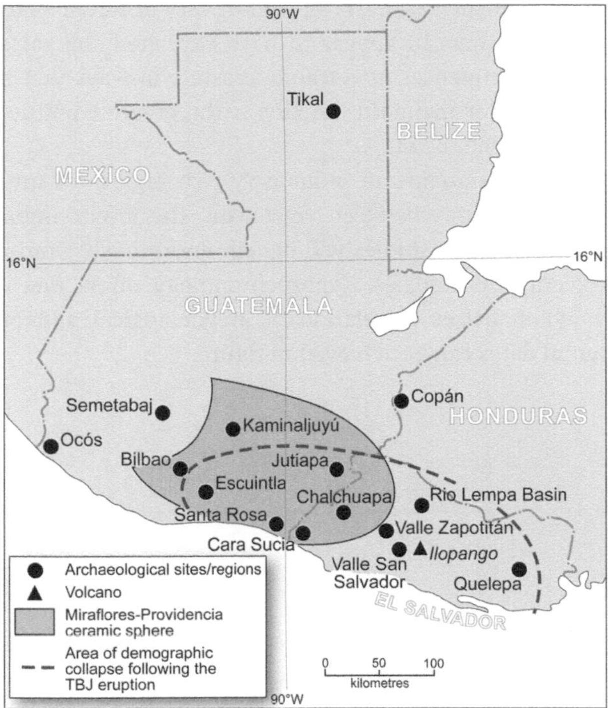

Figure 10.7 Map of region affected by the Tierra Blanca Joven (TBJ) eruption of Ilopango in El Salvador. The eruption had a major impact on ancient society in central and western El Salvador. The approximate area of demographic collapse is shown. The demise of Chalchuapa and collapsing regional economy brought down the formerly powerful mercantile zone known as the Providencia-Miraflores. Modified, with new information from Robert Dull, from reference 187.

widening area of demographic, political and economic authority arising as a consequence of the depletion and exhaustion of the highland Maya due to two volcanic disasters.

10.3 SUMMARY

Environmental change often results in uneven impacts across different societies. Volcanic eruptions represent disasters for immediately afflicted communities – their physical, social and emotional worlds can be turned upside down in the blink of an eye. Large-scale disasters can also represent immense social, political and economic upheaval for communities that are not directly affected by the disaster. How they react in the face of such stimuli can lead to very different outcomes. The rulers of Teotihuacán appear to have exploited the social and demographic consequences of volcanic disasters in ways that helped to establish and later maintain the city as the greatest political and economic force in the region.

Mesoamerica remains an immensely rich region for investigation of the interactions between volcanism, the environment and human society but such studies will benefit enormously from refinements in event chronologies. Tephrochronology offers one of the means by which better correlation of archaeological and palaeoenvironmental dates can be achieved in future.

11

Dark Ages: dark nature?

> It is frequently overlooked that both "climate" and "history" are blanket terms, situated on such a high level of abstraction that relationships between them cannot be investigated in a meaningful way in accordance with the rules of scientific methodology . . . Whether and to what extent climatic factors mattered for social vulnerability needs to be determined through empirical analyses.
>
> C. Pfister, *Climatic Change*, 2010 [188]

The signatures of many famous eruptions have been recognised in the ice-core records. For instance, the sulphate layers sourced from the eruptions of Mount Mazama (~5677 BCE), Vesuvius (79 CE), Huaynaputina (1600), Krakatau (1883) and Pinatubo (1991) have all been reported. But even a perfunctory review of the ice-core sulphate archives indicates that there are many more acid layers from as yet unidentified eruptions. The responsible volcanoes have thus far evaded detection. What is more, tree rings and, for the historical period, assorted documents, give testimony to a number of major explosive volcanic episodes and their associated climatic impacts. In the case of one of these mysteries, taking place around 536 CE, much has been written, including excessive claims that the consequences of a great eruption marked a turning point in history between the 'ancient' and 'modern' worlds.

In truth, it is rather difficult to substantiate some of these more startling or fanciful hypotheses – the 'Dark Ages' did not get their moniker without reason (though since the term is considered so tendentious by medieval historians today I promise not use it again). What is more, it doesn't help that the record of volcanic eruptions is so incomplete. One of the most striking illustrations of this fact is that barely two centuries ago, in 1809, there was an explosive eruption that, based on its sulphate stamp in the polar ice cores, was of comparable magnitude to the 1883 paroxysm of Krakatau (Figure 4.7). And yet we have no idea where it took place! An even more flagrant example of an unidentified sulphate anomaly is that dated to around 1259 CE. In the Greenland GISP2 ice core it represents the largest sulphur fallout of the past 7000 years (Figure 4.6). Nevertheless, the medieval period provides the backdrop for a fascinating collection of volcanological detective stories, where the historical and eruption records start to intersect and come tantalisingly into focus. This chapter aims to reveal the great potential for future work in this arena.

11.1 THE MYSTERY CLOUD OF 536 CE

The period between the fourth and eighth centuries witnessed immense changes in Europe, North Africa and the Near East. The Roman Empire declined (or collapsed according to your point of view), Christianity, and then Islam, spread and clashed. If ever there were a case to avoid looking for a single crisis point, this could be it . . . However, it would be wrong to dismiss out of hand the possible influences of volcanism, albeit with some help from the humble flea.

11.1.1 The aerosol veil in contemporary literature

According to one estimation, a volcanic aerosol veil present in 536 CE was 'the densest and most persistent' on record, and its climate impacts the most exceptional in the past three millennia [189]. Given the paucity of evidence, this is probably overstating the case but a handful of literary sources from the Mediterranean region do attest to an uncommon and prolonged atmospheric disturbance and its associated impacts on crops [79]. One of the sources is actually not contemporary but rather written in the twelfth century by one of the patriarchs of the Syrian Orthodox Church, Michael the Syrian. Of the year 536/7, he says:

> . . . there was a sign in the sun the like of which had never been seen and reported before in the world. If we had not found it recorded in the majority of proved and credible writings and confirmed by trustworthy people, we would not have recorded it; for it is difficult to conceive. So it is said that the sun became dark and its darkness lasted for one and a half years, that is, eighteen months. Each day it shone for about four hours, and still this light was only a feeble shadow. Everyone declared that the

> sun would never recover its original light. The fruits did not ripen, and the wine tasted like sour grapes.

An independent account, referenced to the tenth year of the reign of the Emperor Justinian (Figure 11.1), which dates it to 536/7, comes from the Byzantine historian, Procopius of Caesarea. He lived in Rome during this period and wrote:

> And it came about during this year that a most dread portent took place. For the Sun gave forth its light without brightness, like the Moon, during this whole year, and it seemed exceedingly like the Sun in eclipse, for the beams it shed were not clear nor such as it is accustomed to shed. And from the time when this thing happened men were free neither from war nor pestilence nor any other thing leading to death.

Relating the events surrounding the visit of Pope Agapetus I to Constantinople in 536, a further chronicle (attributed to the bishop Zacharias of Mytilene) tells us:

> The Earth with all that is upon it quaked; and the Sun began to be darkened by day and the Moon by night, while the ocean was tumultuous with spray from the 24th of March in this year till the 24th of June in the following year . . .. And, as the winter [in Mesopotamia] was a severe one, so much so that from the large and

Figure 11.1 Mosaic from San Vitale, Ravenna, dated to *circa* 546 depicting Justinian (centre) and emphasising his integrated military, political and ecclesiastical agenda.

> unwonted quantity of snow the birds perished ... there was distress ... among men ... from the evil things.

Another contemporary resident of Constantinople, John the Lydian, also gives some impression of the aerosol's coverage:

> If the Sun becomes dim because the air is dense from rising moisture – as happened in the course of the recently passed fourteenth indiction [535/36] for nearly a whole year ... so that the produce was destroyed because of the bad time – it predicts heavy trouble in Europe. And this we have seen from the events themselves, when many wars broke out in the west and that tyranny was dissolved, while India, and the Persian realm, and whatever dry land lies toward the rising Sun, were not troubled at all. And it was not even likely that those regions would be affected by the calamity because it was in Europe that the moisture in question evaporated and gathered into clouds dimming the light of the Sun so that it did not come into our sight or pierce this dense substance.

Actually, there is evidence for sighting of the cloud well beyond the Mediterranean. At the equinoxes each year, the ancient Chinese astronomers searched for Canopus, one of the brightest stars in the night-time sky, for astrological divination and to mark the passage of the seasons. But in 536, it was not visible, consistent with the dimming effects of stratospheric aerosol veil. Further records from southern China detail summer frost and snow in 539 that killed crops, resulting in famine.

Although the volcanic origin of this 'Mystery Cloud' has been called into question (an alternative explanation being that the atmospheric dust resulted from the debris caused by a comet impact), a strong sulphate layer in the Greenland ice-core record dated to 533/4 with an uncertainty of two years [190], and another, though less securely dated, anomaly in the Antarctic archive, do corroborate a large eruption around this time. If the bipolar signals are considered to match, then it would point to a low-latitude eruption. One recent suggestion for the volcano's identity is Ilopango in El Salvador, whose M_e 6.9 eruption severely disturbed Mayan society (Section 10.2) [191].

The wider significance of the apparent climate downturn, and its alleged link to harvests, is that the year 541 marked the onset of the first pandemic in Europe of bubonic plague, so-named because of the blister-like swellings (buboes) of the lymph nodes under the armpits or in the groin or neck of victims. Plague is transmitted to humans by the bite of an infected flea but also by air, particularly during pandemics. The root cause – the bacterium *Yersinia pestis* – has been confirmed in the case of the 541 pandemic by studies of DNA extracted from the teeth of three sixth-century victims excavated from a mass grave in Burgundy [192].The so-called Justinian plague went on to ravage the Mediterranean, western Asia and Europe intermittently for two centuries, claiming as many as 100 million lives. The argument goes that poor harvests drove the plague's vector, rats, out of fields and into grain stores where they came into proximity with people.

11.1.2 Plague

At its height in the second century, under the Emperor Trajan, the Roman Empire stretched from Britain to Mesopotamia and included the entire Mediterranean region. But it was not long before it all started to unravel through endless civil wars and attacks on its borders, and in 324 Constantine I, the first Christian Roman emperor, established the city of Constantinople, which became the imperial capital, ultimately dividing the Western and Eastern Roman empires. The Huns migrated into Europe from across the Volga in the late fourth century, displacing the Goths who then fled across the Danube to settle in Roman territory. But hostilities soon broke out between the refugees and the settled authorities. A large contingent of the Eastern Roman army was crushed in combat with the Goths at the Battle of Adrianople in 378 doing much to sap the strength of the Empire. Under Attila, the Huns went on to bleed the resources of the Western Roman Empire, accelerating its demise in the mid-fifth century [193]. Attila died before he could strike at Constantinople, which he had in his sights, and the Eastern Empire enjoyed a period of stability.

However, when Justinian I took the throne in Constantinople in 527 he set about re-uniting the territory, retaking Rome and much of the lost western Mediterranean lands. He also oversaw the construction of the magnificent Hagia Sophia basilica, renowned for its massive dome, and the compilation of the Corpus Juris Civilis, the rewriting of Roman law (which still lies at the foundation of civil law in many parts of the world). Somewhat ironically, one of his acts of religious zealotry had been to abolish worship to the goddess Isis at the temple dedicated to her on Philae, an island in the Nile [194]. Isis was closely associated with spells and ritual used in healing and cure, and her followers may well have secretly rejected the imperial folly as Egypt was ravaged by a new invasion – of *Yersinia pestis*.

The bacterium responsible for plague probably evolved in eastern or central Africa but its outbreak appears to have been first reported in the entrepôt of Pelusium on the Nile delta. From there it spread quickly, following military and trading routes to Alexandria, the rest of Egypt and to Palestine [195]. It reached Constantinople, Antioch, the Balkans and Spain in 542, and a year later struck Italy and Gaul. According to Procopius, even the Emperor himself was infected. Travelling through Syria, John of Ephesus witnessed 'houses and way-stations occupied only by the dead, corpses lying in the fields and along the roadside, and cattle wandering untended into the hills'. Procopius gives an insight into the calamity as it befell Constantinople in 542:

> And when it came about that all the tombs which had existed previously were filled with the dead, then they dug up all the places about the city one after the other, laid the dead there, each one as he could; but later a quarter of those who were making these trenches, no longer able to keep up with the number of dying, mounted the towers of the fortifications in Sycae (Galata) and tearing off the roofs, threw the bodies in there in complete disorder.

These mass burials may explain why there is very limited evidence from funerary inscriptions that might otherwise quantify the lethal effect of the contagion. John of Ephesus even claims that many corpses in Constantinople were thrown into the sea. Evagrius Scholasticus, an ecclesiastical historian writing at the end of the sixth century, related how the plague waxed and waned with a 15-year cycle, decimating successive generations:

> [The plague] made a circuit of the whole world in succession, leaving, as I suppose, no part of the human race unvisited by the disease. Some cities were so severely afflicted as to be altogether depopulated, though in other places the visitation was less violent. . . . the disease occurred, with the almost utter destruction of human beings . . . Thus it happened in my own case . . . I was seized with what are termed buboes, while still a school-boy, and lost by its recurrence at different times several of my children, my wife, and many of my kin, as well as of my domestic and country servants . . . Thus, not quite two years before my writing this, being now in the fifty-eighth year of my age, on its fourth visit to Antioch . . . I lost a daughter and her son, besides those who had died previously. The plague was a complication of diseases: for, in some cases, commencing in the head, and rendering the eyes bloody and the face swollen, it descended into the throat, and then destroyed the patient. In others, there was a flux of the bowels: in others buboes were formed, followed by violent fever; and the sufferers died at the end of two or three days, equally in possession, with the healthy, of their

> mental and bodily powers. Others died in a state of delirium, and some by the breaking out of carbuncles ...
>
> The ways in which the disease was communicated, were various and unaccountable: for some perished by merely living with the infected, others by only touching them, others by having entered their chamber, others by frequenting public places. Some, having fled from the infected cities, escaped themselves, but imparted the disease to the healthy. Some were altogether free from contagion, though they had associated with many who were afflicted, and had touched many not only in their sickness but also when dead. Some, too, who were desirous of death, on account of the utter loss of their children and friends, and with this view placed themselves as much as possible in contact with the diseased, were nevertheless not infected; as if the pestilence struggled against their purpose. This calamity has prevailed, as I have already said, to the present time, for two and fifty years, exceeding all that have preceded it.

11.1.3 Consequences for the Empire

The Eastern Empire was greatly debilitated by depopulation through the ravages of plague and the impacts of endless skirmishes with the Persians as well as bloody internecine disputes and periods of anarchy. The economic impacts would have been severe, not least since the deaths of so many tax-payers diminished the Empire's revenue and, by the middle of the sixth century, the regime was effectively exhausted. Procopius denounces the Emperor in one account for refusing to rescind taxes owed by landowners despite the fact that their agricultural workforce had been decimated. A further reaction of Constantinople during the first decade of the pandemic was to reduce the weight of new issues of gold coinage, while continuing to exact tax payments in the former full-weight *solidi*. Meanwhile, the weight of the copper *follis*, the everyday currency used by the urban poor, artisans and field labourers, was reduced substantially. According to a historian of this period, Peter Sarris, these reflections of a crisis in state finances are most readily explained by the impacts of agrarian depopulation [195]. In a similar move in the early seventh century, the Emperor Heraclius pronounced that the state would make payments in silver coinage but continued to demand taxes in gold. A fiscal crisis translated into military vulnerability since the Byzantine army was the principal beneficiary of state revenue.

Plague spelt ultimate failure for Justinian's reforms and conquests, and the Empire never recovered. Following the establishment of the Caliphate by the disciples of the prophet Muhammad, the seventh century saw the Islamic Conquests capture territories across the Near East and much of North Africa, drawing further peoples into the grip of plague. The associated economic, demographic, political, dynastic and religious shifts have reverberated through history to the present day.

While historians disagree on the scale of impacts of the Justinian plague and its contribution to the descent of the Eastern Roman Empire, there can be even less certitude concerning any role played by volcanism. Nevertheless, if two sulphur-rich eruptions (or even an eruption followed by a bolide impact) occurred shortly before 536 and 541, and were responsible for the climatic swings apparent in the tree-ring records [196], then it is conceivable that they conspired with other internal and external pressures on the Empire to deepen an agrarian crisis. Interestingly, an important Irish chronicle known as the *Annals of Ulster* refers to a 'failure of bread' in 536.

Several volcanoes have been implicated on the basis of very approximately dated tephra, including Krakatau, but the most recent correlation, suggested by Robert Dull and his colleagues, is to the Tierra Blanca Joven eruption of Ilopango volcano in El Salvador (Section 10.2). This is a fascinating hypothesis since it would link a major known eruption with clear evidence for local-, regional- and hemispheric-scale social change. Long before this link was suggested, Mesoamerican specialists had recognised a period known as the Hiatus in the middle to latter half of the sixth century CE, which is associated with depopulation and abandonment across large areas of the lowland Maya region. If Ilopango volcano in El Salvador were responsible for the 536 CE event, it could link the Hiatus to the regional disturbances of the eruption discussed in Section 10.2 and the upheavals in the Old World. The Tierra Blanca Joven eruption of Ilopango would then rank as one to have had the greatest and most widespread human impacts in the historic period.

Another major volcanic eruption appears to have taken place around 626, judging by ice-core records, together with Byzantine and Chinese chronicles. Its climatic effects have been linked to the collapse of the Eastern Turkic Empire [197]. But with little else to elaborate on, we fast-forward to the mid-thirteenth century . . .

11.2 VEILS AND WHIPS

Without exception, all relevant ice cores from the polar regions reveal a striking layer of sulphate fallout and associated tephra around 1259

CE. In the chemical stratigraphy of the GISP2 Greenland core it is by far the most prominent horizon of the last 7000 years (Figure 4.6). The deposition of ash and sulphate aerosol at both ends of the planet points strongly to a massive, sulphur-abounding, equatorial explosive eruption. As we shall see, some rather extraordinary claims have also been made for the human consequences of the eruption.

11.2.1 Ice cores

The signature of this mystery eruption was first recognised three decades ago in the acidity record of the Crête ice core from Greenland. Since then, it has shown up in many more polar cores. Significantly, in both the Arctic and Antarctic records, the sulphate layer contains very fine ash particles of similar intermediate to silicic composition [198]. Dating accuracy is reported variously as ±1 or 2 years, or 0.5%. Allowing for up to a year time delay for polar fallout of a stratospheric aerosol veil, and for the quoted age uncertainty of one or two years, the ice cores provide clear evidence for a major explosive eruption in the second half of the 1250s. In the GISP2 record, which extends back 110,000 years, events of this magnitude - in terms of sulphur yield - recur roughly once in 5000 years. (By this ice-core-based measure, the Tambora 1815 (Chapter 13) and Krakatau 1883 eruptions are approximately 1-in-500- and 1-in-100-year events, respectively.) Based on the calibration obtained from ice-core measurements of radionuclides from 1960s atmospheric weapons tests (Section 4.2.2), the stratospheric sulphur mass of the mystery eruption cloud was more than 60 megatonnes (roughly ten times greater than that generated by the Pinatubo eruption cloud in 1991). The eruption also appears to be revealed in concentrations of the trace metal thallium in the ice core recovered at the breathless elevation of 6350 metres above sea level near the summit of Nevado Illimani, Bolivia [199].

11.2.2 Identifying the mystery eruption

I have previously tried to estimate the size of the responsible eruption from the relationship between magnitude and sulphur output for known eruptions. This approach suggests an eruption of massive size: M_e 7.7–8.3. An alternative scenario is that the 1259 ice-core layer resulted from a much smaller eruption, greatly enriched in sulphur. If it had been as sulphur-rich, say, as Pinatubo, then it could have been as small as M_e 7.3 (accepting that the ice-core-based estimates of sulphur yield for the mid-thirteenth-century eruption are valid). For the time being, we cannot distinguish between these two scenarios – i.e. a 'typical' M_e 8 eruption or a sulphur-rich M_e 7 eruption. But even if the mystery eruption is not *the* largest of the historic period, it is almost certainly of the magnitude of the 1815 eruption of Tambora (Chapter 13).

One relevant consideration in identifying such mystery eruptions is the size of hole in the ground they must have formed. The statistics for crater and eruption sizes, again for known events, can give a rough guide to the size of crater to expect for our unknown episode (though some large eruptions such as that of Huaynaputina, Perú, in 1600 barely left a scratch on the volcano). This tells us that a typical M_e 7 eruption should have an associated caldera more than10 kilometres across, while an M_e 8 event will likely leave a crater 30 kilometres in diameter. Something that size is rather difficult to hide from modern mapping techniques, even if it is on the seabed.

Following these arguments, one suggestion was that the mystery eruption could have taken place at El Chichón volcano in Mexico, renowned for its exceptionally high-sulphur-yield explosive eruptions in March–April 1982, though the chronological and volcanological evidence is scant. A proposed link to an eruption of Okataina volcano in New Zealand can also be discounted, as its age is demonstrably younger. A more recent proposition is the volcano Quilotoa in Ecuador. Though of respectable magnitude (M_e 6.6), the eruption seems to have been too small to account for the mystery cloud under discussion and the radiocarbon dates allow a lot of room for manoeuvre [200].

So, what to do? We can reasonably assume that the volcano is located within the tropics or subtropics given the global dispersion of the aerosol cloud. We would also expect an eruption of this size to have revealed various sensible phenomena, the nature of which can be gauged from contemporary records of eruptions such as Tambora in 1815 (Chapter 13) and Krakatau in 1883 (Section 5.3). These include audible detonations up to 2000 kilometres away, daytime darkness beneath ash clouds (within a radius of 500 kilometres), pumice rafts (thousands of kilometres away) and tsunami (fatalities from the Krakatau 1883 tsunami were recorded up to 800 kilometres from the volcano). Nevertheless, the thirteenth century literary record for most of the tropics is sparse, and it is difficult to exclude any volcanic regions as candidates for the eruption on this basis. The strongest candidate provinces are less-well-studied volcanic island arcs, including remoter parts of Indonesia and the western Pacific. A small island or submarine eruption would be among the most effective ways to camouflage a large caldera and associated voluminous tephra. In such a case, the deep-sea sediment core record may eventually help to solve this mid-thirteenth century puzzle.

11.2.3 Machine-gun volcanism and the Little Ice Age

The literary record unearthed thus far is not rich enough to represent overwhelming evidence for abrupt environmental change associated with the mid-thirteenth century mystery eruption, but passages in various contemporary chronicles suggest the occurrence of a major stratospheric aerosol veil in 1258–9 [201]. In addition, several composite records of tree-ring data and derived climate reconstructions covering the period have are also consistent with a volcanic cooling signal (Figure 4.10). The northern-hemisphere summers of 1258 and 1259 stand out amongst the coldest in one millennial-scale record, with temperature anomalies of −0.46 and −0.31 °C, respectively. The southern-hemisphere summer of 1257/8 is also one of the very coldest in the tree-ring reconstructions [202] and 1258/9 is not far behind: −0.69 and −0.39 °C, respectively. These compare with anomalies of −0.64 and −0.57 °C in the northern and southern hemisphere records, respectively, for the year following the 1600 Huaynaputina eruption (the coldest northern-hemisphere summer in the entire record). Unseasonal conditions in 1259 are also attested by frost-damaged tree rings in larches from northwest Siberia and reports of summer frost in a Russian 'millennial chronicle of extraordinary natural events'.

Considering references to pronounced temperature decreases in England in contemporary chronicles, one estimate suggests the eruption occurred in January 1258. However, if the southern-hemisphere summer temperature reconstructions do indeed register the same event (the possibility that the anomaly is just part of the random noise in the annual climate signal cannot be discounted), then the eruption may have occurred in 1257. Interestingly, a similar temperature pattern is apparent from spaceborne microwave measurements of lower-tropospheric temperatures following the Pinatubo 1991 eruption. These reveal southern-hemisphere summer cooling of Australasia and southern South America in the austral summer of 1991/2. In other words, the first summer cooling occurred in the southern hemisphere. A mid-1257 eruption is therefore consistent with the summer-cooling patterns apparent in both hemispheres, and the optical phenomena witnessed in Europe in 1258 that suggest the presence of a stratospheric aerosol veil (including a dark lunar eclipse in May 1258). The stratospheric volcanic aerosol may also have favoured initiation of an El Niño episode through its forcing on the tropical Pacific [203].

The Little Ice Age is conventionally considered to have got underway in the sixteenth century but more recent evidence has pushed back the initial glacial advances to the second half of the thirteenth century. The idea that waxing and waning glaciers during the historic period might be related to volcanism is not new. One possible explanation to link volcanism to more extended climate change during this period is through the accumulative effects of repeated sulphur-rich eruptions. From an ice-core perspective, the thirteenth century witnessed the most volcanically perturbed upper atmosphere in the past 1500 years (Figure 4.6). In addition to the 1257/8 event, there were major eruptions around 1227, 1232, 1268, 1275 and 1284. Over the whole period of the Little Ice Age, there appears to be a remarkable correspondence between episodes of particularly marked cooling and volcanic eruptions [204].

David Schneider from the National Center for Atmospheric Research in the USA and his colleagues ran a 50-year duration simulation of the climate response to this serial volcanism [205]. For the initial 1257/8 event, the largest of the ensemble, the model suggested up to 4 °C of cooling in the mid latitudes, with up to 1 °C of summertime cooling persisting for four years after the eruption. Cooling of more than 0.5 °C persisted throughout the half-century model run. Furthermore, the results indicated a major perturbation to precipitation with a decrease of around 7% of summer rainfall averaged for the year across the northern-hemisphere mid-latitude landmasses. The reduction was twice as large for the tropics. Of potential significance for longer-term climate change, the model reproduced considerable expansion of sea ice in the northern hemisphere. By the end of the 50-year simulation, there remained an excess of more than one million square kilometres of sea ice (compared with a simulation lacking the volcanism). This effect could be responsible for a longer-term cooling, especially in the Arctic, through the increased scattering of light due to the reflective nature of ice compared with open water.

A further interesting finding of the computer modelling was that the very large mid-thirteenth-century eruption did not result in the typical winter warming pattern seen over continental landmasses of the northern hemisphere after eruptions such as that of Pinatubo. This effect was reproduced in the cases of the successive thirteenth-century events, suggesting that the magnitude of the aerosol veil was sufficient to override the circulatory response of the atmosphere that generates the winter warming pattern seen after Pinatubo's 1991 eruption (Section 3.2.2). This might be consistent with harsh winter conditions documented in Iceland, England and Alsace in 1260–1261.

One important caveat to climate simulations of very large sulphur-rich eruption clouds is the treatment of aerosol physics in the models. Claudia Timmreck, from the Max-Planck Institute for Meteorology in Hamburg, and her colleagues have shown that larger sulphur yields from explosive eruptions do not result in commensurately large temperature responses at the Earth's surface [29]. This is because the increased concentrations of sulphur in the stratosphere result in generation of larger individual particles (due to condensation of gases onto existing particles, and collisions between particles that then stick together). The resulting aerosol is less effective at scattering incoming sunlight and also drops from the sky more rapidly because of its aerodynamic properties. Nevertheless, in the favoured model scenario with rather large aerosol size, the globally averaged surface cooling still exceeds 0.5 °C some three years after the eruption.

11.2.4 Religious fervour and regime change

Richard Stothers highlighted various demographic, political and cultural disturbances that afflicted Europe and the Middle East in the period immediately following the mid-thirteenth-century eruption. One of his more intriguing arguments is that the climate deterioration accompanying the aerosol veil brought on yet more outbreaks of famine and pestilence in parts of Europe. These, he argued, inspired one of the most bizarre cults of the period: the flagellants (Figure 11.2).

The first recorded public manifestation of this movement took place in Perugia, Italy, in 1260, following an outbreak of epidemic disease. The cult spread across northern Italy and beyond the Alps into Alsace, Bavaria and Poland. The demonstrations involved up to 10,000 penitents who processed through the streets, chanting hymns and canticles. Meanwhile, stripped to the waist, they scourged their backs with leather goads until the blood ran, apparently to atone for the sins of the world. But the practice awoke the consternation of the Pope who promptly forbade the processions, and the movement dissipated as quickly as it had arisen. Though Stothers does concede that 'some reasonable doubt seems to exist as to a possible connection between those natural disasters and the flagellant movement', he

Figure 11.2 This procession of mid-fourteenth-century flagellants at Doornik in Belgium gives an impression of events that took place in northern Italy in 1260. Original from the chronicle of Aegidius Li Muisis (1272–1327).

nevertheless suggests that flagellants' mortification of their own flesh might symbolise the lethal depredations of famine and pestilence. In fact, the flagellant movement did re-emerge during the Black Death, a century later (Figure 11.2).

A further geopolitical coincidence of the era is that the year 1260 has been seen by some to mark the 'dissolution of the Mongol Empire' [206], recently expanded through the sacking of Baghdad. Although the maximum extent of the Empire was achieved some 20 years later, the internal divisions that were ultimately to lead to its downfall followed the death of Möngke Khan (one of Genghis' innumerable grandsons) in August 1259, and the election of his brother Ariq Böke. Most of the Mongol Empire's difficulties concerned internal feuding and battles for succession rather than the external enmities faced by the Romans. However, to maintain internal order over the world's largest-ever contiguous empire required 'above all the maintenance of internal order, a sufficiently functioning government and in particular an adequate income from taxation revenue' [207].

Frost-damaged rings in the Mongolian dendrochronological record indicate climatic stress in 1258 and 1259, while ring widths point to a decade of disturbed growth [208]. The 1260s also witnessed grain shortages in Mongolia, though these have been thought to be because Ariq Böke's brother and rival, Kublai Khan, had disrupted grain imports into Mongolia from his power base in northern China. Nevertheless, since the steppes sustained the semi-nomadic, subsistence lifestyles of most of the Mongolian Empire's subjects, climatic deterioration that affected pasture could have had significant impacts: 'steppe society rested on animal wealth' [209]. Given that military superiority was based on the horsemanship and archery skills of an otherwise pastoralist population, if famine did visit the steppes, exacerbated by volcano-induced climate change then, combined with warmongering sibling rivalry at the top, it may have contributed to the Empire's implosion.

11.3 SUMMARY

There are many good reasons to be wary of anything that whiffs of environmental determinism. There have also been few, if any, periods of history that were not turbulent in some way or another. Given a five-year range selected at random, the task of finding something bad that happened during it is unlikely to pose much of a challenge. Thus, linking a penitent movement in Italy to an unknown eruption on the other side of the planet on the assumption that want begets self-mortification is probably pushing the boat out too far. (On the other hand, it was far too good a story for me to overlook it completely!) And, for that matter, how could we ever prove that the Justinian plague would not have happened without the 536 mystery eruption?

Nevertheless, it is clear that ice cores and tree rings concur that several large (M_e 6 and 7), and possibly one very large, as yet unidentified volcanic eruptions occurred over the past two millennia. These events were likely responsible for global climatic swings that went beyond those of more common phenomena such as the El Niño – La Niña cycle. Because of the potential for associated climate stress (short growing seasons and reduced precipitation) on both agrarian production and the grasslands that supported pastoralist societies, it is justified to explore the coincidences. What is required to do this rigorously, however, is to pool the expertise of scholars from diverse fields including philology, medieval history, historical climatology, geochronology and volcanology. The historical climatologist, Christian Pfister, from the University of Bern expresses this well [188]:

> On a very general level, it could be said that beneficial climatic effects tend to enlarge the scope of human action, whereas climatic shocks tend to restrict it. Which sequences of climatic situations mattered depends upon the impacted unit and the environmental, cultural and historical context. This statement needs even to be restricted in the sense that the

> term "climatic shock" itself is ambiguous, as it is well known that some of the people and groups involved always take advantage of situations of general distress, both economically and politically. It needs to be emphasized that investigating past climate and its significance for, and its perception by, humans does not imply as a matter of course that climate is considered to be a determinant factor. Rather, it is assumed that climate is among those conditions – as are population growth and wars – which may be significant in accounting for a given situation.

In this chapter, I have collated a few facts and figures but there is tremendous scope for future detective work along the lines discussed here.

12

The haze famine

Fires from beneath, and meteors from above,
Portentous, unexampled, unexplain'd,
Have kindled beacons in the skies, and th' old
And crazy Earth has had her shaking fits
More frequent, and forgone her usual rest.

From *The Task* by William Cowper (written in late August 1783)

The eruption of Eyjafjallajökull volcano in 2010 demonstrated just how disruptive Icelandic volcanoes can be. The effects were all the more remarkable since, in volcanological terms, the 2010 activity was relatively trivial in terms of magnitude or intensity. Imagine then scaling up the 2010 episode by several hundred times. This begins to help in understanding the severe consequences of the Laki eruption, which began in June 1783, also in southern Iceland. This had a magnitude, M_e, of about 6.6. It is the largest-known lava-flow eruption in the past millennium - nothing close to it has been witnessed in recorded history - and it led to the deaths of around a quarter of the Icelandic population. It also precipitated an environmental crisis in Europe. Tens of thousands of people perished in England and France, possibly as a repercussion of the eruption.

In this chapter, we examine the nature of these phenomena. One of the most important issues is to explain the loss of life. Some have argued for even greater socio-political reverberations, claiming that the eruption's climatic aftermath helped to fuel the uprising that culminated in the French Revolution. Does this claim hold up to scrutiny?

From volcanological and atmospheric science perspectives, the 1783–4 eruption is also of great significance. It might seem that effusive eruptions would not have much impact on hemispheric- to global-scale climate because their plumes do not climb so high in the atmosphere. There are compensating factors, however. Firstly, lava-flow eruptions are predominantly of basaltic magma composition, and so tend to be sulphur-rich. Also, a large-magnitude lava eruption can persist for days, weeks, months and even years - much longer than the several-hour bursts of an eruption like that of Mt Pinatubo in 1991. The response of climate to a forcing agent depends not just on its magnitude but how long it is applied for.

Although the Laki eruption discharged vast quantities of sulphur gases, its predominantly effusive character and high-latitude situation contrast with characteristics we have come to associate with volcano-induced climate change, i.e. the summer-cooling effects of tropical, Plinian eruptions. The 1783–4 eruption coincides with very different patterns of climate change that atmospheric scientists are still trying to understand. Lastly, as a paradigm for volcanogenic pollution in the lower atmosphere, there is much to learn from Laki concerning human, animal and environmental health effects of chronic volcanic gas emission.

12.1 THE ERUPTION

Iceland is a volcanologist's paradise: it rises over three kilometres (on average) above the seabed and is essentially built entirely from basalt with a modicum of rhyolite thrown in. It also has its attractions for glaciologists, puffin and whale watchers, and enthusiasts of enduring rain-swept hikes during which not a single other human soul will be encountered (though plenty of trolls will be met with). Its volcanic history is blessed with innumerable eruptions (more than 200 are documented in historic times [210]), and thanks to a millennium of human occupation, reciting of sagas and meticulous tephrostratigraphy, much is known about them. Furthermore, the acts of volcanism have taken many forms, from submarine to subglacial, phreatomagmatic to fissural, lending their own vocabulary to the lexicon of volcanology. Elsewhere in the world, Surtseyan eruptions and jökullhlaups can be recognised. All of this activity stems from Iceland's situation both on the mid-Atlantic boundary between the North American and Eurasian plates, and above a mantle hotspot (though not all geologists accept the latter point). In the past 1200 years, there have been four episodes of Icelandic volcanism that exceeded the magnitude of the 1991 Pinatubo eruption. One or another of Iceland's volcanoes erupts every four or five years or so.

The Laki eruption of 1783–4 (also known as the Lakagigar and Skaftá or Síða Fires eruption) is by far the best known to volcanologists, and, to Icelanders, the most notorious. We know a good deal about it thanks to an abundance of contemporary letters, weather logs, diaries, scientific commentaries and newspaper reports. No one has done more than Þorvaldur Þórðarson from the University of Edinburgh to bring these to light [211], in combination with forensic geological research of the lava and tephra produced by the eruption. Much of the material in this chapter is based on his research.

One of the most illuminating contemporary accounts is that of Pastor Jón Steingrímsson, who served at the church at Kirkjubæjarklaustur, and we shall turn often to his testimony. His parish in the Síða district was one of the most severely affected by the disaster, and this is how it all started:

> Around midmorn on Whitsun, June 8th of 1783, in clear and calm weather, a black haze of sand appeared to the north of the mountains nearest the farms of the Síða area. The cloud was so extensive that in a short time it had spread over the entire Síða area and part of Fljótshverfi as well, and so thick that it caused darkness indoors and coated the earth so that tracks could be seen.

The eruption took place along a 27-kilometre-long fissure in the Síða highlands, home to the upland pastures where farmers grazed their sheep in summer. It followed a few weeks of increasing earthquake activity. The fissure is not a single rent in the crust but made up of ten shorter tears [212]. These collectively host more than 140 vents and scoria, and lava spatter cones and ramparts developed at several foci along the fissure system, some of them reaching over 100 metres in height (Figure 12.1).

The magma erupted vigorously: fire fountains climbed a kilometre or two into the sky, and immense torrents of lava gushed from the bases of cinder and lava spatter cones constructed on the fissure. At night, Steingrímsson's imperilled parishioners could see the lava glow behind the uplands.

Steingrímsson tells us that on 12 June – day five – 'the flood of lava spilled out of the canyon of the River Skaftá and poured forth with frightening speed, crashing, roaring and thundering. When the molten lava ran into wetlands or streams of water the explosions were as loud as if many cannon were fired at one time.' Meanwhile, ash continued to fall: 'It was calm on the 14th and the entire area around here was covered by a fall of cinders with even more of them shaped like threads

Figure 12.1 Part of the Lakagigar (Laki) crater row, showing the 1783 eruptive fissure and rampart of cinders built up on each side. The site can be reached by bus from Kirkjubæjarklaustur in summer.

than in the previous downpour, on the 9th. They were blue-black and shiny, as long and thick around as a seal's hair.'

Five days later, Steingrímsson travelled to the canyon itself and witnessed that: 'The flood of fire flowed with the speed of a great river swollen with melt water on a spring day. In the middle ... great cliffs and slabs of rock were swept along, tumbling about like large whales swimming, red-hot and flowing. ... they cast up such great sparks and bursts of flames hither and thither that it was terrifying to watch'.

Over the next days, several farms were engulfed and, on 22 June, a newly constructed church was set ablaze by lava, destroying the building and all its contents. Steingrímsson particularly lamented the loss of a magnificent 240-pound church bell and ecclesiastical books and ornaments, blaming the loss on the failure of the haughty incumbent priest to recognise the threat.

The lava flows stretched from the uplands down the Skaftá valley (Figures 12.2, 12.3) towards gentle sand plains (composed of older volcanic debris). Where they crossed swampy ground, steam explosions built up innumerable cinder cones, now covered in grass. The lavas partly buried older lava flows, some of which were erupted in the tenth century from Eldgjá, a fissure extending from Katla volcano. There were at least five surges of lava down the valley.

For Steingrímsson, 'the tumult reached its peak' on 18 July, and two days later he was gathering with his terrified congregation

Figure 12.2 Lava flows in the Skaftá valley erupted in 1783. The flows appear as the lighter ground just the other side of the stream. In most places, the lavas are now covered in moss and form a seemingly endless expanse of light-green–brown hummocks.

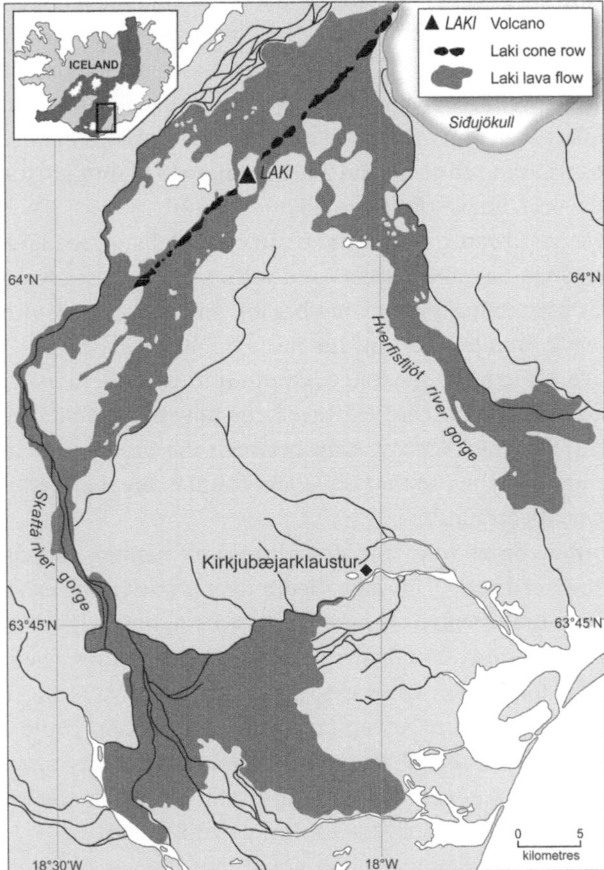

Figure 12.3 A map of the lava flows based on reference 211 and used with permission of Þorvaldur Þórðarson.

for what he imagined might be his last service before the church was consumed in the advancing lava flows.

> As we approached, the clouds of hot vapours and fog coming from the fire farther down the river channel were so thick that the church could hardly be seen . . . from the doors of the cloister building. Claps of thunder were followed by such great flashes of lightning . . . that they lit up the inside of the church and the bells echoed the sound, while the earth tremors continued unabated.

It was now that Pastor Steingrímsson read the 'Fire Mass', still celebrated by Icelanders today, in which he beseeched God to stem the advance of lava towards Kirkjubæjarklaustur. 'Both myself and all the others in the church were completely unafraid there inside its walls. No one showed any signs of fleeing or leaving during the service, which I made slightly longer than usual.' His intercession seems to have worked as the lava came to a halt, there at least, just a hair's breadth from the first dwellings. Instead, the lava had 'collected and piled up in the same place, layer upon layer, in a downward-sloping channel some 70 fathoms wide and 20 deep, and will rest there in plain sight until the end of the world, unless transformed once again'.

However, towards the coastal plain of Meðalland, the lava flows continued to fan out across farms and valued tracts of sea-lyme grass (which was used to make gruel and bread), and people fled with their possessions. It even looked as if the lavas might reach the sea, which would have effectively separated communities and, in particular, cut off those lying to the east. However, the flow fronts stalled around 20 July [213] 'leaving a narrow strip between the fire and the impassable pools and marshes, so that travellers could make their way to and fro on necessary errands'.

But then, on 29 July, 'the first thuds and rushing sounds were heard to the northeast'. This signalled a new phase in the eruption – episode six in the accepted chronology – as the fissure migrated to the northeast. By 31 July, lava had reached the Hverfisfljót river gorge and, over the following two weeks, lava destroyed two farms, one of them 'among the best of sheep farms'. Steingrímsson could barely describe the terrors visited upon the population at this time, with lava flows seeming to capture them in a pincer movement between the Hverfisfljót and Skaftá river valleys (Figure 12.3).

On 25 October, Steingrímsson noted that 'a great spout of flame shot upwards into the air' from the ice sheet 'accompanied by a terrible surge or fire' from the Laki fissure. He described these lava flows, which again followed the Hverfisfljót gorge, as the most threatening and the most powerful. However, hemmed in by the margins of the prior lavas, they did little further damage, except to engulf the grazing lands and hay meadows of three farms. (Steingrímsson remarked that the tenants of these farms had had it coming to them, anyway, on account of their constant squabbling over property. This bickering and litigation 'was now, fortunately, brought to an end'.) The activity continued through November and it was not until December 'that all the flames and glare in the sky began to decrease'. Then, on Christmas Eve:

> A while before sunset a thick cloud piled up here above the cloister, or above the edge of the slope beyond it . . . It was not unlike a work of sculpture, forming a wreath . . . The bulge in the middle was light blue, with branches, curls and spheres, extending out into the wreath itself. These were coloured dark red, bright red, black, reddish black, yellow, pink and saffron, with other colours mixed in as well, which I know not words to describe. A great number of people observed this strange cloud or portent which hung there in the sky without moving, until it disappeared instantaneously just before sunset. Although I can well imagine that it may have been caused by a collection of the various mineral vapours arising from the new lava, which practically surrounded us, it occurred to me as well as others that it might be an indication of the famine and death which was to follow.

The effusion of lava seems to have ended by 7 February 1784, eight months after the whole episode began. But how much lava and tephra was actually produced? The estimates are staggering though subject to uncertainty. Although the area of lavas is known accurately, their depths are generally poorly constrained since there are no reliable topographic maps of the inundated area made before 1783. The widely quoted figure for the magnitude of the eruption is thus based on estimations of the original depths of the Hverfisfljót and Skaftá gorges, and indeed of the thicknesses of lava on the plains downstream and around the fissures themselves. For the gorges, approximate dimensions (widths of around 0.5 kilometres, and lengths of 35 kilometres for the Skaftá and 15 kilometres for the Hverfisfljót) and a lava depth of about 150 metres amount to some four cubic kilometres of lava. A large part of the area of the flow field is contained in the regions furthest from the vent where lava spread out across the sandy plains. Based on assumed thicknesses of 21 metres for the Skaftá and 14 metres for the Hverfisfljót segments, they constitute nearly six cubic kilometres of lava. Add to these figures an overspill of lava from the Skaftá gorge and the total adds up to an astonishing 14.7 cubic kilometres of lava. In addition, around 0.4 cubic kilometres of tephra (converted to volumes of dense magma) were also ejected by the fire fountains along the fissure [214]. This is equivalent to an eruption magnitude, M_e, of about 6.6.

According to Þórðarson, six cubic kilometres of lava were erupted in the first dozen days of the eruption, equating to an average discharge rate equivalent to tipping out the water from two Olympic swimming pools every second. This is over 1500 times the mean eruption rate of Kīlauea since 1983; more than four times more lava was erupted in a matter of a few months in Iceland than has accumulated in nearly three decades on Hawai`i! Presumably, peak outpourings at Laki in June 1783 must have been even higher than 6000 cubic metres per second. The extent of lava inundation is likewise staggering – lava flows covered an area of 600 square kilometres.

12.2 GAS EMISSIONS AND AEROSOL VEIL

There were two main sources of volcanic gases and particles from the eruption: one at the vents associated with the violent fire fountains and convective plumes that soared high into the atmosphere, the other due to the more limited degassing from the spreading lava flows. This latter source had no explosive impetus behind it and contributed mainly to a local volcanic fog. One of the most crucial aspects for understanding the longer-range impacts of the volcanism is the height to which thermal convection carried eruptive plumes above the vents. In this respect, opinions and evidence differ. Based on eyewitness accounts and simple trigonometry it has been suggested that convective columns of gases, aerosol and fine ash rose up to 10–13 kilometres altitude, into the region of the upper troposphere and lower stratosphere. Stratospheric transport of some of the volcanic emissions also seems to be consistent with the pattern of fallout of sulphuric acid aerosol and tephra preserved in the Greenland ice core.

The 'petrological technique' to estimate the sulphur emission during the Laki eruption appears to work well (based on differencing the sulphur contents of melt inclusions and the matrix glass of the lava, and the product of this quantity and the total mass of lava erupted; Section 4.1.5). The total release of sulphur dioxide thus estimated comes to 122 megatonnes (80% of which was emitted at the vents and the remainder from the lava flows) [215]. This total represents at least six times the sulphur emission from the 1991 eruption of Pinatubo. The most intense emission of sulphur dioxide occurred at the start of each surge in activity, with around 5–20 megatonnes of the gas being released per episode. In addition, the eruption discharged into the atmosphere 235 megatonnes of water vapour, 349 megatonnes of carbon dioxide, 15 megatonnes of hydrogen fluoride and 7 megatonnes of hydrogen chloride. Of course, all these figures rest on the validity of the estimated size of the eruption.

12.2.1 Spread of the volcanic cloud

Europe's Age of Enlightenment witnessed a proliferation of scientific experimentation, founding of academies and of systematic meteorological observations. Amateur naturalists across the continent got through many a quill pen scribbling notes on the weather and other phenomena in their diaries. Pre-eminent among the latter in late-eighteenth-century England was Gilbert White. In his *Natural History and Antiquities of Selbourne*, published in 1789, and based on his observations in Hampshire in England, he records:

> The summer of the year 1783 was an amazing and portentous one . . . the peculiar haze, or smokey fog, that prevailed for many weeks in this island, and in every part of Europe, and even beyond its limits, was a most extraordinary appearance, unlike anything known within the memory of man. By my journal I find that I had noticed this strange occurrence from June 23 to July 20 inclusive . . . The sun, at noon, looked as blank as a clouded moon, and shed a rust-coloured ferruginous light on the ground, and floors of rooms; but was particularly lurid and blood-coloured at rising and setting.

The haze appears to have reached Europe a week or two after each pulse of activity in Iceland. In Sweden, it was referred to as *sol-röken* (sun smoke), in Germany *Höhenrauch* (lofty smoke) and in Iceland *móða* (haze). From the first sighting in Copenhagen on 24 June, it spread to central Germany, Switzerland, Upper Bavaria and Poland. It appeared all across France and as far south as central Italy. In Paris, rumours spread that the pollution might bring plague. The city's physicians advised people to sniff some vinegar before going outdoors and not to go out on an empty stomach. Meanwhile, scientists from the Paris Observatory launched kites from which they suspended chunks of meat to investigate the corrupting effects of the extraordinary haze.

Another observer in Paris was Benjamin Franklin, one of the founding fathers of the United States of America. In September of 1783, he was a signatory to the Treaty of Paris, which formally ended the

American War of Independence. Despite being immersed in these momentous negotiations, he found time to exercise his polymathic tendencies by making his own observations of the haze [216]:

> This fog was of a permanent nature; it was dry, and the rays of the sun seemed to have little effect towards dissipating it, as they easily do a moist fog, arising from water. They were indeed rendered so faint in passing through it, that when collected in the focus of a burning glass they would scarce kindle brown paper.

The haze reached Moscow on 25 June and Lisbon on the following day, Syria on 30 June and Baghdad and the Altai mountains of western China on 1 July.

Further perspicacious affirmation of the haze is owed to Jan Hendrik van Swinden, a professor of physics and pioneering meteorologist at the former University of Franeker in Friesland (now part of the Netherlands). There, the haze was first seen on 19 June and was discernible from typical cloud 'by its constancy, density, and especially by very great dryness' [217]. The volcanic cloud reached ground level on 24 June when 'it brought with it as a companion a sulphurous odour very readily perceived by the senses, crawling through everything, even closed houses. Men with delicate lungs . . . were unable to contain a cough, as soon as they were exposed to air. I myself experienced this, and many others, first in the city, then in the country'. The grounding of the plume caused visible damage to vegetation. 'In the morning of 25 June the leaves of many trees were discovered drooping; grass and leguminous plants were drooping; the leaves fell as in autumn and were followed little later by a number of fruits. The appearance of the fields was very sad.' Later he noticed that 'the green colour of the trees and plants had disappeared' and that brass pillars showed signs of corrosion.

A colleague of van Swinden's, living in the neighbouring city of Groningen, noted that the haze 'not only affected the sense of smell, but even taste.' He went on to report that many people experienced 'very troublesome headaches and respiratory difficulties' and that asthma sufferers were particularly afflicted. These health manifestations, both human and vegetative, strongly suggest fumigation at the surface by a mixture of sulphur dioxide and acid particles. In fact, this is one of the puzzles not yet resolved through computer-modelling efforts. Qualitatively, it seems plausible that a slow-moving high-pressure system that established itself over Europe at the time may have acted to funnel aerosol down to ground level.

Also of note, during the first two months of the eruption, ash fell out across most of Iceland and large parts of northern Europe. In Caithness in Scotland, 1783 became known as the 'Year of the Ashie' on account of the frequent accumulation of ash.

12.3 WEATHER AND CLIMATE

Many accounts of the period, such as the following from the *Natural History and Antiquities of Selbourne*, collectively reveal that western and central Europe experienced a blazing summer in 1783:

> All the time the heat was so intense that butchers' meat could hardly be eaten on the day after it was killed; and the flies swarmed so in the lanes and hedges that they rendered the horses half frantic, and riding irksome. The country people began to look with a superstitious awe, at the red, louring aspect of the sun; and indeed there was reason for the most enlightened person to be apprehensive; . . . the minds of men are always impressed by such strange and unusual phaenomena.

This picture is corroborated by detailed reconstructions of monthly temperatures across Europe, based on a range of tree-ring, ice-core and documentary evidence, which show that the heat wave reached a peak in July 1783, spanning France, the Low Countries, Germany, Southern Scandinavia and England (Figure 12.4). A long time-series of temperature observations for central England dating back to 1659 [218] reveals July 1783 as the hottest July on record until recently (1983 and 2006). The heat wave experienced in Europe in the summer of 2003 was more extreme but gives an idea of the 1783 conditions.

Beyond Europe, tree rings tell a contrasting story of summer cooling. In white spruce from Alaska, the last, thicker-walled and flatter-celled ring, which marks the late growing season, records the second-coldest summer in four centuries [219]. This may coincide with oral history of the native inhabitants of 'the time when the summer did not come'.

A further climatic extreme coinciding with the Laki eruption was a very severe winter (1783/4) in Europe and beyond. Benjamin Franklin was quick to make the link between the two phenomena:

> [The] summer effect [of the Sun's rays] in heating the earth was exceedingly diminished.
>
> Hence the surface was early frozen.
>
> Hence the first snows remained on it unmelted, and received continual additions.

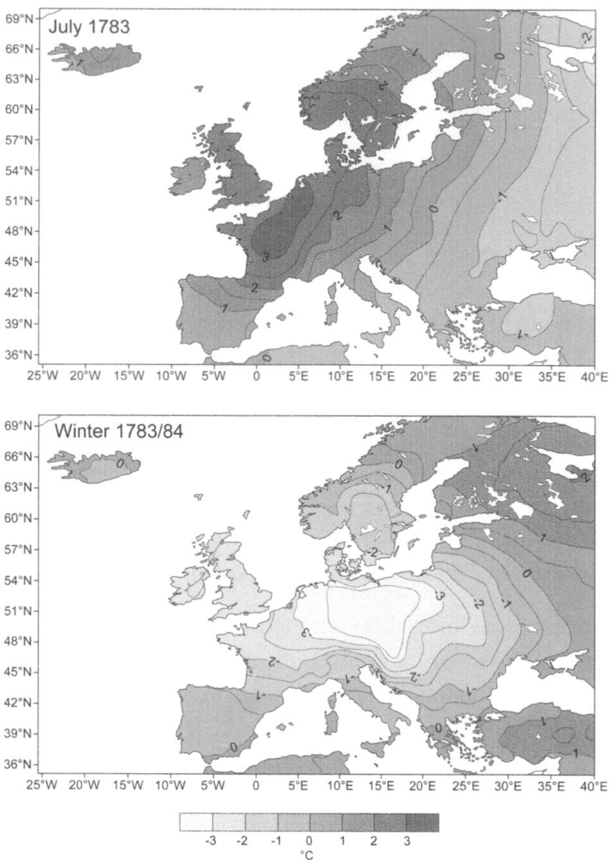

Figure 12.4 Maps of temperature anomalies (referenced to the relevant 31-year-mean values) for July 1783 and winter 1783/1784. Modified from figures prepared by Luke Oman.

> Hence the air was more chilled, and the winds more severely cold. Hence perhaps the winter of 1783–4 was more severe, than any that had happened for many years.

Across Europe, rivers froze. Traffic crossed the iced surface of the Thames in London; in Rotterdam, the 'largest fair on the ice ever known' was held with 'play houses' and other amusements [220]. A letter from Edinburgh (dated 21 February 1784) illustrates the severe conditions in Scotland and the suffering they caused:

> The river Clyde is so filled with ice that there has been no communication between Glasgow and Greenock by water these last

Figure 12.5 Restoration of the Charles Bridge in Prague, damaged during the flood on 28 February 1784. More than a metre of ice had formed on the River Vltava and the level eventually rose more than five metres. K. Salzer, copperplate engraving, courtesy of the Museum of the City of Prague, catalogue no. 1.324.

> seven weeks. By accounts from the North we learn, that the storms of snow are more severe than ever, so that the poor people are in a very distressed condition for want of meal, and many of the sheep and cattle are dying. In many places very liberal contributions have been made for the relief of the poor.

The freezing of waterways as well as river catchment soils, and the accumulation of vast amounts of snow was, in many places, followed by sudden thawing and intense rainfall on as many as three occasions during the winter of 1783/4. This led to some of the most devastating floods ever experienced in many European towns and cities, including Vienna, Paris, Würzburg, Bratislava and Prague (Figure 12.5), evident from documentary sources and the high water marks chiselled into many bridges and buildings [221]. A church diary from Počaply, on the River Elbe in Bohemia, records that the freeze had set in during December of 1783 but at midnight on 28 February 1784 'the ice started to break, with a terrible crack'. The water rose fast and flooded the church and services had to be cancelled the following day. Downstream, in Dresden, the water level rose more than 8.5 metres causing much damage. In Prague, the Vltava (which feeds into the Elbe) rose four metres in just 12 hours [222], a record that was not even surpassed during the catastrophic flooding there in 2002.

In Ely, just a short distance from my home city, floods resulting from thaw caused much damage in January 1784 with the townsfolk forced to live upstairs due to the rise in the level of the Ouse. The Seine froze in mid December of 1783 accompanied by heavy snowfall. A sudden thaw in early January saw the river level rise by three metres. Several *quartiers* of Paris were inundated. Meanwhile, the Loire flooded in Orléans, carrying away boats loaded with wood and wine. The Île de la Cité in Paris was submerged again in late February, and by early March people most at risk had to be evacuated. The official response to the disasters across Europe had significant political repercussions, reminiscent of the situation following Hurricane Katrina in the USA in 2005. In December of 1783, Marie-Antoinette, Queen of France, had let it be known she wished the snow to remain on the streets of Paris as long as possible so she could continue to enjoy sledging. When she realised her *faux pas*, she donated 500 gold coins to support Parisians stricken by the flooding. And on 14 March 1784, King Louis XV himself issued a decree providing for compensation of three million pounds to victims of the flooding. Though this represented a trivial 1% of the annual royal revenue, it was nevertheless the first time the monarchy had extended such financial assistance to the kingdom.

Also in Paris at the time, Benjamin Franklin speculated whether previous hard winters had been preceded by 'extended summer fog'. If so, he suggested that future apparitions of such fogs would serve as warning of an impending harsh winter and potential flooding, enabling practical measures to be taken to mitigate their effects.

A key observation is that available climate reconstructions for the northern hemisphere at the time of the Laki eruption reveal a generally contrary pattern to that observed after high-intensity explosive eruptions in the tropics. Instead of the summer cooling and winter warming witnessed, for instance, following Pinatubo's 1991 eruption, the Laki episode coincides with a warm summer and cold winter. This opposing pattern has yet to be fully reconciled but it is likely relevant that the Laki eruption occurred at relatively high latitude and was prolonged.

Over the past decade, atmosphere modellers have explored the climate effects of the Laki eruption. Most efforts have taken the sulphur dioxide emission inventory outlined in Section 12.2; some have prescribed the state of the atmosphere according to expectations for the pre-industrial world; others have simply worked with modern weather patterns. Accordingly, the results differ somewhat but all yield a massive atmospheric perturbation with peak loadings of 80 megatonnes or more of sulphate aerosol; reproduce its hemispheric dispersal; and demonstrate significant backscatter of solar radiation into space, with its associated cooling effect at the surface. However, according to the models, the climate-forcing effect should have peaked in the autumn of 1783, reflecting the chronology of sulphur dioxide emissions at the vent; the oxidation timescale to form sulphate aerosol; and its faster removal with increased winter precipitation. Thus, the models predict only modest extra cooling in the boreal winter [223].

More startlingly, model simulations infer summer cooling across Europe in 1783 [224], whereas the documentary evidence for an exceptionally hot summer seems irrefutable. These discrepancies suggest plenty of scope for higher-resolution modelling efforts, perhaps focused on the effects of a sustained tropospheric gas and aerosol source to see if it reproduces the heating over Europe apparent in the summer of 1783 (sulphur dioxide is a greenhouse gas). In particular, it will be worthwhile to investigate the interrelationships between the synoptic weather pattern prevailing in late June through early July of 1783 (a high-pressure front firmly established over northeast Europe) and the volcanogenic pollution from Laki. Perhaps radiative and circulation anomalies arising from the dispersed cloud helped to stabilise the high-pressure system. Although the coincidence of the climatic anomalies with Iceland's biggest eruption for a millennium is very hard to dismiss, it is just conceivable that the extreme weather witnessed in 1783 and 1784 was part of natural inter-annual variability unrelated to the volcanic emissions.

12.4 THE HAZE FAMINE

To understand the calamity that befell Iceland during and immediately after the Laki eruption (Figure 12.6), it is helpful to examine its premodern pastoralist society, which effectively lived under serfdom, and had languished under the disinterested rule of the Danish crown since the fourteenth century. Farms were centred on scattered homesteads and the nearby hay fields. They were run by tenant farmers, many of whom even had to rent their livestock. Sheep and cattle were the main source of food and there was little to supplement them by way of garden crops. Rye flour and barley, tools and timber were imported from Denmark and paid for (at high prices due to very limited access to traders) by selling hides or, in good years, some meat and fish. Horses were used for transport and hay was cut by hand. Wild foods provided an important contribution to the diet. These included seals, seabirds and their eggs, crowberries and blueberries, sea-lyme grass, Icelandic moss (a lichen containing a little starch) and some seaweed.

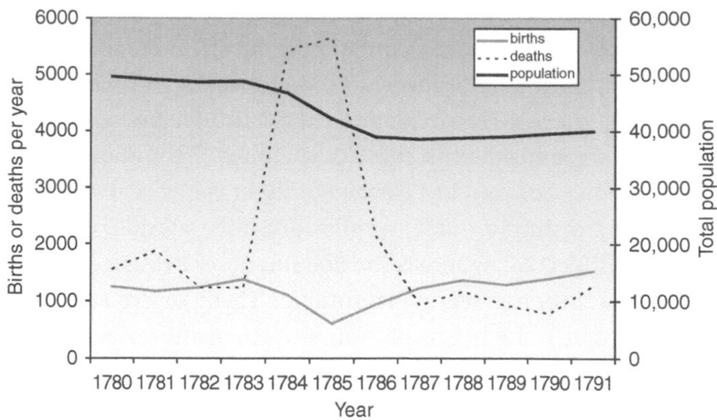

Figure 12.6 Annual figures for births, deaths and total population in Iceland before, during and after the Laki eruption.

Iceland was, thus, at the 'margin of the habitable world' and, especially in the eighteenth century, the community has been described as being 'close to extinction' [225]. The country's climate, particularly through the Little Ice Age, combined with the nature and poor management of its soils made agriculture extremely borderline and famine was a constant spectre for farmers. Sheep were put out to roam the higher ground during summer but the rest of the year grazed the precious 'home fields' or were kept indoors. The main crop was hay, which was used to sustain the animals over the winter when much of the ground was covered with ice and snow. Peat was used as fuel. Summers were barely warm or long enough to provide adequate fodder for the animals. And because stores of hay were generally limited, a late frost during the growing season or a cool, short summer could prove calamitous. The Laki episode needs, therefore, to be seen in the wider demographic context in pre-modern Iceland. Many hundreds of people died in the famine of 1756–1758 (following a large subglacial and explosive eruption of Katla volcano in 1755), and a smallpox epidemic in 1707–1709 represents Iceland's worst mortality crisis of the eighteenth century.

Autumn was a crucial season for Icelandic farmers. It was the time when they might sell some meat, skins and woollens, and estimate their reserves of hay so as to make contingency plans for the impending subarctic winter. A key decision was whether to insure against harsh winters by storing hay and killing some animals, or to avoid the associated costs but risk losing livestock. Evidence suggests that

Icelandic farmers were risk takers in this respect, and that time and again they were caught out by particularly severe winters. Even the sagas relate the perils. According to the *Landnámabók* (Book of Settlement), Flóki Vilgerðarson, one of the first Norsemen to settle in Iceland, skimped in preparing hay for his flock. His animals perished during the first winter, prompting him to return to Norway.

Another tale that vividly applies to pre-modern Iceland comes from the seventeenth century Faroe Islands, whose customs were in many ways similar. A farmer wanted to keep a ram over the winter and disregarded the advice of his experienced hired hand, Oli, who had urged culling the animal to ensure sufficient hay for the cows. Oli disobeyed orders and killed the ram secretly (and ate it!). The spring proved to be cold and the farmer was resigned to killing the cows, and said to Oli that the hay consumed by the ram would not have made much difference anyway. Then, Oli confessed but showed the farmer that he had put aside all the hay it would have eaten, asking wryly if it still made no difference. There was enough hay to save all the cows that year.

In Iceland, it was not until 1884 that legislation known as the Starvation Act was passed to ensure that farmers would not keep more animals than they could feed in a hard year. Even so, many farmers ignored the new law, perhaps because it went against their deeply ingrained social norms. It is also surprising that Iceland did not fully develop its fisheries until the late nineteenth century; Icelanders 'only attended part-time to one of the world's most valuable fisheries that surrounded their country' [226]. Their exploitation could have greatly enhanced the socioeconomic wellbeing of the country and provided a means for many to escape abject poverty. Perversely, social and legal institutions actively discouraged fishing to the benefit of landlords – for instance, a Medieval Icelandic law forbade someone to set up a household without livestock unless the local community to which it belonged guaranteed to insure its welfare. Nor were the Icelanders' small, open boats suitable for serious winter fishing in the stormy seas of the North Atlantic. Lastly, Danish colonial policy isolated Iceland from the wider European economy: only in 1787 did crown rule finally abolish trade monopolies, allowing Danish merchants to operate freely in Iceland. A reflection of Iceland's chronic subsistence economy and social practices is reflected in its demography: between the thirteenth and nineteenth centuries, the population never grew much above 50,000.

Returning to the Laki eruption, the timing of its onset could probably not have been worse for Icelanders. In the first week of

June, Pastor Steingrímsson tells that the grass, so crucial to the farmers' survival through the winter, was 'green and luscious'. But the grazing lands and home fields were about to be ruined:

> ... more poison fell from the sky than words can describe: ash, volcanic hairs, rain full of sulphur and saltpetre, all of it mixed with sand. The snouts, nostrils and feet of livestock grazing or walking on the grass turned bright yellow and raw. All water went tepid and light blue in colour ... All the earth's plants burned, withered and turned grey ... The first to wither were those plants which bore leaves, then sedges were checked, and the horsetails were the last to go, and would later be the first to return.

The people frantically cut the ash-plastered grass and rinsed it with water before feeding it to their cattle 'but it was all in vain unless they still had some older hay to mix with it'. Once the uncontaminated hay was used up, the livestock 'withered away'. With hindsight, Steingrímsson realised 'it would have been for the best to slaughter them all while they still had flesh on their bones and could be rounded up, and thus have food for ourselves'. His descriptions of crippling bone and joint deformities and muscle wasting of the animals point firmly to fluorine poisoning. Horses lost all flesh, their hair rotted and 'hard, swollen lumps appeared at joints'. Their jaws became so weak they could barely eat. Sheep were worse affected: 'large growths appeared on their rib joins at the chest, on their hips and legs, bowing the legs ... Both bones and gristle were as soft as if they had been chewed'. And cattle fared no better, their hips and ribs disfigured, fused and disintegrated, their hooves fell off and they sprouted growths on jaws and shoulders.

According to a nineteenth century source, the Laki eruption resulted in the loss of 76% of horses, 79% of sheep and 50% of cattle in the country [227]. Horses were vital in Iceland for moving food around the country. Much of the estimated eight megatonnes of fluorine emitted during the eruption was likely to have been adsorbed onto ash at the vents and dissolved by rainfall, contaminating the land. The 2010 Eyjafjallajökull eruption also emitted abundant fluorine, which was carried on ash fallout. The threat of fluorosis was an immediate concern and farmers were advised to keep all livestock indoors and to ensure they were only given uncontaminated feed.

Throughout the rest of 1783 '... all water, whether flowing in surface streams or from underground sources, had been scarcely drinkable due to its bad flavour and bitter taste in the mouth.' Several of Steingrímsson's remarks are even suggestive of acute fluorosis afflicting people: 'Ridges, growths and bristle appeared on their rib joins, ribs, the backs of their hands, their feet, legs and joints. Their bodies became bloated, the insides of their mouths and their gums swelled and cracked, causing excruciating pains and toothaches'. His account also suggests many experienced severe respiratory discomfort, heart palpitations and inflammation:

> The foul smell of the air, bitter as seaweed and reeking of rot for days on end, was such that many people, especially those with chest ailments, could no more than half-fill their lungs with this air . . . it was astonishing that anyone should live another week.

As an aside, in 2004, I joined a small team of Icelandic archaeologists and a volcano medicine specialist looking for pathological support of Steingrímsson's remarks on bone changes. This involved exhuming a number of early-nineteenth-century tombs in the Síða district to retrieve human skeletal remains. I hasten to add that the research was fully approved of by all relevant authorities. Further, the farming communities on whose land the abandoned cemeteries lay took a keen interest in the work (as did the farm dog). Ultimately, the three skeletons exhumed were found not to show any signs of bone abnormality or excessive fluoride in teeth. However, the graves were unmarked, and therefore only dated approximately by their juxtaposition between identifiable tephra layers. Two of the individuals were aged about 30 and 40, and were probably born after the Laki eruption. Although nothing conclusive emerged from the study, the various signs of trauma that were identified from the remains, and the young age at death, nevertheless highlight the harsh life led by Icelanders up to the middle of the nineteenth century.

Steingrímsson considered that the combined effects of poor air quality and contamination of water and food were responsible for the ill health and deaths it occasioned. 'What passed for meat was both foul-smelling and bitter and full of poison, so that many a person died as the result of eating it.' The deficiencies in the diet may also have caused some cases of scurvy, consistent with Steingrímsson's observations of swelling and cracking of the gums. He also documents diarrhoea, dysentery, worms and sore growths on necks and thighs, and both young and old were especially plagued by loss of hair. Despite these indications of deteriorating health, parish burial records across Iceland, and even Steingrímsson's testimony concerning the most directly affected district, indicate that relatively few people died in

1783. It was the hardship and starvation that followed that claimed most lives.

The sulphurous gases, acid particles and fluoride not only ravaged the hay crop, they poisoned the Icelandic moss, wild berries and sea-lyme grass, as well as inland and coastal waters, eradicating trout, salmon and cod. Nesting birds were likewise decimated and migratory birds did not return. Off the northern shores of the country, the volcanic smog was so thick that fishing boats could not go out to sea. In any case, most of the catch was exported to Copenhagen.

Steingrímsson remarked that the crisis could have been ameliorated by more decisive action from the authorities: 'The fact that the Intendant was so slow with his decrees regarding grain rations from the trading towns for the people here did much to increase the famine and number of deaths ... [Had it come sooner], less would even have been enough, if it had been used sparingly'. At New Year of 1784, a pamphlet describing the plight of the Icelanders was read out in all the churches of Copenhagen, and money was collected to support the imperilled outlier of the Danish monarchy. However, when financial aid did arrive it did little good since the money given to farmers to buy livestock was 'taken back for the payment of rents and other debts in arrears.' An increasingly desperate community took flight or took to crime:

> As famine and death oppressed us more and more, people began to flee westward to save their lives, some attempting to find a plot of land somewhere. They had to leave almost all their possessions behind and anything which was not placed in the custody of honest men was ruthlessly consumed or stolen, houses and locked stores broken into, so that it is painful even to think of it.

Over the course of the first months of 1784, once their last supplies of uncontaminated meat, cheese and dried fish had expired, thousands of people died in the eastern three-quarters of the country [228]. Reported deaths at Thykkvabaejarklaustur, just 60 kilometres southwest of Laki, soared from a handful per year to 54 in just the first half of 1784. The north of Iceland suffered too, particularly in the grip of its typically colder winters.

In the Síða district, some respite came in the summer of 1784 when a ship grounded offshore. This enabled Steingrímsson and others with any money left to buy quantities of flour, hemp, iron and other items 'which were enough to bring us through the worst of the time of dearth and high prices.' October 1785 brought another windfall in the form of 190 seals that alighted on the beach where they were promptly clubbed by a farmer and two boys.

Many who did make it through 1784 were probably so weakened that they could not work their farms as productively as they would have in normal times, and this may have contributed to the next wave of human mortality, which peaked in March 1785, now particularly in the west of Iceland in the districts surrounding Reykjavik. Many who fled the worst-affected areas heading for the west and southwest of Iceland simply died on the road. The crisis hit all age groups but especially infants, people in their teens and twenties, and the over-fifties. In the parishes that Pastor Steingrímsson reported on, a third of the populace succumbed. (Though his evaluation of the demography of the health crisis seems rather cruel: 'The 215 who died in this area were in such condition that they were not felt to be too great a loss, with the exception of 12 to 20, according to the subsequent deliberation and reckoning of upright men, and with the exclusion of children, as no one could say what sort of person they would have become as adults.')

Combined with a substantial drop in the birth rate (a typical consequence of famine), the country's population decline was just over 22% [228]. A smallpox epidemic struck Iceland in late 1785 and might be seen as evidence for the adage 'pestilence follows famine'. The Vestmannaeyjar (Westman Islands), off the south coast of Iceland, were one of the few parts of the country that escaped relatively unscathed. Their tradition of cod fishing saw them through the disaster.

12.5 LONG REACH OF THE ERUPTION

The evidence for immense suffering in Iceland during and immediately following the Laki eruption is unambiguous, even if the precise aetiology of the demographic crisis is more ambiguous. But Iceland's subsistence pastoralist society already rendered it vulnerable to such calamity. What is more surprising is that the eruption may be linked to even greater loss of life beyond Iceland's shores.

12.5.1 Mortality crisis in England

The year 1981 saw the first publication of a recognised classic on demography in England [229]. At its heart was a statistical analysis of reams of parish records on births, burials and marriages. Here, it is burials that concern us, since (one hopes that) interment is a reliable proxy for mortality. One of the patterns that emerged in the original

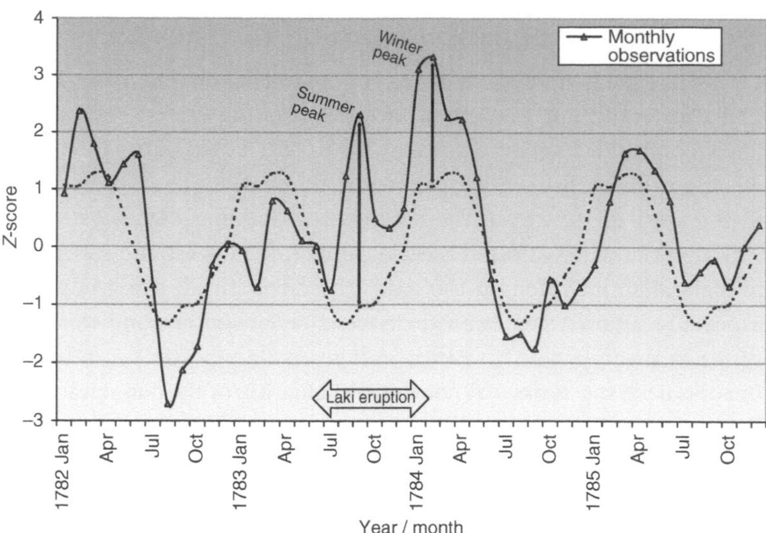

Figure 12.7 Mortality crises in England in summer 1783 and winter 1783/4 evident in burial records for 39 counties. The Z-score measures how far a particular month's mortality figures differ from what would be expected if there were no seasonal effects. The dashed line reveals the seasonal mortality trend based on the 50-year-period 1759–1808. Its sinusoidal pattern reveals the typical mortality pattern evident even today – the death rate is higher in winter than summer. The solid line then shows the month-by-month Z-scores from the start of 1782 to the end of 1785 and shows significantly enhanced death rates in late summer 1783 and late winter 1784. Data courtesy of Claire Witham.

study was of 'mortality crises' of significantly higher than expected death rates. One of these crises occurred in the 'harvest year' from July 1783 to June 1784. The raw data were actually broken down by month, enabling a more time-resolved investigation of the period encompassing the Laki eruption [230]. The first step was to calculate the average death rate for the second half of the eighteenth century. Then, it is possible to calculate, month by month, how far any given month's burial rate differs from that baseline. The average monthly trend that emerges looks like a sine-wave, with a peak in deaths around March and a low around August. In fact, a similar pattern prevails today: winter kills.

However, the pattern looks very different shortly following the onset of the Laki eruption (Figure 12.7). There is a sharp and unseasonal increase in the death rate in August and September of 1783, corresponding to 40% higher mortality than expected and equivalent to 11,500 'excess' deaths. The curve returns to the average trend in the autumn but then shoots up again in January and February of 1784, representing a further 8200 deaths. Curiously, in both episodes, eastern parts of England were particularly affected suggesting a strong regional element to the crisis. In the county of Bedfordshire (which neighbours my home county of Cambridgeshire in the east of England), mortality in August to September of 1783 was five times the average, with the month of September recording the highest mortality in the entire eighteenth century. A similar picture is emerging for parts of France [231]. But how can we explain it?

The first consideration is temperature. Extremes of heat, cold, drought, storms and associated flooding, and changes in air quality can all directly affect human health, and indirectly through their effects on the ecology of infectious diseases. One of the most alarming recent demonstrations of this in Europe was the heat wave of August 2003, when many temperature records were broken. Indeed, it is widely recognised as the hottest summer in at least 500 years. It resulted in around 15,000 deaths in France alone, and many more thousands in England, Spain, Germany and Italy, with a total toll across Europe of as many as 45,000. Heat stress was the main contributing factor but increased levels of ozone and particles in the lower atmosphere likely also claimed lives.

Today, the effects of extreme heat are exacerbated by the growth of cities and their urban heat island effects, but if this degree of vulnerability exists in modern Western Europe today with its access to ventilation systems and modern medicine, it becomes easier to imagine how extremes of temperature might have stricken vulnerable groups centuries ago. Returning to England, we know that July of 1783 was particularly sweltering. In the extended central England record, average daily temperature peaked on 11 and 28 July at 22.6 and 21.5 °C (compared with peaks of 23.1 and 23.9 °C in August 2003). However, the mortality rate peaked in August and September 1783. It is difficult to reconcile this time lag with the direct effects of heat stress. During the 2003 heat wave, the statistics show an essentially instantaneous correlation between high-temperature days and increased death rate. For example, in Paris, the average number of deaths per day is five but, in mid August 2003, more than 30 deaths per day were recorded. It is also clear from modern cases of heat stress that it is especially the elderly and those with pre-existing respiratory and cardiovascular conditions that are at risk from dehydration and heat stroke and exhaustion.

What is more, the most intense fumigation at ground level in England in summer 1783 occurred from mid June to mid July. So, it is difficult to appeal to poor air quality as the cause of the later summer peak in mortality. In any case, although Pastor Steingrímsson's parishioners were certainly experiencing tremendous discomfort from the volcanogenic pollution – and they were very considerably closer to its source – the death toll there did not mount until the winter. This leaves the possibility that epidemic disease was the root cause of the August and September deaths in England. This could still be linked to the very hot July, which might have enabled certain disease vectors, including flies and mosquitoes, to flourish. At this time in England, diseases such as malaria, typhoid and dysentery were common, and hot summer temperatures would plausibly have heightened their transmission.

Some evidence for the prevalence of a fever epidemic comes from the memoirs of Reverend Charles Simeon of Cambridge who travelled through central England in August and September 1783. On his return to his Cambridge parish on 19 September, he wrote '… many whom I left in my parish well are dead, and many dying; this fever rages wherever I have been' [232]. Of note, the memoir implies that the illness affected previously healthy people. The diary of the Reverend James Woodforde (written while he was in the eastern county of Norfolk) and letters of the poet William Cowper (then in Bedfordshire) also allude to a lethal fever. On 2 September, Woodforde recorded that: 'almost all the house ill in the present disorder … It is almost in every house in the village …' Cowper, writing on 8 September, stated that 'The epidemic begins to be more mortal as the autumn comes on'.

The winter hike in death rate in England is easier to understand as a direct consequence of the severe temperatures experienced, since the two anomalies coincide. Extremes of cold also particularly affect the elderly who are more vulnerable to hypothermia and respiratory and circulatory disease, as well as to accidents on icy pavements.

12.5.2 Africa and Asia

Taking a broader view, climate fluctuations remain strongly linked to health outcomes across the globe but especially in Africa and India. While Europe was to experience extreme levels of damaging flooding from December 1783 to March of 1784, Egypt suffered from a lack of flooding. The French orientalist Constantin-François de Volney travelled through Egypt and Syria at this time [233] and provides us with a graphic portrayal of its consequences:

> The inundation of 1783 was not sufficient; a great part of the lands therefore could not be sown for want of being watered, and another part was in the same predicament for want of seed. In 1784, the Nile again did not rise to the favourable height, and the dearth immediately became excessive. Soon after the end of November, the famine carried off, at Cairo, nearly as many as the plague; the streets, which before were full of beggars, now afforded not a single one: all had perished or deserted the city. Nor were its ravages less dreadful in the villages; an infinite number of wretches, who attempted to escape death, were scattered over the adjacent countries. . . . it was the received opinion, that the country had lost one-sixth of its inhabitants.

This inefficacious flooding of the Nile, so vital to the agriculture of Egypt, has been attributed to summer warming of the Sahel region of Africa due to Laki's aerosol veil [223]. Its presence reduced the temperature difference between Asia and Europe, and between the Indian and Atlantic Oceans, which had the effect of weakening the Indian and African monsoon circulation. Reduced cloud and rain over the Nile source regions in Sudan and Ethiopia that are simulated for the summer of 1783 are consistent with the documented very low flow at the mouth of the river. Interestingly, the almanacs for Nile river levels indicate that 1782 was also a low-flow year. This event is unexplained but the coincidence does not detract from the evidence linking Laki's aerosol to the 1783 dearth.

The climate model also simulated comparable warming and reduced rainfall over parts of India linked to the diminished monsoon. This is a striking coincidence since 1783 brought starvation to almost all of India (the *Chalisa* famine), claiming the lives of up to 11 million people during a severe drought [234]. The same year saw the terrible *Tenmei* famine in Japan, though it should be noted that there was a substantial M_e 5.1 explosive eruption of Asama in central Honshū (Japan), less than 200 kilometres from Tōkyō, between May and July of 1783 [235]. This claimed around 20,000 lives and destroyed rice crops, resulting in angry demonstrations in Edo (Tōkyō).

12.5.3 The French Revolution

It is worth touching briefly on a popular supposition that the climatic upheavals caused by the Laki eruption contributed in some way to the French Revolution of 1789. There is not much to commend the idea, however. The first real signs of public unrest in France occurred in 1788, and there had been intermittent good and bad weather and associated harvests between then and 1783–4. In particular, there was drought in 1788, whose consequences have been suggested to have destabilised public order in France. However, given that around 90% of the French population lived at or below the subsistence level and were hit hard by fluctuating grain and bread prices, and given that the French aristocracy and church enjoyed privileges that were only multiplied under the reign of Louis XVI, one might avoid making 'some absurd "explanation" of the French Revolution and related events in terms of climate, in another attempt to unravel over-determination and multifaceted causality' [236]!

The final word on this matter can be left with John Kington of the University of East Anglia's Climatic Research Unit. On examining the synoptic weather map reconstructed for 14 July 1789, he observed that with 'a ridge of high pressure extending east over France, the weather was partly cloudy with occasional showers, there was a light to moderate westerly wind and a temperature of 22 °C. The weather in Paris would not have deterred participants in a popular disturbance such as the Storming of the Bastille: it was a perfect day for outdoor activity!' [237].

12.6 SUMMARY

If the 2010 Icelandic eruption taught us anything in Europe, it is how easy it is to be under-prepared. There was, in fact, a lava eruption in Iceland even larger than the 1783–4 Laki episode within the historic period. The Eldgjá eruption took place in the 930s CE not long after the island was first settled by Norsemen. It also emitted more sulphur dioxide (though probably over a longer time period) [238] than Laki, and there is evidence that it, too, had profound climatic and human impacts across Eurasia. Other great eruptions occurred in Iceland in 1362 (Öræfajökull) and 1477 (Bárdabunga–Veiðivötn volcanic system). The point is that the Laki eruption should not be seen as a one-off. Similar-magnitude eruptions will occur in Iceland in the future. What the case of Laki demonstrates is that the effects can be severe both home and abroad. Much work remains to be done, however, to clarify and understand both the environmental and human impacts of the Laki episode, and to be able to analyse the risks of future eruptions of this character.

13

The last great subsistence crisis in the Western world

Its noise reverberated loudly
Torrents of water and ash descended
Children and mothers yelling and screaming
Believing the world had turned to ash
At first it seemed Allah was angry
At the deed of the King of Tambora
In shedding the blood of a worthy pilgrim,
Rashly and thoughtlessly

From an epic poem (*syair*) from Sumbawa, compiled in Malay around 1830 [239]

In Western Europe, 1816 was a time of optimism. The Napoleonic wars, which had ravaged the continent, were over. Unfortunately, a little known volcano on the island of Sumbawa in Indonesia had erupted the previous year and its impacts were about to be felt across the northern hemisphere . . .

Tambora, on the Indonesian island of Sumbawa, may once have been the highest peak of the East Indies as they were known in the colonial period. Sailing eastwards past Bali, it appeared as high on the horizon, despite being further away, as the 3726-metre-high Mt Rinjani on Lombok Island, suggesting it may have reached over 4000 metres above sea level. We shall never know for certain because the cone was toppled in April 1815 by the largest eruption of recorded history. The events also resulted in the greatest-known death toll attributable to a volcanic eruption [18], and the associated climate change has been implicated in 'The last great subsistence crisis in the Western world' (see Further Reading). This chapter recalls the events and consequences of the 1815 eruption, and evaluates the arguments for long-range impacts in North America and Europe.

13.1 SUMBAWA BEFORE THE DISASTER

By the beginning of the nineteenth century, settlements were established across Sumbawa, all belonging to one of six princedoms or sultanates: Sumbawa, Bima, Dompo, Sanggar, Pekat and Tambora. The villages were generally located close to rivers and teak forests, and their inhabitants cultivated and harvested rice, honey and beeswax, birds' nests (used to make the greatly prized soup), horses, salt, cotton and sappanwood (a yellow-flowering tree cultivated as a source of dye and for its medicinal properties). One account of the island before the eruption paints a particularly idyllic picture [240]:

> Nature had poured its bountiful blessings on this island, [which,] no matter how mountainous, is the proud possessor of the most extensive of plains and the loveliest of verdant valleys. Rice, beans and maize were plentiful, the forests provided wax and excellent timber, in particular sappanwood, the quality of which is second to none in the entire archipelago. Coffee, pepper and more especially cotton were grown, which latter crop constituted an important source of income for the inhabitants. There are birds' nests, including some of the most excellent quality, and the island has been proved to contain gold resources, even though these have never been exploited. In a bay in Dompo on the south coast there are pearls, some of which are very large, although pearl fishing has never been properly supervised. There are saltpans in Bima, which supplies the entire coast and part of Boneratte, Manggarai, Saleijer and Bonie with salt from these. Finally, who has never heard of the fine horses from this island, which are surpassed in quality by none?

On the other hand, the district of Tambora itself – closest to the volcano – was described by a visitor [241] in 1786 in somewhat less favourable terms, as:

> a tiny, barren, rocky district, where nothing grows in the mountains but a little paddy, hardly enough to feed its inhabitants, who therefore obtain this from traders in exchange for the products which are found in abundance in the forests and are available in their purest form here, and by which, as well as by horse breeding, the king, nobles and subjects are compensated annually for the barrenness of their country.

13.2 THE ERUPTION

In 1811, just a few years before Tambora exploded, Java was the scene of several military campaigns between the Europeans: one of the global consequences of the Napoleonic wars. The Kingdom of Holland had been defeated by Napoleon, and thus the Dutch East Indies (Indonesia) came under French administration. This resulted in a British Navy assault against a weak Dutch and French army, which only held out for a matter of weeks. Sir Thomas Stamford Raffles, who had been working for the British East India Company in Penang (then called Prince of Wales Island and now part of Malaysia) until joining the military campaign, was appointed Lieutenant Governor of Java. Raffles took a keen interest in the culture and natural history of the island and the Indonesian archipelago, and much of what we know about the Tambora eruption is handed down in his *History of Java* [242] and his memoirs [243]. He was so impressed with the flora and fauna of southeast Asia that he went on to become a founder and the first president (in 1826) of London Zoo. Raffles' accounts, along with other documents and letters published in the *Asiatic Journal* (a compendium of intelligence information, literary reviews and news items on the region, published by the East India Company from 1816–1829), provide fascinating insights into the nature and immediate impacts of the eruption.

13.2.1 Initial blasts

Tambora forms the Sanggar (formerly Sanggir) peninsula on Sumbawa, one of the fleet of predominantly volcanic islands rising from the Flores Sea in eastern Indonesia (Figure 13.1). It appears to have been considered extinct (perhaps even non-volcanic) until 1812, when it began rumbling and emitting small ash clouds. Magma, which had accumulated in a chamber three or four kilometres beneath the crater, was on its way to the surface. One eyewitness of these premonitory signs was John Crawfurd [244], who sailed past the Sumbawa coast in 1814:

> . . . even then the volcano of Tomboro was in a state of great activity. At a distance, the clouds of ashes which it threw out blackened one side of the horizon in such a manner as to convey the appearance of a threatening tropical squall (in fact, it was mistaken for one and the commander of the ship in which I was, took in sail, and prepared to encounter it). As we approached, the real nature of the phenomenon became apparent, and ashes even fell on the deck.

Then, on the evening of Wednesday 5 April 1815, the first serious eruption began, lasting for around two hours. Raffles describes the surrounding events thus:

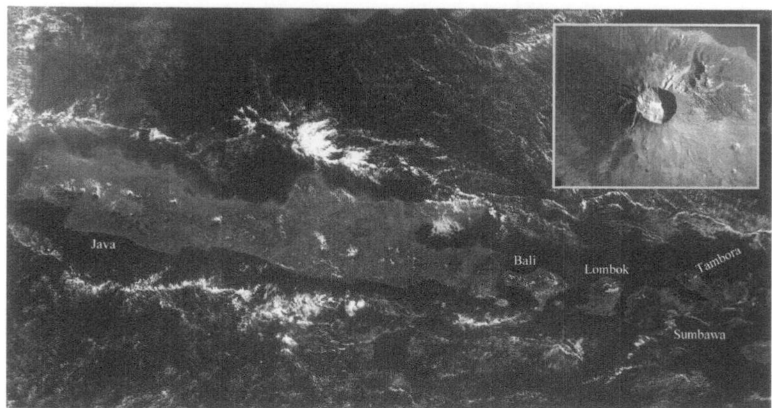

Figure 13.1 Satellite image stretching from Java to Sumbawa. Inset shows a photograph of Tambora taken from the International Space Station in 2009. The caldera has a diameter of six kilometres. Credit: Jacques Descloitres, MODIS Land Rapid Response Team, NASA/GSFC; inset from NASA-JSC.

> The first explosions were heard on this Island in the evening of the 5th of April, they were noticed in every quarter, and continued at intervals until the following day. The noise was, in the first instance, almost universally attributed to distant cannon; so much so, that a detachment of troops were marched from Djocjocarta, in the expectation that a neighbouring post was attacked, and along the coast boats were in two instances dispatched in quest of a supposed ship in distress.
>
> On the following morning, however, a slight fall of ashes removed all doubt as to the cause of the sound, and it is worthy of remark, that as the eruption continued, the sound appeared to be so close, that in each district it seemed near at hand, it was attributed to an eruption from the Marapi, the Gunung Kloot or the Gunung Bromo.

The East India Company's cruiser *Benares* was at Makassar on Celebes Island (now called Sulawesi), 350 kilometres north-northeast of Tambora on 5 April. An extract from a letter published in the *Asiatic Journal* written by the commander reads:

> On the fifth of April a firing of cannon was heard at Macassar: the sound appeared to come from the southward, and continued at intervals all the afternoon. Towards sun-set the reports seemed to approach much nearer, and sounded like heavy guns occasionally, with slighter reports between.

From the dispersal pattern of lithic fragments (rocks ripped out from conduit and vent erosion during eruption) in the tephra layer resulting from this event, the eruption magnitude (M_e) has been estimated at 5.0 with an intensity of 11 propelling the plume to a maximum altitude of 33 kilometres [245]. The alarmed local population petitioned the colonial Resident in Bima, the main town in eastern Sumbawa (around 80 kilometres away), for help. As a result, a man by the name of Israël was dispatched to assess the situation, reaching the scene on 9 April.

13.2.2 Cataclysmic eruption

After a lull in activity, a second major eruption began around 19:00 local time on Monday 10 April. This event lasted less than three hours but was one of the most powerful Plinian eruptions yet analysed, with an estimated intensity of 11.5 and maximum column height of 43 kilometres. A valuable eyewitness account of this phase of the eruption was recorded by Owen Philipps, an officer dispatched by Raffles to Sumbawa with quantities of rice to 'proceed and adjust the delivery thereof, with instruction, at the same time, to ascertain, as nearly as possible, the local effects of the volcano'. While staying in Dompu, a town 60 kilometres to the southeast of Tambora, he met with the Rajah (local chief) of Sanggar who had miraculously survived the eruption:

> As the Rajah was himself a spectator of the later eruption, the following account which he gave me is perhaps more to be depended upon than any other I can possibly obtain. About seven p.m. on the 10th April, three distinct columns of flame burst forth near the top of the Tomboro mountain (all of them apparently within the verge of the crater), and after ascending separately to a very great height, their tops united in the air in a troubled and confused manner.

Up to this point, just 1.8 cubic kilometres of dense magma had been erupted in total. But during this powerful episode, the vent was eroding back and, at the same time, the water content in the magma was decreasing. These factors led to catastrophic collapse of the eruption column shortly before 20:00. The Rajah of Sanggar described the events thus:

> In a short time, the whole mountain next Sang'ir appeared like a body of liquid fire, extending itself in every direction. The fire and columns of flame continued to rage with unabated fury, until the darkness caused by the quantity of falling matter obscured it at about 8 p.m. Stones, at

> this time, fell very thick at Sang'ir; some of them as large as two fists, but generally not larger than walnuts.
>
> Between 9 and 10 p.m. ashes began to fall, and soon after a violent whirlwind ensued, which blew down nearly every house in the village of Sang'ir, carrying the ataps or roofs, and light parts away with it. In the part of Sang'ir adjoining Tomboro its effects were much more violent, tearing up by the roots the largest trees and carrying them into the air, together with men, horses, cattle, and whatever else came within its influence.
>
> The whirlwind lasted about an hour. No explosions were heard till the whirlwind had ceased, at about 11 a.m. From midnight till the evening of the 11th, they continued without intermission; after that time their violence moderated, and they were only heard at intervals, but the explosions did not cease entirely until the 15th July.

The 'whirlwind' surely describes pyroclastic currents, which razed the village of Sanggar. Over the next three or four days, several tens of cubic kilometres of magma gushed from the vent, much cascading downhill as pyroclastic currents, destroying Tambora village, and generating immense phoenix clouds that covered the region in ash (Figure 13.2). Israël, the official emissary from Bima, was among those killed. Ash fallout from the phoenix clouds accounts for around 40% of the mass of the resulting deposit [246]. In terms of physics, the eruption was not dissimilar from that of Pinatubo in 1991, though the more prolonged magma discharge of Tambora established a giant, stratospheric umbrella cloud covering an area of up to one million square kilometres (three times larger than Pinatubo's). The total

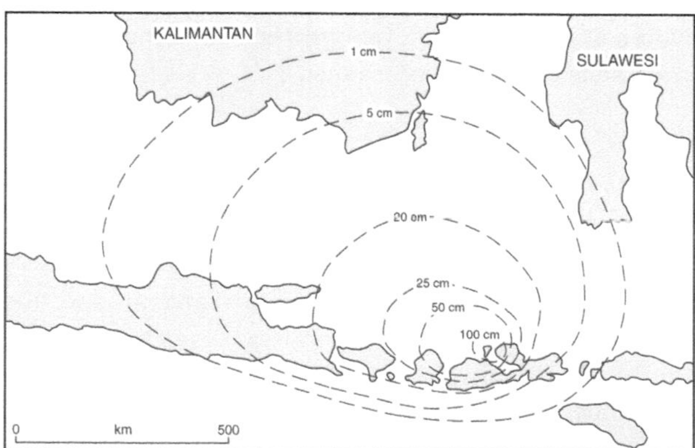

Figure 13.2 Isopach map for the fallout from the Tambora 1815 phoenix cloud based on contemporary reports. Modified from reference 246.

magnitude of the eruption (whose bulk magma composition is intermediate-trachyandesite) is estimated as M_e 6.9 [247].

The crew of a vessel sailing from Timor also appears to have witnessed the paroxysmal phase of the eruption. They reported that the foot of the volcano was 'engulfed in flames', and the summit encircled by 'dark clouds', with 'fire and flames' shooting out. Bima remained in complete darkness until noon on 12 April. Ash fall there was so heavy that most roofs collapsed, including that of the Resident's house. Curiously, the Resident appears to have singled out 14 April as the night of the most terrific explosions, which were like 'a heavy mortar' fired close to his ear. This could suggest the climactic phase of the eruption, though this seems unlikely in view of the geographically widespread reports of concussions heard on the night of 10 April. After leaving Bima, the *Benares* approached Tambora on 23 April, its crew witnessing the volcano following its convulsions:

> In passing it at the distance of about six miles, the summit was not visible, being enveloped in clouds of smoke and ashes, the sides smoking in several places, apparently from the lava which has flowed down it not being cooled; several streams have reached the sea; a very considerable one to the N.N.W. of the mountain, the course of which was plainly discernible, both from the black colour of the lava, contrasted with the ashes on each side of it, and the smoke which arose from every part of it.

The paroxysmal phase of the eruption drained so much magma from its reservoir that the volcano toppled in on itself, forming a six-kilometre-wide, one-kilometre-deep caldera. Today, Tambora's crater rim reaches only 2850 metres above sea level, easily surpassed by Rinjani volcano on Lombok island.

13.2.3 Darkness at noon

The distant effects of the eruption were astonishing. Explosions were heard through the night of 10–11 April in Bengkulu (1800 kilometres away), Mukomuko (2000 kilometres) and perhaps Trumon (2600 kilometres) on Sumatra. A report from the British garrisoned at Fort Marlborough in Bengkulu reads thus (revealing their general level of enlightenment about the country):

> A somewhat remarkable instance has occurred recently on this coast. A noise, as if firing of guns, has been heard, nearly at the same time, at different stations, lying between 2° 30' and 5° 30' of south latitude.
>
> The noise was heard by some individuals in this settlement, on the morning of the 11th April. In the course of that day, some deputies

> (or head men) of villages situated at a considerable distance towards the hills, came down, and reported that they had heard a continual heavy firing since the earliest dawn. It was feared that some feud had broken out into actual hostility, between villages in the interior. People were sent to make inquiries; but all was found tranquil.
>
> Our chiefs here, immediately decided, that it was only a contest between Jin (the very devil), with some of his awkward squad, and the manes of their departed ancestors, who had passed their period of probation in the mountain, and were in progress towards paradise.
>
> The most natural method of solving the difficulty, is, possibly, by supposing, that there must have been a violent eruption from some one of the numerous volcanoes amidst our stupendous mountains, centrally situated between Moco-Moco and Semanco. If so, we shall not, perhaps, ever learn the particulars; for we have very little communications with, and still less knowledge of, the mountaineers (though some of them are said to be Lord Monboddo's men, and have tails) or of the country they inhabit.

Across eastern Java the explosions were strong enough to shake houses. The Resident of Surakarta (Solo), quoted by Raffles, noted that:

> On Tuesday the 11th the reports were more frequent and violent through the whole day: one of the most powerful occurred in the afternoon about 2 o'clock, this was succeeded, for nearly an hour by a tremulous motion of the earth, distinctly indicated by the tremor of large window frames; another comparatively violent explosion occurred late in the afternoon, but the fall of dust was scarcely perceptible. The atmosphere appeared to be loaded with a thick vapour: the Sun was rarely visible, and only at short intervals appearing very obscurely behind a semitransparent substance.

A correspondent in Gresik (west of Surabaya) records that:

> In travelling through the district on the 13th, the appearances were described with very little variation from my account, and I am universally told that no one remembers, nor does their tradition record so tremendous an eruption. Some look upon it as typical of a change, of the re-establishment of the former government; others account for it in an easy way, by reference to the superstitious notions of their legendary tales, and say that the celebrated Nyai Loroh Kidul has been marrying one of her children on which occasion she has been firing salutes from her supernatural artillery. They call the ashes the dregs of her ammunition.

Effects were severe in the town of Banyuwangi on the eastern tip of Java, where ash accumulated to a thickness of 23 centimetres. Many places within a 600-kilometre radius remained pitch black for a day or two, accompanied by dramatic chilling of the air. Presumably, Raffles himself experienced the event, since it is recorded that ash fell at Buitenzorg (Bogor), the residence of the Governor, 60 kilometres south of Batavia (Jakarta). The commander of the *Benares*, at that time still anchored at Makassar, recorded that:

> During the night of the eleventh the firing was again heard but much louder; and towards morning the reports were in quick succession, and sometimes like three or four guns fired together, and so heavy, that they shook the ships, as they did the houses in the fort. Some of the reports seemed so near that I sent people to the mast-head to look out for the flashes, and immediately the day dawned, I weighed and stood to the southward, with a view to ascertaining the cause.

Spotting a ship approaching from the south around daybreak, the captain sent a party to meet it to glean further intelligence on the events. The Dutchman commanding that vessel had also heard the firing all night, and also on 5 April when he had been at Salajar Island (at the southeast tip of Sulawesi and 360 kilometres northeast from Tambora). Initially mistaking the concussions for an attack by pirates, soldiers had taken up battle positions in the fort! The commander of the *Benares* went ashore to meet the Resident of Makassar, Captain Wood, whose house had been shaken by some of the detonations. By 08:00, the sky was increasingly gloomy, and the ship's crew soon had to contend with heavy ash fallout:

> The face of the heavens to the southward and westward had assumed the most dismal and lowering aspect . . . At first it had the appearance of a very heavy squall or storm approaching, but as it came nearer it assumed a dusky red appearance, and continued to spread very fast . . . By ten it was so dark that I could scarcely discern the ship from the shore, though not a mile distant. I then returned on board.
>
> It was now evident that an eruption had taken place from some volcano, and that the air was filled with ashes or volcanic dust, which already began to fall on the decks. . . . the appearance altogether was truly awful and alarming. By noon, the light that had remained in the eastern part of the horizon disappeared, and complete darkness had covered the face of day . . .
>
> The darkness was so profound throughout the remainder of the day, that I never saw any thing equal to it in the darkest night; it was impossible to see your hand when held up close to the eye. The ashes continued to fall without intermission through the night. At six in the morning, when the

> sun ought to have been seen, it still continued as dark as ever; but at half past seven I had the satisfaction to perceive that the darkness evidently decreased, and by eight I could faintly discern objects on deck. From this time it began to get lighter very fast, and by half past nine the shore was distinguishable; the ashes falling in considerable quantities, though not so heavily as before. The appearance of the ship, when daylight returned, was most extraordinary; the masts, rigging, decks, and every part being covered with the falling matter; it had the appearance of a calcined pumice stone, nearly the colour of wood ashes; it lay in heaps of a foot in depth in many parts of the deck, and I am convinced several tons weight were thrown over board; for although a perfect impalpable powder or dust when it fell, it was, when compressed, of considerable weight; a pint measure filled with it weighed 12¼ oz.; it was perfectly tasteless, and did not affect the eyes with any painful sensations; it had a faint burning smell, but nothing like sulphur.

(Note that this report gives a measure of the density of the uncompacted ash of around 0.6 kilogrammes per cubic metre, a typical value for fresh ash deposits.) The Sun finally reappeared by noon on 12 April, though only faintly penetrating a stagnant, dusky atmosphere still charged with fine ash. These conditions prevailed up to 15 April. Meanwhile, the crew of the *Benares* was preparing the ship for passage:

> It took several days to clear the ship of the ashes; when mixed with water they formed a tenacious mud, difficult to be washed off. My chronometer stopped, owing, I imagine, to some particles of dust having penetrated into it.

From the estimated magnitude and duration of the major phase of the eruption, the mean intensity must have been around 11.7, launching the phoenix clouds into the lower stratosphere.

13.2.4 Tsunami

The pyroclastic currents engulfed the Sanggar peninsula and crossed the sea, reaching the small island of Moyo due west of the volcano. Tsunami were generated as the pyroclastic currents hit the water, and were observed across the region on the night of 10 April. The waves inundated Sanggar with a peak height of four metres around 22:00:

> The sea rose nearly twelve feet higher than it had ever been known to do before, and completely spoiled the only small spots of rice land in Sang'ir, sweeping away houses and every thing within its reach. [Lt. Philipps]

Considerable damage was also evident in Bima, as related to the commander of the *Benares* by the Resident:

> The wind was still during the whole time, but the sea uncommonly agitated. The waves rolled in upon the shore, and filled the lower part of the houses a foot deep; every prow and boat was forced from the anchorage, and driven on shore; several large prows are now lying a considerable distance above high water mark.

The tsunami hit Besuki in eastern Java (roughly 500 kilometres distant) by midnight (thus travelling around 250 kilometres per hour), and Surabaya, with a height of up to two metres, hurling boats inland.

Pyroclastic-current deposits enlarged the Sanggar peninsula by building deltas of tephra. Abundant carbonised tree trunks found preserved in the deposits testify to their high temperatures; inland the pumice deposits were hot enough to weld. In coastal areas, the deposits contain depressions several hundred metres across, which are probably explosion craters formed as the hot currents mixed with seawater. Since the volcano is almost completely surrounded by the sea, this interaction may have resulted in a nearly circular 40-kilometre-diameter curtain of rising ash.

13.2.5 Pumice rafts

Massive rafts of pumice, up to five kilometres across, and tree trunks floated across the Flores Sea. They still hindered navigation between Moyo and Sanggar three years after the eruption. The *Benares* encountered them when it first approached Sumbawa, and soon ran into difficulties:

> . . . on the eighteenth made the island of Sumbawa. On approaching the coast, passed through great quantities of pumice-stone floating on the sea, which at first had the appearance of shoals; so much so, that I hove to, and sent a boat to examine one, which at the distance of less than a mile I took for a dry sand bank, upwards of three miles in length, with black rocks upon several parts of it, concluding it to have been thrown up during the eruption. It proved to be a complete mass of pumice floating on the sea, with great numbers of large trunks of trees and logs among it, that appeared to be burnt and shivered as if blasted by lightning. The boat had much difficulty in pulling through it; and until we got into the entrance of Bima bay, the sea was literally covered with shoals of pumice and floating timber.

> On the nineteenth arrived in Bima bay . . . The shores of the bay had a most dreary appearance, being entirely covered with ashes, even up to the summit of the mountains. The perpendicular depth of the ashes, as measured in the vicinity of Bima town, I found to be three inches and three quarters.

The East India Company's ship *Fairlie*, crossing the Indian Ocean en route to Calcutta, encountered rafts between 1 and 3 October 1815, about 3600 kilometres west of Tambora, though the crew mistakenly attributed them to the eruption of a nearby submarine volcano:

> On the 1st of October our latitude at noon was 13 deg. 25 min. S. longitude 84 deg. 0 min. E. we observed quantities of stuff floating on the surface of the water, which had, to us, the appearance of seaweed, but were quite astonished to find it burnt cinders, evidently volcanic. The sea was covered with it during the two next days.

Tambora continued rumbling intermittently at least up to August 1819. A small cone and lava flow, called Doro Afi Toi, erupted within the caldera sometime between 1847 and 1913. A strong earthquake recorded on 13 January 1909 may be related to this activity. Since then all appears to have been quiet.

13.3 ATMOSPHERIC AND CLIMATE IMPACTS

The sulphur mass injected into the stratosphere by the eruption has been estimated by several independent methods including modelling of polar ice-core sulphate concentrations (Figure 4.7), petrological measurements of 1815 tephra and analysis of atmospheric optical phenomena. The results vary by an order of magnitude but the most recent estimates have revised the sulphur output down to around 30 megatonnes, three times as much as the eruption of Pinatubo in 1991; Section 3.1). It is difficult to partition this total between the different phases of the eruption and, in particular, to determine the fractions derived from Plinian versus phoenix clouds. Nevertheless, we would expect this much sulphur to have had strong impacts on regional and global climate.

Spectacular sunsets and twilights were observed in London in the summer of 1815. The twilight glows appeared orange or red near the horizon, purple to pink above, and were sometimes streaked with diverging dark bands. In the spring and summer of 1816 a persistent 'dry fog' was observed on the east coast of the United States. According to a report from New York, the fog reddened and dimmed the Sun such that sunspots were visible to the naked eye. Neither wind nor rain dispersed the 'fog', indicating its stratospheric height. It is claimed that some of the painter J. M. W. Turner's work in this period, characterised by lurid orange and red skies, was inspired by the volcanically induced stratospheric optics (see Section 5.3.1).

There is also abundant evidence for extreme weather in 1816, especially in the spring and summer in much of Europe and the Eastern Seaboard, where folkloric memories of 'the year without a summer' apparently live on. Contemporary meteorological data for the region, including measurements of temperature, precipitation and wind direction, paint a literally chilling picture [248]. On 4 June 1816, frosts were reported in Connecticut and, by the following day, most of New England was gripped by a cold front. On 6 June, snow fell in Albany, New York, and Dennysville, Maine, and there were killing frosts at Fairfield, Connecticut. Severe frosts had spread as far south as Trenton, New Jersey, the next day.

Many contemporary diaries document personal experiences of the extraordinary weather of 1816. Chauncey Jerome of Plymouth, Connecticut, writing in 1860, recalled 7 June dressed all day in thick woollens and an overcoat [249]:

> My hands got so cold that I was obliged to lay down my tools and put on a pair of mittens . . . On the 10th of June, my wife brought in some clothes that had been spread on the ground the night before, which were frozen stiff as in winter.

Weather logs for Williamstown, Massachusetts, compiled by Chester Dewey, then a professor at Williams college, also bear witness to the harsh conditions [250]:

> Frosts are extremely rare here in either of the summer months; but this year there was frost in each of them . . . June 6th the temperature about 44° through the day – snowed several times . . . June 7th no frost, but the ground frozen, and water frozen in many places . . . Moist earth was frozen half an inch thick, and could be raised from round Indian corn, the corn slipping through and standing unhurt. June 8th, some ice was seen in the morning . . .

Canada also experienced this remarkable cold wave in summer, with substantial snow fall settling on Montreal and Quebec City.

Contemporary meteorological observations and tree-ring chronologies have been used to reconstruct 1816 temperature, pressure and precipitation patterns across Europe. They reveal an exceptionally cold summer in an already cold decade [60]. Summer temperatures across much of western and central Europe were 1–2 °C cooler than the average for the period 1810–1819 and up to 3 °C cooler than the mean during 1951–1970. Rainfall was also high across most of Europe except the eastern Mediterranean during the summer of 1816. Tree rings confirm summer cooling both sides of the Atlantic (Figure 4.11). In one reconstruction of northern-hemisphere summer temperatures, 1816 is one of the very coldest of the past six centuries. The northern-hemisphere summers of 1817 and 1818 are also anomalously cold (5th and 22nd coldest in the 600-year record). The tree-ring-based mean northern-hemisphere (land and marine) surface-temperature anomalies in the summers of 1816, 1817 and 1818 are −0.51, −0.44 and −0.29 °C, respectively. Broadly, the northern-hemisphere patterns are consistent with those observed following Pinatubo: summer cooling, winter warming.

13.4 HUMAN TRAGEDY

The local impacts of such a devastating eruption are hard to imagine. We lack detailed evidence of the nature and numbers of casualties but do have some important contemporary sources that illustrate the terrible circumstances that befell not only Sumbawa but the neighbouring island of Lombok, and possibly Bali and beyond. The consequences for the local population were not more widely perceived until ships reached ports on Sumbawa. The *Benares* dropped anchor at Bima on 19 April, followed by the *Dispatch*, which was en route from Ambon in the Moluccan islands (Maluku), reaching Bima on 22 April.

The damage at Sanggar is recorded first by the crew of the *Dispatch*. They had mistaken Sanggar Bay for Bima and had, with difficulty, anchored and sent a boat ashore. A ship's officer met the Rajah of Sanggar and learnt that:

> . . . the greater part of the town and a number of people had been destroyed by the eruption; that the whole of his country was entirely desolate, and the crops destroyed. . . . a considerable distance from the shore being completely filled up with pumice-stones, ashes, and logs of timber; the houses appeared beaten down and covered with ashes.

Further destruction was witnessed in Sumbawa Besar (the present-day administrative capital of Sumbawa) by another ship's crew putting ashore there for water. They came across boats strewn inland by tsunami, and many corpses. Sailing away on 12 April, they were trapped all night by a pumice raft.

Raffles' agent, Lieutenant Philipps, then in Sumbawa on his mercy- and fact-finding mission, further details the desperate conditions for the survivors of the tragedy:

> On my trip towards the western part of the island, I passed through nearly the whole of Dompo and a considerable part of Bima. The extreme misery to which the inhabitants have been reduced is shocking to behold. There were still on the road side the remains of several corpses, and the marks of where many others had been interred: the villages almost entirely deserted and the houses fallen down, the surviving inhabitants having dispersed in search of food.

Poor hygiene, contaminated water and malnutrition quickly led to the spread of disease:

> In Dompo, the sole subsistence of the inhabitants for some time past has been the heads of the different species of palm, and the stalks of the papaya and plantain . . .
>
> Since the eruption, a violent diarrhoea has prevailed in Bima, Dompo, and Sang'ir, which has carried off a great number of people. It is supposed by the natives to have been caused by drinking water which has been impregnated with ashes; and horses have also died, in great numbers, from a similar complaint.
>
> The Rajah of Sang'ir came to wait on me at Dompo . . . The suffering of the people there appears, from his account, to be still greater than in Dompo. The famine has been so severe that even one of his own daughters died from hunger. I presented him with three coyangs of rice in your name, for which he appeared most truly thankful.
>
> A messenger who returned yesterday from Sambawa, relates that the fall of ashes has been heavier at Sambawa than on this side of the Gulf, and that an immense number of people have been starved: they are now parting with their horses and buffaloes for a half or quarter rupee's worth of rice or corn. The distress has, however, I trust, been alleviated by this time, as the brig, with sixty-three coyangs of rice, from Java, arrived there the day he was leaving.

The eruption had also released around 18 megatonnes of fluorine [247]. Much of this would have ended up on the ground adsorbed to tephra and it is likely that fluorosis afflicted any surviving livestock and potentially even the human population. Inhalation of fine ash is also likely to have caused widespread respiratory distress. Following the

Pinatubo eruption, several hundred evacuees died from diseases while in refugee camps. Deaths were primarily caused by measles (31%), diarrhoea (29%) and respiratory infections (22%), and reached 349 in the three months following the eruption [251]. Similar ailments must have afflicted many of Tambora's survivors, and probably with much higher mortality rates given the more limited public health response. Estimates of the number of casualties vary considerably and there are no really reliable data available. Philipps tried to collate some statistics:

> Of the whole villages of Tomboro, Tempo, containing about forty inhabitants, is the only one remaining. In Pekáté no vestige of a house is left: twenty-six of the people, who were at Sumbawa at the time, are the whole of the population who have escaped. From the most particular inquiries I have been able to make, there were certainly not fewer than twelve thousand individuals in Tomboro and Pekáté at the time of the eruption, of whom only five or six survive. The trees and herbage of every description, along the whole of the north and west sides of the peninsula, have been completely destroyed, with the exception of a high point of land near the spot where the village of Tomboro stood . . .

The naturalist Heinrich Zollinger, who visited Sumbawa in 1847, concluded that about 10,000 people were killed during the eruption, probably by pyroclastic currents, and that a further 38,000 died from starvation on Sumbawa. He estimated another 10,000 deaths from disease and hunger on Lombok. The most widely quoted statistics today are derived from a report published in 1949 [252], which calculated that there were 48,000 victims on Sumbawa and 44,000 on Lombok, around a third and a quarter of the islands' populations, respectively. Based on these estimates and adding a probable 25,000 victims on Bali, the historian Bernice de Jong Boers estimated a total death toll of 117,000. However, the estimates for Lombok have been described as 'entirely unfounded' [18]. Nevertheless, there is evidence for casualties as far as the east of Java (reported in the *Asiatic Journal*):

> The roofs of the houses at Bangeewanzee [Banyuwangi on the eastern tip of Java] fell in from the weight of the ashes . . . Most of the inhabitants of Sumbawa, who are not buried, must be starved, and as the crops in Bali and the east end of Java have been destroyed, they will also suffer considerably.

Damage was also considerable around Makassar. The commander of the *Benares* recorded . . .

> On going on shore . . . I found the face of the country covered to the depth of an inch and a quarter. Great fears were entertained for the crop of paddy that was on the ground, the young plants being completely beaten down and covered by [ash]; the fish, in the ponds . . . were killed, and floating on the surface, and many small birds lying dead on the ground.

Further afield in Java, crops were spared because 'the cultivators every where took the precaution to shake off the ashes from the growing paddy as they fell' demonstrating positive steps taken by the community to limit the effects of the disaster. Heavy rainfall on 17 April also helped to wash away the ash. According to Raffles, this 'prevented much injury to crops, and removed an appearance of epidemic disease, which was beginning to prevail. This was especially the case at Batavia, where, for two or three days preceding the rain, many persons were attacked with fever'.

We shall never know for sure but considering that there were certainly victims of famine and disease in Bali, and quite likely even in eastern Java, the death toll in Indonesia must have been of the order of 60,000–120,000. This makes the Tambora eruption the deadliest known in history.

The impacts on Sumbawa were prolonged and complex. Life must have been desperately harsh for the survivors. Apparently, thousands sold their children into slavery just to afford a small quantity of rice [253]. Thirty-two years after the eruption, Zollinger found the western end of the island still knee-high in ash. Along the western edge of Saleh Bay, once fertile lands were sterile, while inland, rice terraces and villages lay abandoned. Many refugees resettled in the interior rugged parts of the island where the ash washed away more rapidly (only to entomb further the lowlands). Many present-day villages in the interior of western Sumbawa are thought to date from the first decades after the eruption. The migrants first had to change their manner of rice cultivation from the wet *sawah* techniques they had employed for lowland paddies to the dry *ladang* swidden agriculture suited to their new environment. Later, the Sultan of Sumbawa embarked on a deliberate programme to resettle the lowland and coastal areas through immigration. Many of the new settlers came from southern Sulawesi as slaves. Today, the imprint of this assimilation is still apparent in the distinct customs, dialects and village names of the lowland communities.

These significant demographic changes had long-lasting political repercussions. In particular, the traditional village-focused ties of authority came apart in the traumatised, itinerant and politically fragmented post-disaster society. The Sultan reasserted his authority to some extent through his resettlement policy but he still faced many challenges both from within his clan and from rival states, resulting in enduring political turmoil and change. The catastrophe is still remembered in Sumbawa today: in one folk story, the eruption is seen as an act of divine retribution after the Rajah of Tambora had fed a pious Muslim dog meat and then killed him (referred to in the stanza at the front of the chapter).

13.5 GLOBAL REACH OF THE ERUPTION

The years 1816–1817 witnessed the worst famine in Europe and North America in over a century. The killing frosts in New England all but destroyed the main crops in 1816. Further south in North Carolina, towards the American grain-producing heartlands, the outcome of the harvest was summed up on 1 September as follows [254]:

> The very cool and dry weather in spring and summer hurt our grain fields badly, and it was with sorrowful and troubled hearts that we gathered our second crop of hay and our corn crop, which were so scanty that we reaped only a third of what we usually get, and wondered how we could subsist until next year's harvest.

The recurring cold conditions drastically shortened the growing season (Figure 13.3), ruining crops. Professor Dewey in Williamstown, Massachusetts reported:

> Cucumbers and other vegetables nearly destroyed. . . . June 10th, severe frost in the morning. . . . Ten days after the frost, the trees on the sides of the hills presented for miles the appearance of having been scorched. June 29th and 30th some frost. July 9th, frost, which killed parts of cucumbers. August 22nd, cucumbers killed by the frost. August 29th, severe frost. Some fields of Indian corn were killed on the low grounds, while that on the higher was unhurt. Very little Indian corn became ripe in the region.

Many livestock in New England died due to the ensuing lack of feed during the winter 1816–1817. The region was particularly vulnerable to disaster since the land was in any case climatologically marginal at that time for farming, especially with increasing competition from the midwestern United States and central Canada. In fact, Canada's population seems to have avoided serious social distress thanks to an

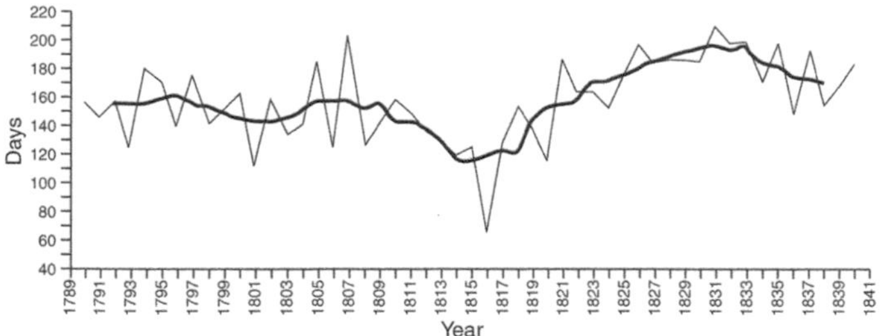

Figure 13.3 Length of growing season in southern Maine between 1790 and 1840. 1816 stands out as by far the shortest. Based on data in reference 249.

embargo on grain exports imposed between July and September 1816, as well as to lower demand on resources.

Meanwhile, in Europe, the summer of 1816 was also miserable, reflected in Lord Byron's oft-quoted poem, *Darkness*, an extract from which follows:

> I had a dream, which was not all a dream.
> The bright sun was extinguish'd, and the stars
> Did wander darkling in the eternal space,
> Rayless, and pathless, and the icy earth
> Swung blind and blackening in the moonless air;
> Morn came and went – and came, and brought no day,
> And men forgot their passions in the dread
> Of this their desolation; and all hearts
> Were chill'd into a selfish prayer for light

Byron wrote this whilst staying in Geneva, brooding gothically on the bloody turn of European history during the Napoleonic wars, and presumably complaining about the weather. The general mood and 'wet, ungenial summer' seems to have struck an even stronger chord with one of his companions in Geneva, Mary Godwin (soon to become Shelley). It was then that she conceived of the monster *Frankenstein*.

The cool temperatures and heavy rains resulted in failed harvests in parts of the western British Isles. Welsh families took to the road as refugees, begging for food. Famine was also prevalent in the north and southwest of Ireland following failures of wheat, oat and potato harvests. A report in the *Times* of London on 20 July 1816 stated bleakly that:

> Should the present wet weather continue, the corn will inevitably be laid, and the effects of such a calamity at such a time cannot be otherwise than ruinous to the farmers and even to the people at large.

In parts of continental Europe the situation was just as bad according to a report in the *Norfolk Chronicle* on the same day:

> Melancholy accounts have been received from all parts of the Continent of the unusual wetness of the season; property in consequence swept away by inundation and irretrievable injuries done to the vine yards and corn crops. In several provinces of Holland, the rich grass lands are all under water, and scarcity and high prices are naturally apprehended and dreaded. In France the interior of the country has suffered greatly from the floods and heavy rains.

The crisis was especially severe in the German lands, especially in the countryside and southwest, where food prices exceeded those in urban and northern areas. When Carl von Clausewitz toured the Prussian Rhineland in the spring of 1817, he found a bleak scene [255]:

> The author, who traveled on horseback through the Eifel region in spring 1817, where he passed the night in villages and little towns, often had a heartrending view of this misery, because these areas belong to the poorest classes in the land. He saw ruined figures, scarcely resembling men, prowling around the fields searching for food among the unharvested and already half rotten potatoes that never grew to maturity.

Popular reaction to the dire circumstances included demonstrations in grain markets and in front of bakeries, and in some regions riots, looting and arson. In May 1816, riots broke out in various parts of East Anglia in England, including Norfolk, Suffolk, Huntingdon and my home town of Cambridge. Acts of protest included destruction of threshing machines, and torching of barns and grain sheds. The insurrection culminated in the formation of marauding groups of rioters armed with heavy sticks studded with iron spikes and carrying flags proclaiming 'Bread or Blood'. The riot act was read, carrying with it the death penalty. Not far from Cambridge, angry mobs were also on the loose in the towns of Littleport and Ely. Dozens of the rioters were arrested and five charged with burglary and theft were hanged on 28 June following a short trial.

The historian John Post characterised these extraordinary upheavals as the 'last great subsistence crisis' to affect the Western world (see suggested reading in Appendix B). To quantify the effects, he collated reports of grain prices from both sides of the Atlantic. These show clearly the immediate impacts of the deteriorating harvests: as crops failed through the cold summers, costs at the market doubled between 1815 and 1817 (Figure 13.4). The highest price increase was four and a half times the 1815 index in Saint Gallen, Switzerland, in

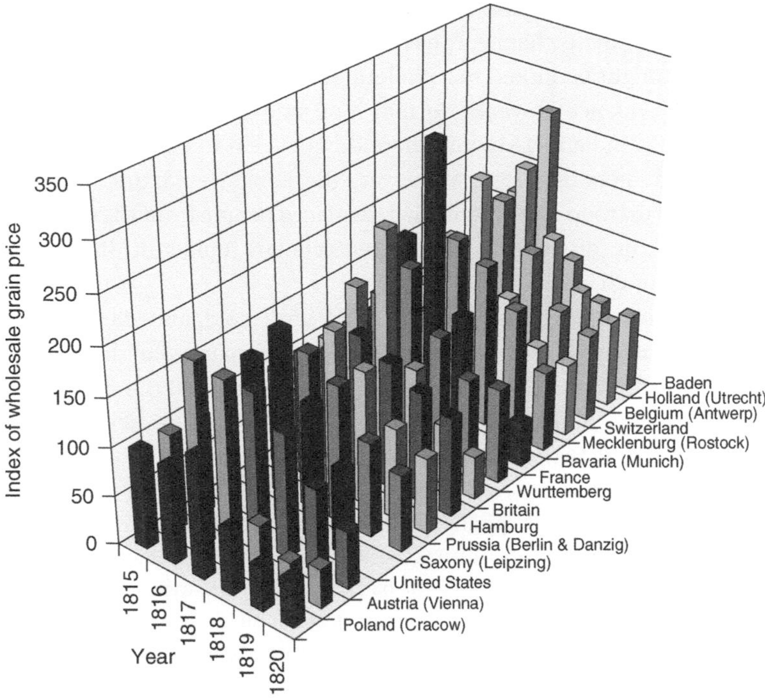

Figure 13.4 Indices of wholesale grain prices in Europe and North America from 1815–1820 based on national, regional or market averages. The figures are standardised to an index of 100 for all areas in 1815 and represent wheat prices except for Austria, Bavaria, Wurttemberg, Saxony and Poland, for which rye prices are quoted. They present a picture of deficient harvests and increased demand. Data from J. D. Post, 1977, see Appendix B.

June 1817. These rises should be viewed in the contemporary context: expenditure on cereals (especially bread) accounted for over half of a typical worker's budget. This economic earthquake, coming hot on the heels of the massive political dislocation and depression in manufacturing and trade following from the Napoleonic wars, affected millions of people. It drove the price of bread – the principal food in popular budgets – beyond the reach even of the majority employed at customary wage levels.

13.5.1 Disease, demography, economics and politics

Between 1816 and 1819, there were serious outbreaks of typhus and plague in various regions in Europe and the eastern Mediterranean –

there is a reasonable case to be made that these had a lot to do with Tambora's climate change. It has also been suggested that Tambora's eruption might be linked to a cholera outbreak in Bengal in 1816 but here the evidence is weak. Normal monsoon rainfall seems to have prevailed in India in 1816 and there do not seem to be any reports of drought or flooding [256]. Moreover, the Bengal epidemic is better attributed to troop movements that displaced people from the endemic source of the disease. The first great cholera epidemic in Europe occurred much later, in 1831–1832.

Typhus epidemics are associated with cold and damp, poor hygiene and louse infestations. They often accompany war and famine. A doctor working in the Belfast Fever Hospital detailed the epidemiology of typhus outbreaks in Ireland in a strikingly prescient letter written in April 1818 as follows [257]:

> I consider the predisposing causes of the present Epidemic to have been the great and universal distress occasioned among the poorer classes, by the scarcity which followed the bad harvest of 1816, together with the depressed state of trade and manufactures of all kinds. The low condition of bodily health arising from the deficiency and bad quality of the food; the want of cleanliness both in the persons and dwellings of the poor . . . I consider the contagion to have been rapidly spread by the numbers wandering about in search of subsistence, and also by the establishments for the distribution of soup and other provisions among the poor where multitudes were crowded together, many of whom must have come from infected houses, or were perhaps even laboring under the early stages of the disease.

Harty estimated that around 800,000 people were infected during the epidemic in Ireland, and that 44,300 'perished from the joint ravages of Famine, Dysentery, and Fever'.

The typhus epidemic visited almost every town and village in England, and was reported in many cities in Scotland. Europe in 1816 was already mired in social and economic distress at the end of a quarter century of war, not least due to the sudden appearance of several million men on the labour market after their demobilisation from the military. These disorders created vulnerability, and the combination of colder weather, migration due to famine and a malnourished population would have represented conditions conducive to an epidemic. Post acknowledged the role of the Napoleonic wars in the economic, demographic and political convulsions:

> The harvest failures of 1816 came at an inopportune time, superimposing a subsistence crisis not only on a stagnating economy but also on a Western society still unsettled as a residue of the war years.

Across Europe, Post identified increased death rates, lower birth rates and delayed marriages in the grip of the crisis. Mortality rates between 1816 and 1817 increased by 4% in France, 6% in Prussia, 10% in Bohemia, 22% in Switzerland and 25% in Tuscany. He concluded that actual starvation (usually attended by dysentery and enteric disease) was the cause in the countries with some of the highest mortality-rate increases.

He also examined 'whether famine progressed automatically to mortality . . . or whether famine and epidemic disease operated essentially independently of one another'. Firstly, to meet the demand for grain, traders increasingly sourced supplies from ports where plague-carrying rats or fleas were endemic. Thus, hunger ushered in disease. But Post also found that the places with the highest food costs were not necessarily those with highest mortality rates. What mattered more, he argued, was how efficient the authorities were in distributing relief to the poor and administering quarantines.

According to Post, more 'rationalised' and modernised city and state administrations in France, Germany, Britain and the Netherlands coped better with the crisis through improved communications, hygiene and through taking direct steps to combat famine (and the associated political unrest, which perhaps was their greater concern). In France, vigorous measures were introduced to control prices and the government succeeded, at least in the cities, in keeping bread affordable. Meanwhile, in Britain, the parish poor relief system set up during the Napoleonic wars distributed means-tested wage supplements to alleviate the plight of the rural population. Both countries raised funds to import food from Russia and the Baltic states. One long-term outcome was the introduction of welfare reforms and renewed recognition of the need for cooperation at the state level [258].

A knock-on effect of the high food prices was to put people out of work: the demand for manufactured goods plummeted as people spent what little money they had on expensive food. This in turn shifted purchasing power from small to large grain producers, who were themselves unable or unwilling to buy textiles or other goods. Farmers reacted to the high prices by expanding production, which led to excessive supplies and a collapse in prices in 1818–1819. The misery prompted many thousands of Europeans to emigrate to North America, while thousands of Americans abandoned New England and the Carolinas in search of a better life in the West.

Post's boldest claim, perhaps, was to suggest that the series of social and economic reversals experienced in these years initiated a political shift to the right, especially in France and Germany, as authorities reacted to the popular unrest.

> The few weak liberal compromises that made a timid appearance in 1815–16 vanished in a climate of mistrust and fear in 1819–20.

Anti-Semitic violence erupted in August 1819 in Würzburg in northern Bavaria in the so-called hep-hep riots, triggered during an academic ceremony when an aged professor who had recently come out in favour of civic rights for Jews had to run for his life as students assaulted him. The violence quickly spread to Frankfurt, Hamburg and Heidelberg, and as far as Copenhagen, Amsterdam and Kraków. Post concluded that the political reaction of European governments to the epidemics, commercial depression, unemployment, hunger, rioting, widespread begging and vagrancy, and large-scale emigration 'can be seen as the last link in a connected sequence of events that began with the meteorological effects of the volcanic dust clouds of 1815'.

13.6 SUMMARY

The M_e 6.9 Tambora 1815 eruption represents a fascinating case study. Even as we approach the two hundredth anniversary of the event, it still offers important lessons for the future, some of which are explored in the final chapter. The eruption happened just at the conclusion of what some would regard the 'first world war'. Close to the sailing routes between the Spice Islands and the centres of colonial power in southeast Asia, and with its own natural resources traded widely, it was not 'off the map'. Thus, even though there wasn't anything like the kind of scientific enquiry that followed the Krakatau eruption of 1883, we know a good deal about Tambora from vivid eyewitness testimony. There are good accounts, too, of the various attempts of central and local authorities to alleviate the misery of a deeply traumatised island community and of more sinister machinations to restructure and rebuild the society. In the short term, the aid response was totally inadequate to meet the scale of the disaster, and most of the casualties died from a combination of starvation and pestilence. Around 60,000 lives were lost in Sumbawa and Lombok alone, and perhaps, in reality, more than 100,000 perished as a result of the eruption.

The eruption pumped around 30 megatonnes of sulphur into the stratosphere. This generated up to 100 megatonnes of sulphuric acid aerosol, noted across the globe in the guise of various atmospheric optical phenomena. Evidence linking the eruption to recorded extreme weather events in Europe and North America is found in the overall hemispheric pattern of summer cooling and winter warming and storminess. The cold climate that prevailed prior to 1815 may even have been partly induced by a Krakatau-sized stratospheric aerosol veil cast over the Earth by an unknown equatorial eruption in 1809 (Figure 4.7). This raises the issue of the potential of repeated high-sulphur-yield eruptions being able to cause more prolonged climate change than seen after isolated eruptions like those of Krakatau in 1883 or Pinatubo in 1991 (see also Section 11.2.3).

The global climatic perturbations arising from the Tambora aerosol veil led to poor harvests in 1816 and 1817. According to the historian John Post, the ensuing food insecurity, combined with serious political and social dislocation at the conclusion of the Napoleonic wars, contributed to outbreaks of typhus, dysentery and other ailments in many parts of Europe. While the use of documentary evidence for reconstructing past climate is not without problems, Post's arguments are compelling and perhaps, if history could be re-run from April 1815 without Tambora going off, the world might be a different place today.

14

Volcanic catastrophe risk

One can make two diametrically opposed kinds of assertion: global risks inspire paralyzing terror, or: global risks create new room for action.

U. Beck, *World Risk Society* (2009) [259]

... the diversity of individuals and institutions that draw on reservoirs of practices, knowledge, values, and worldviews ... is crucial for preparing ... for change, building resilience, and for coping with surprises.

W. N. Adger *et al.*, *Science* (2005) [260]

Each earthquake, flood and famine disaster tragically re-confirms how limited are the achievements that humankind has made in reducing catastrophic losses of life, especially in poorer countries. While actuaries do succeed in turning uncertainty into profit, their industry is rendered impotent in the face of the most severe global risks. Nevertheless, our exploration of some of the most significant volcanic eruptions, and the evidence for human resilience in terms of both the anticipation and reality of disaster, provide important lessons for understanding and managing volcanic risk in the future. It would be extremely difficult, for instance, to envision the impacts of a future super-eruption without knowledge of the pyroclastic deposits of Yellowstone, Toba or Taupo. Meanwhile, both archaeology and history offer rich insights into human ecology in the context of volcanic threat and impact. In this chapter, we draw these themes together to consider:

- What are the worst-case eruption scenarios for the future?
- How likely are they to happen?
- Are they really as apocalyptic as some would have us believe?
- Should we be concerned enough to take action to mitigate their adverse consequences and, if so, what kind of action?

The dangers of global climate change and their association with the fossil-fuel economy are widely accepted to be among the gravest concerns facing humanity. One proposal to combat global warming is the seeding of aerosol in the stratosphere via sustained release of sulphur dioxide: in effect, a permanent explosive volcanic eruption (but without the ash). This would cool the Earth's surface by deflecting some sunlight back to space, much as Pinatubo's aerosol did. Another 'geo-engineering' concept considers fertilising the oceans with iron. This would mimic the capacity of volcanic ash fallout to stimulate algal growth (thereby extracting carbon dioxide from the atmosphere). Put both of these endeavours together and you really have something a lot like an intense and permanent volcanic eruption!

If greenhouse gas emissions do proceed unabated without carbon capture or geo-engineering solutions applied, global warming is expected to manifest some of its strongest effects in the polar regions. In extreme scenarios, vast bodies of ice such as the West Antarctic Ice Sheet could disintegrate, dramatically raising global sea level. Both sea-level change and deglaciation alter the magnitude and distribution of stresses on the Earth's crust and upper mantle. Beneath some volcanoes, these changes could enhance rates of melting at the magmatic source and modify conditions in magma reservoirs, acting overall to increase global eruption rates.

Given what the record of volcanism and atmospheric change has revealed concerning the feedbacks between volcanism and climate, we also ask in this final chapter:

- Are geo-engineering projects to combat global warming a good idea?
- Could future climate change trigger real volcanic eruptions?

This chapter thus covers considerable ground and necessarily compromises on detail and sophistication. However, the aim is to highlight the value of understanding the past record of volcanism and its diverse consequences for climate and human ecology as we shape and formulate our anticipation of, and response to, future volcanic extremes.

14.1 THREE CATASTROPHE SCENARIOS

Chapter 2 reviewed the wide range of volcanic phenomena and their associated hazards. We have seen how the largest-scale impacts of eruptions arise in two ways: heavy ash fall (which, in the case of a super-eruption, can cover much of a continent); and global climate change arising from generation of a sulphuric acid aerosol veil in the stratosphere. What translates these phenomena into risks is their potential juxtaposition in space and time with lives and livelihoods, infrastructure, the economy and environment. By one count, around 10% of the global population lives within 100 kilometres of an active or dormant volcano.

Drawing on what we have understood from the past record of volcanism, this section identifies three eruption scenarios capable of massive human consequences in the future: a Plinian or caldera-forming eruption directly affecting a major city; a very large lava eruption precipitating a volcanic pollution crisis; and a super-eruption (Table 14.1). These are by no means the only volcanic catastrophe scenarios, nor will all their potential ramifications be considered; rather the key political and planning implications for risk management will be framed for subsequent discussion.

14.1.1 Explosive eruption near a major city

Rapid and unregulated urbanisation in many countries has proliferated exposure to a wide range of environmental and technological hazards – over half the world's people now live in cities. Continued population growth in coastal regions, in particular, has aggravated volcanic jeopardy because of the preponderance of explosive volcanoes near oceanic plate boundaries, as well as the long-range hazard of volcanogenic tsunami. While smaller and sparser communities possess their own vulnerabilities and must not be side-lined in risk management, cities can be uniquely disaster-prone because of their population density and the complex interdependencies of urban environments and lifelines (power and water utilities, communications, transportation and public health needs). Hundreds of thousands of people can be imperilled within a few days if a handful of city water-treatment and pumping facilities power down.

At least a dozen major cities and metropolitan areas worldwide lie within range (according to wind directions) of Plinian eruptions of nearby volcanoes. These include some of the largest metropolises and capitals in the world, including México City, Jakarta, Manila and Tōkyō. For example, the last lies just 100 kilometres from the iconic volcano, Mt Fuji (Figure 14.1). The Tōkyō Metropolitan Area has a population exceeding 30 million and ranks top in the global list of economic hotspots (its GDP is around US$1.5 trillion). Mt Fuji's last eruptions took place in 1707–1708 and involved Plinian events of magnitude

Table 14.1 *Volcanic catastrophe risk scenarios*

Scenario	Where could it happen?	Risks	Precedents and analogies	Maximum protection
Plinian eruption (M_e 6–7) of volcano near major urban areas	Tōkyō, Manila, Naples ...	Very heavy loss of life; economic and production losses; breakdown of civil order; evacuation panic	St Pierre 1902, Pompeii 79 CE, Armero 1985, Hiroshima and Nagasaki 1945, New Orleans 2005, Port-au-Prince 2010, Lisbon 1755; great fires (e.g. Edo 1657; San Francisco 1906)	Warning and evacuation of city
Energetic tsunami triggered by explosive or caldera-forming eruption (M_e 6–7) or through large volcanic landslide	Pacific Ocean, Mediterranean Sea, Indian Ocean, Caribbean Sea, Atlantic Ocean	Very heavy losses of life, property and economy in coastal areas	Santorini Minoan eruption (Chapter 9), Indian Ocean tsunami 2004	Warning and evacuation of areas susceptible to inundation

Table 14.1 *(cont.)*

Scenario	Where could it happen?	Risks	Precedents and analogies	Maximum protection
Super-eruption (M_e 8–9)	USA, Indonesia, New Zealand, central Andes . . .	Very heavy loss of life; global climate change (drought, famine); reduced agricultural capacity; greatly impaired transport and communications at continental scale; geopolitical instability and breakdown of civil order	Younger Toba Tuff eruption (Chapter 8), 'Little Ice Age', Tambora 1815 (Chapter 13); war and famine; floods in China 1931	Warning and evacuation on scale of >> 100 km radius around volcano; total cooperation at international level
Large lava eruption (or massive intrusion) ($M_e > 6$)	Cities and metropolitan areas in Europe, Japan, México . . .	Public health crisis; air-transportation chaos, heavy economic losses; agricultural and farming losses from acid and fluoride deposition	Laki 1783 (Chapter 12), Miyake-jima 2000, London 1952, Moscow 2010	Management of anthropogenic pollution sources, public health response

Figure 14.1 The Nihon-bashi bridge in Edo (Tōkyō) seen in one of Hokusai's *Thirty-six Views of Mount Fuji* (*circa* 1827).

M_e 5.3. During a powerful basaltic phase of the eruption the prevailing winds carried ash clouds towards Edo (the historical name of Tōkyō, which was probably even then the largest city in the world despite the devastating impacts of the magnitude 8.2 'Genroku' earthquake of 1703). Tephra accumulated to a depth of about ten centimetres in Edo, sufficient to cause serious disruption. A similar episode today could blanket the Tōkyō Metropolitan Area in around a cubic kilometre of ash.

Thankfully, Tōkyō is not threatened by pyroclastic currents (except in the most extreme imaginable circumstances). A densely populated region around Vesuvius is, however, and much effort has been applied to evaluating and managing the associated risk (Figure 14.2). In this case, an immense body of historical and geological evidence helps in imagining future eruption scenarios, beginning with Pliny the Younger's account of the events in August of 79 CE (Figure 14.3). This is being increasingly supplemented by evaluation of risks by structural engineers and public health experts, and implementation of risk reduction measures by political and civil protection authorities in the region. Major challenges include: the multiplicity and combination of hazards that can be expected during a single eruptive episode (tephra fallout, pyroclastic currents, mudflows and earthquakes); the fact that Vesuvius is not the only volcano that poses a threat to the region; and the size of the population at risk (nearly six million).

One approach adopted by the civil protection authorities and volcanologists concerned with risk assessment for Vesuvius itself is to focus on a reference event, believed to represent the worst probable

Figure 14.2 Aerial view of the Neapolitan region. Part of the scalloped caldera of Campi Flegrei is visible in the lower part of the photograph (neatly infilled by urban sprawl), and in the centre distance rises Vesuvius (recognisable from its double peak formed by the Somma caldera wall and the historical Vesuvius cone). Naples lies on the coast between Campi Flegrei and Vesuvius.

Figure 14.3 The archetypes for the impacts of volcanism on urban areas are Pompeii and Herculaneum, of course (Plymouth on Montserrat is the latest to be added to their ranks). This view today from the western end of the excavations at Herculaneum takes in the ancient alleyways, homes and public buildings, evoking urbanite vignettes *circa* 79 CE. In the foreground are the boat houses wherein dozens of people hoping for escape were killed by pyroclastic currents that swept through the town. In the distance – just seven kilometres away – is the cone of Vesuvius, dormant since 1944. But in between, and rising precisely from the edge of the excavated area, is a mass of multi-storey dwellings and an urban zone whose population density rivals that of Hong Kong. If ever a lesson from history is stared in the face it is here!

eruption. This hypothetical scenario is based on the largest eruption of the volcano since 79 CE, which took place in 1631, claiming up to one thousand lives. Since all has been quiet on the mountain since 1944, renewed activity would very likely involve fracturing of rock as pathways re-open between subterranean magma bodies and the surface. This would generate earthquakes, some of which could be damaging. The cumulative effects of seismic shaking would increasingly take their toll on the building stock within a few kilometres of the volcano. Eventually, hydrovolcanic explosions would be likely once the conduit reached the water table; these would also be associated with strong earthquakes. With a fully opened magma conduit, sustained eruption would follow, with a tephra column climbing up to 20 kilometres above sea level. Tephra fallout would result in extensive building damage but the location of the worst-affected zones would depend strongly on prevailing wind directions at various heights in the atmosphere. While it is possible to model expected building damage statistically, the fact that one cannot know until the day of eruption which way the wind will carry tephra represents a major uncertainty in risk control (this point was illustrated vividly in the case of the Eyjafjallajökull eruption in 2010). A total area of around 1500 square kilometres, home to nearly two million people, could be affected to varying degrees by the immediate impacts of a future eruption of Vesuvius.

As occurred during the 1631 event, the eruption column would likely founder at some juncture, generating pyroclastic currents that would rush down the flanks of the volcano, potentially affecting an area of more than two hundred square kilometres. They would even be funnelled to some extent by streets and buildings as has been deduced from studies of the deposits that buried Pompeii. The dynamic force of the flows and their entrained large chunks of volcanic and building debris, combined with their searing temperatures, would devastate the building stock and threaten the lives of all caught either inside or outside. There would be a further wider area at risk from firestorms as hot tephra would ignite buildings and fuel tanks and supplies.

Of course, there will be wider impacts and other hazards of future activity, including major disruption to aviation and shipping. Structural failure of the volcanic edifice could also occur, triggering landslides and, potentially, caldera formation. In the waning stages of an eruption, mixing of ground water and magma within the conduit could produce hydrovolcanic explosions, while rainfall would redistribute the fresh tephra deposits, generating destructive mudflows. And it is feasible that the next eruption culminates in something bigger than a re-run of 1631. Prehistoric eruptions of Vesuvius including two major Plinian events (the Pomici di Base and Avellino Pumice eruptions, 22,000 and 4300 years ago, respectively) were accompanied by debris avalanches that reached the coast. Such evidence raises implications for tsunami hazard associated with any future large-scale flank failure of Vesuvius.

No one knows when Vesuvius will reawaken, reflecting the difficulty of reconciling purely statistical analyses of eruption frequency and clustering with what the geological record tells us about magma-reservoir processes. There have been numerous estimates, some just a matter of 20 years hence, but the focus now seems to be on using whatever quiet period is available to make good preparations rather than endlessly re-iterate the prognostication.

14.1.2 Volcanogenic pollution crisis

The threats posed by volcanogenic pollution (Section 2.8) have been encountered in several of the case studies in this book. They include the effects of sulphurous gases and particles on respiratory and cardiovascular health; fluoride contamination of pasture; acid deposition on soils and vegetation; the disruption of air transport by aerosol and ash clouds; and associated perturbations of regional climate. Arguably, the most notable example from the historical period is the 1783–4 eruption of Laki in Iceland and its long-range impacts across Europe (Chapter 12). The 2010 Eyjafjallajökull eruption provided just a taste of what Icelandic volcanism is capable of. Given that the Laki episode was preceded by the even larger Eldgjá eruption in the late tenth century, there is a chance (very crudely) of order 1–10% of a comparable (e.g. $M_e > 6$) lava flood in the next century. Notwithstanding the achievements of the modern world, the next Laki-type event could still have severe ramifications for public health in Europe.

In fact, vulnerability to such a crisis may well be greater today than it was in the late eighteenth century because of modern levels of air pollution (mostly from road traffic) in major European cities, especially in summer when air-quality standards are often exceeded. Adding volcanic fumes on top of air already burdened with high levels of fine particles, ozone and nitrogen oxides would certainly aggravate public health risk. For instance, emissions from the chronically degassing volcanoes Popocatépetl (México) and Miyake-jima (Japan) have sporadically reached México City and Tōkyō, respectively, increasing levels of acid aerosol and sulphur dioxide.

The association between air pollution and health has been firmly established through many robust statistical studies of both chronic and acute exposures. Differences between the composition of anthropogenic air pollutants and the constituents of volcanic plume make it difficult to extrapolate risks of a volcanogenic pollution crisis from the conventional studies. Variations in the general health, lifestyles and behaviours of exposed populations, and the likely complex spatial and temporal patterns of exposure to air pollution from a distant eruption, also hamper evaluations of the associated health risk. Nevertheless, a scenario combining volcanogenic and urban pollution with a summer heat wave (recalling the prolonged hot weather that coincided with the intense phase of the Laki eruption in the summer of 1783) would plausibly result in considerable loss of life.

Acid deposition can also result in significant damage to farmland and natural ecosystems. Evidence for the damaging effects of Laki's acid deposition on vegetation is found in many contemporary anecdotes and diaries from various sources across Europe. Leaves were variously described as withered, scorched, burnt or dried and premature shedding was observed (Section 12.2.1). Some cereal crops appear to have been damaged in Britain while fish kills were reported in Scotland. However, we lack robust estimates of the timing or spatial distribution of acid loadings across Europe during the Laki episode. In respect of future eruptions of similar magnitude, we can only suggest that the uptake of volcanic pollutants by foliage and via soil acidification will depend on many factors including season, levels and duration of exposure to pollutants, topography and meteorology. Whether doses could exceed critical loads for agriculture is uncertain.

Without question, however, a future eruption on the scale of Laki's would significantly affect aviation across the North Atlantic. Air travel has become so much a part of leisure and commerce that relying on alternative transportation harks back to the era of steam ships and trains. In this light, it is easier to understand why the intersection between geology, meteorology and aviation witnessed during the Eyjafjallajökull eruption in 2010 inspired such chaos. At the height of the ash-cloud crisis in April of that year, it was indeed ships that came to the rescue! Thousands of stranded passengers were repatriated from Spain by a British Royal Navy warship and a luxury cruise liner (which had been diverted ahead of its inaugural voyage).

Eyjafjallajökull began erupting on 20 March 2010. The initial activity consisted of basaltic fire fountains and lava flows whose accessibility spawned a vigorous but ephemeral volcano tourism industry. Then, on 14 April, a swarm of earthquakes presaged a dramatic switch in eruptive style. Magma of intermediate composition had encroached on the glaciated summit crater of Eyjafjallajökull and its explosive interaction with the ice cap yielded extremely fine-grained ash that was blasted violently into the atmosphere by explosions of steam and magmatic gases (Figure 1.6). The prevailing weather system dispersed the floury tephra in the direction of Europe. As civil-aviation administrations closed airspace, people living near airports such as London's Heathrow enjoyed the sound of birdsong in their gardens for the first time.

Following the near disaster in Alaskan airspace during the 1989 Redoubt eruption (Section 2.1.1), a worldwide network of Volcanic Ash Advisory Centres (VAACs) was established and tasked with monitoring the threat. But while geologists, meteorologists and the International Civil Aviation Organization took the issue seriously, and despite experience of prior flight-corridor restrictions that were imposed due to Icelandic eruptions, it appears that both the aviation industry and political authorities were taken by surprise by Eyjafjallajökull's outburst. The aviation administrations adopted the precautionary principle and closed down vast tracts of airspace. Around a hundred thousand flights were cancelled, stranding ten million passengers. The economic cost is almost impossible to gauge accurately but estimates range up to US$2–3 billion. Given the limited knowledge available at the time concerning the intensity of the eruption and altitude of ash clouds, the atmospheric concentrations of ash downwind and the threshold tolerances of jet-engines to ash, the official response is understandable.

There are important differences in products and eruptive style between Laki's 1783 fissure eruption and Eyjafjallajökull's summit explosions. These include the higher altitude of Eyjafjallajökull's crater, the magma–ice interaction and intermediate-composition magma – all these factors (combined with the contemporary meteorological patterns) contributed to the aviation hazard experienced in spring of 2010. On the other hand, the eighteenth-century Laki eruption involved multiple episodes over a period of eight months, and appears to have been capable of lofting fine material into the lower stratosphere (i.e. at cruising altitude of trans-Atlantic air traffic). A future Laki-style episode would, therefore, almost certainly have major repercussions for aviation.

14.1.3 'Super-eruption' scenario

Super-eruptions have been described as 'the ultimate geologic hazard' [261]. The last one (that we know of) took place 26,000 years ago at Taupo volcano on the North Island of New Zealand. Before that (73,000 years ago) there was Toba (Chapter 8). We have no useful constraints on when the next eruption of such great magnitude will take place, though such events may have a return period as short as 50,000 years [5]. We can also speculate where the next super-eruption might occur based on prior track record. Thus, Toba, Yellowstone or Taupo all remain possible candidates, raising the possibility of future super-eruptions in the tropics or mid latitudes of both north and south hemispheres. However, there is nothing to rule out the appearance of a debutant on the super-volcano scene. (Recent seismological investigation at Yellowstone revealed a shallow magma reservoir of more than 4300 cubic kilometres in total volume, a third of which is molten [262]. Whether this magma is eruptible is another question.)

But how would a super-eruption in the not-too-distant future affect us? No detailed scholastic attempts have yet been made to model risk scenarios of a super-eruption (though the BBC (British Broadcasting Corporation) rose to the task in a 'factual drama' titled *Supervolcano*, shown in 2005). We can speculate, however, on the generic effects based on what is known from the deposits of past super-eruptions and experiences of smaller, recent eruptions. Pyroclastic currents from an M_e 8 or 9 event would extend up to 100 kilometres or more radially from the volcano. These would bury an area of a few tens of thousands of square kilometres in incandescent pumice to a depth of up to 200 metres. Even the history of the relatively small 1902 eruption of Mont Pelée demonstrates that the chances of surviving exposure to pyroclastic currents can be vanishingly small (Section 2.3.1). Beyond the fringes of a super-eruption's ignimbrite deposits, there may be some chances of initial survival, though many would subsequently die from exposure, burns and other injuries. Additionally, many structures and facilities not physically buried in pyroclastic material will nevertheless have been incinerated as a result of hot pumice igniting flammable material.

A much wider zone will be affected by substantial ash fallout. Where more than half a metre or so of ash accumulates, there would be substantial damage to buildings, likely claiming many victims. Those caught out in the open during ash fallout would fare no better as it would be difficult to avoid inhaling large quantities of ash. Movement through thick ash falls in absolute darkness beneath the ash clouds would be all but impossible and probably futile unless shelter was very close at hand. Power lines would be brought down and wireless telecommunications compromised by the effects of airborne ash.

The duration of past super-eruptions is poorly constrained. Estimates based on meticulous field studies of ignimbrites and associated Plinian tephra deposits that constitute the 760,000-year-old, M_e 8.1, Bishop Tuff (the eruption of which formed Long Valley caldera in eastern California) suggest the whole episode lasted just a matter of days [263]. Other super-eruptions seem to have involved intermittent bursts of activity that spanned a few years. Either way, air quality and visibility during, and for a long time after, a future super-eruption would be poor across a vast area due to windblown ash. Close to ground zero, this would compound the difficulties of mounting search-and-rescue operations. Meanwhile, many roads and railways would be impassable due to ash fallout, and aviation hazardous due to re-suspended ash. Valleys would be inundated by mudflows, bringing devastation to communities that had otherwise escaped the worst of the ash fallout.

Farming and agriculture would be severely affected where ash has accumulated to depths of more than a few centimetres. Past eruptions of Yellowstone blanketed much of North America in tephra. Looting would be rife and food and water resources would become increasingly scarce and contaminated. More vulnerable members of the community, such as the elderly and infirm, would be particularly at risk. Hospitals would be crippled by power cuts and shortages of medicines. In arid regions, water could become scarce very quickly and, even where rain does fall, fluoride and other chemicals leached from the ash might contaminate surface water. Meanwhile, pumped water supplies would dwindle due to power shortages. A monumental effort would be required to respond to such a disaster and mitigate loss of life in the most affected regions, especially to stem outbreaks of infectious disease.

For a super-eruption outside the tropics, the human impacts would likely vary according to the season of eruption – a winter-time scenario might prove deadlier as freezing temperatures would compound the exposure of millions of people. On the other hand, a summer eruption would probably result in a deeper hemispheric climate response and immediate impacts on crops and livestock, presenting a greater threat to food security. Agriculture in the zone affected by ash fallout would likely be curtailed for years, and potentially decades, due to deficits in rainfall that can be expected after a super-eruption. This would perturb regional climate for much longer than the typical few years' residence of sulphuric acid aerosol in the stratosphere. Silicic ash is as bright as snow and hence reflects sunlight that would otherwise be absorbed by vegetation and soils. One climate model for the effects of ash cover from an eruption of Yellowstone predicted surface cooling of around 5 °C throughout the year for North America [264].

The most extreme scenarios for super-eruptions include the demise of technological civilisation and have led Michael Rampino (an expert on geological catastrophes) to suggest that volcanism might represent a universal constraint on the number of extra-terrestrial civilisations [265]. He argues that the impacts of such a large eruption on worldwide climate would severely reduce global agricultural yields:

> The major effect on civilization would be through collapse of agriculture as a result of the loss of one or more growing seasons . . . The result could be widespread starvation, famine, disease, social unrest, financial collapse, and severe damage to the underpinnings of civilization.

Just a glimpse of the potential political and economic instability that could arise from worldwide reductions in grain supply can be gauged from the impacts of the 2007–2008 global food crisis. More recently, in 2010, Russia (one of the world's largest producers of wheat, rye and barley) banned exports of grain after a period of drought and a succession of wildfires devastated crops (sparking panic in the international grain market). Rampino concludes that 'volcanic supereruptions pose a real threat to civilization, and efforts to predict and mitigate volcanic climatic disasters should be contemplated seriously'.

Among the mitigation measures he discusses is the establishment of a reserve of foodstuffs adequate to see the global population through years of agricultural crisis. Another more extreme solution is the establishment of an 'interplanetary repository for terrestrial civilization'. This would involve 'the transfer of human civilization, along with all technological and cultural information, to other places in the Solar System for safekeeping'. Such a repository would provide 'a backup system for the planet, fostering recovery of terrestrial civilization in the wake of global disasters such as asteroid collisions or volcanic catastrophes'. Studies of many terrestrial and aquatic ecosystems have shown that lowered resilience can contribute to catastrophic disturbance, and may provide some analogy to the vulnerability of human civilisation. On the other hand, Rampino's arguments for the existential catastrophe are strongly based on inferred climatic consequences that are far more extreme than suggested by recent modelling efforts (Section 8.3).

14.2 RISK CONTROL

Understanding and managing risk - especially towards the 'very low probability–very high impact' end of the spectrum - is profoundly complex. In the case of volcanic risk, this is true enough for M_e 5 and 6 events let alone 7 and above. Does the precautionary principle of 'better safe than sorry' have any value in managing such a statistically extremely unlikely occurrence as a super-eruption? In this section, I raise these philosophical, economic and perhaps even existential matters in giving some thought to mitigating the adverse effects of the eruption scenarios described in the preceding section. But before considering specific control measures to reduce the harmful impacts of volcanic catastrophes, it may be helpful to review a few key concepts in risk analysis and management. A good starting point is a definition of 'risk' itself (in the context of either geophysical or meteorological hazards). This one is taken from the UN International Strategy for Disaster Reduction:

> Risk is a description and measure of potential harmful consequences to life and health, livelihoods, property, the economy or environment. It results from the interactions between natural hazards and human conditions for a given area and reference period.

Implicit in this statement is that risk is the anticipation of a hazard event by a threatened community. It also reflects the community's ability to withstand the consequences of the hazard event. Furthermore, the references to 'measure', 'given area' and 'reference period' introduce a probabilistic dimension to risk assessment. For example, we can ask what the chances are that the specified damaging phenomenon (hazard) will affect a given region, within a stated time period (the next century, say). Such an approach frames risk within a context that can guide policy formulation. It can also support cost–benefit analyses regarding, for instance, measures that might be invested in to reduce risk (economic risk projections may help to motivate political authorities) or for decision-making in respect of evacuation announcements (Section 14.2.3).

At the very least, conceptualising risk in terms of contributory and compensating factors provides an initial step towards risk reduction. Risk increases if the hazard phenomenon is of greater magnitude or occurs more frequently, or if the number of people exposed to the hazard is larger. It also increases if the threatened community is more vulnerable to the effects of the hazard phenomenon. For instance, people might suffer more in the event of a volcanic disaster if their health or the durability of their homes are already compromised by chronic poverty. On the other hand, we have seen repeatedly in this book how societies can be resilient and even opportunistic in the face of volcanic crisis – ancient communities in Costa Rica and New Britain re-established and adapted their social organisation and material culture once natural-habitat recovery allowed stricken areas to be resettled. This is not meant in any way to diminish recognition of the suffering inflicted during volcanic disasters but it is certainly true that, at the broader scale, geographically and temporally, societies can respond in more or less favourable ways. Risk can represent both a threat and an opportunity (though often in an antagonistic way, in which those who reap the rewards are not necessarily those who suffer the consequences).

14.2.1 Vulnerability versus resilience

Over the last three decades, there has been an acceleration of research into assessment of the physical, social, political, economic and environmental conditions that make a community more or less susceptible to the impacts of hazards, and the application of that knowledge to policy formulation for risk reduction. Among the first to recognise the need for social vulnerability to be considered in volcanic disaster preparedness were the eminent geographer and hazards researcher Gilbert White and his colleague Eugene Haas. In the mid 1970s, and before any stirrings at Mt St Helens, they considered that mitigation of the threat of volcanic risks in the USA required greater attention to be paid 'to the social implications of great eruptions so that public policies can be designed with confidence. Special studies on human response ... should be expanded ... by building upon previous geological studies' [266].

The first-order correlation between hazard occurrence and impact is that disasters disproportionately affect the poorest, least powerful people. Poverty goes hand-in-hand with living and working in the most exposed places, scarce access to education or public health services, political exclusion and the least means for escape from catastrophe (for instance a car and a place to go) or for recovery after a disaster (such as financial savings or insurance). Disasters and extremes of inequality are, thus, chronically and deplorably part of a vicious circle. In poorer countries prone to frequent hazard occurrences, so much resource is put into disaster response that little is left over for social and economic development or risk management. There is even evidence that a kind of perverse disaster economy underlies the way both donor and recipient countries behave [267]. For instance, while the humanitarian crisis mounted in Pakistan after the 2010 floods, the West's aid response at times seemed motivated more by politics and 'the war on terror' than moral imperative. That said, a badly managed volcanic crisis that impinges on a poorly prepared society can exact an egalitarian calamity, as was the case in St Pierre, Martinique, in 1902 (when pyroclastic currents killed all but two of the city's residents).

The reciprocal to vulnerability is 'resilience', a term that owes much to the field of ecology, and which broadly refers to the capacity of a social–ecological system to cope with disturbance (disaster) so as to recover its essential behaviours [260]:

> Resilience [is] the capacity of linked social–ecological systems to absorb recurrent disturbances such as hurricanes or floods so as to retain essential structures, processes, and feedbacks ... Resilience reflects the degree to which a complex adaptive system is capable of self-organization (versus lack of organization or organization forced by external factors) and the degree to which the system can build capacity for learning and adaptation ... Part of this capacity lies in the regenerative ability of ecosystems and their capability in the face of change to continue to deliver resources and ecosystem services that are essential for human livelihoods and societal development. The concept of resilience is a profound shift in traditional perspectives, which attempt to control changes in systems that are assumed to be stable, to a more realistic viewpoint aimed at sustaining and enhancing the capacity of social–ecological systems to adapt to uncertainty and surprise.

No one has yet proposed a viable means to switch a volcano off or to prevent it erupting, but there are measures that can be taken to mitigate certain volcanic risks. These include the diversion of lava or mudflows using various forms of earth barriers, or the strengthening of roofs to withstand tephra fallout. However, for larger-scale volcanic threats, substantial gains in risk mitigation are likely to accrue from fostering community resilience and interdependence, and by facilitating means and support for mass evacuation when circumstances demand this last resort. This reflects, in part, the recognition that provincial, national and international authorities can be left essentially powerless to help citizens in worst-case scenarios – they may even have collapsed themselves. When capital cities are devastated by earthquakes such as that which struck Haiti in 2010, and hospitals, police stations, government palaces and water treatment, power and communications facilities are in ruins, the survivors may be left to cope on their own for days before help arrives. Even when evacuation measures are in place, they may not apply to all members of a community. This dichotomy was shockingly apparent in the aftermath of Hurricane Katrina's 2005 landfall in Louisiana.

14.2.2 Risk analysis and the problem of extremes

Over the past decade or so, volcanic risk analysis has shifted away from deterministic methods to what has been termed an 'evidence-based' or 'science-based' approach [268], a concept which owes much to the fields of medicine and law. At its core is the evaluation of volcanic hazards by experts, the probabilistic modelling of plausible risk scenarios (accounting for the exposure and vulnerability of threatened communities) and communication of the forecasts and their associated uncertainties to the authorities responsible for civil protection and to those at risk. The 'evidence' can come from many sources including volcano monitoring networks, petrological investigations, tephrostratigraphy, tephrochronology and meteorology (to determine, for instance, ash hazard scenarios or the triggering of events by rainfall), though providing measures for vulnerability can be far more challenging. Such an approach was followed in the case of Pinatubo's 1991 reawakening and during the prolonged volcanic crisis on Montserrat, and it is being applied to planning for a re-awakening of Vesuvius, Campi Flegrei, Teide (Tenerife) and the Auckland Volcanic Field. However, for the events capable of the regional- to global-scale impacts (such as major volcanogenic pollution episodes, volcanogenic tsunami and supereruptions) an international approach to mitigation is demanded.

The problem of modelling extreme events is that we don't always know about them until they happen and, in any case, lack the information to constrain the 'tails' of probability distributions or the frequency of 'outliers' hiding within them [269]. One branch of statistics that attempts to address these issues is 'extreme value theory' and it has many applications from managing risk in financial markets to flood-defence engineering. Rather than worry about the 'normal' values in a dataset, extreme value theory models the tail of a distribution. This can help to describe the relationship between magnitude and frequency or characterise the size of the largest-possible events. To provide a simplistic example of the inadequacy of conventional statistics, and in particular the failure of the normal (or Gaussian) distribution to capture extremes, consider the distribution of counts of volcano fatalities in the twentieth century (for which more or less complete data are available) [6]. For this period, 259 eruptions resulted in recorded fatalities and the average number of deaths per eruption is 354. If the normal distribution is applied to this dataset, it would suggest a vanishingly small probability of events claiming more than, say, 10,000 lives. And yet, in the twentieth century, two eruptions alone (of Mont Pelée in 1902 and Nevado del Ruíz in 1985) accounted for 50,000 deaths (over half those recorded).

A related problem concerns the return period (average recurrence interval) of very large eruptions. Firstly, only a handful of M_e 7 eruptions have been documented for the last 10,000 years – are these adequate to define the return period of events of this size when we don't know how many comparable cases lie undetected? As for eruptions of M_e 8 and up, five are known to have occurred in the last million years. Here, again, extreme value theory can help to constrain recurrence rates given the strong likelihood that the record is incomplete. By one calculation, M_e 8 and greater eruptions may recur as often as every 50,000 years, on average [5]. That yields a crude estimate of the chances of a M_e 8 eruption occurring in the next century of as high as one in 500. Given that one would have to buy a weekly lottery ticket for the best part of a quarter of a million years to expect (on average) to take home the jackpot once, the odds of getting to experience a super-eruption are, perhaps, not as low as one might have guessed.

But should we do something about such a possibility? The greatest problem in tackling catastrophe risk towards the most extreme end of the scale is that the extent of expected damage starts to escalate faster than the improbability of the events. It then matters greatly what kind of extreme value distribution is employed in the modelling (there are several to choose from) and the exact numerical coefficients used to parameterise it. In the realm of volcanology, we really have little clue how M_e 8 and larger events are distributed in space and time, except that it appears likely that different mechanisms come into play for explosive eruptions larger than M_e 7 (perhaps linked to the formation of calderas) [270]. Given the difficulties in modelling the events themselves, we should not expect to deduce with any ease generalised models for the human response to such extremes. Nevertheless, it is possible and important to make a start.

Among the closest analogies to the threat of a super-eruption that have been examined in detail are the collision of an asteroid with the Earth's surface, global climate change and associated 'tipping points', and nuclear conflict. There is an important and relevant distinction, of course, between these events, since one is uncontrolled (possibly uncontrollable), another enhanced by human activities and controllable, and the third unquestionably self-inflicted. In the case of climate change, considerable efforts have gone into modelling different greenhouse-gas-emissions trajectories and the associated climatic, ecological and socioeconomic impacts. It is instructive to follow the climate-change analogy further, not least since artificial volcanism has been proposed as a means to combat the effects of greenhouse gas emissions (Section 14.3.1).

Forecasts of the impacts of greenhouse-gas emissions and climate change are based on 'integrated assessment models' that combine economic and Earth-system sectors. The economic part of such models considers capital, labour, gross domestic product and so on, with associated carbon energy costs and greenhouse-emission projections. The geophysical part, on the other hand, captures relationships between greenhouse-gas emissions, radiative forcing, and climate-change induced damage such as flooding due to sea-level rise. The integration of economic and Earth-system components allows modelling of different climate-change policies to find optimal balances between damage limitation and its associated costs in terms of restricted carbon emissions.

An important limitation to such investigations of climate change and carbon policy is that they rely on information that is currently unavailable, and which may only become available when it is too late. Two central uncertainties, in particular, pose challenges to quantitative economic analysis. Firstly, how bad will climate get – that is, how much will temperatures rise as a result of increasing atmospheric abundances of greenhouse gases? Secondly, how bad will the worsening climate be for the economy – that is, how much economic damage will be caused by increased temperatures and associated physical impacts of climate change? Notwithstanding these difficulties, such models remain useful for examining the effects of different policies such as market-based emissions trading or carbon taxes, and also for exploring the circumstances that can lead to catastrophic economic outcomes. Given the existing capacity to model the climate change wrought by super-eruptions (Section 8.3.1), it might be feasible to adapt the integrated assessment model approach to consider impact scenarios for future Toba- or Yellowstone-like events. At the very least, there are parallels between aspects of climate economics and the global risk of a super-eruption that merit exploration.

14.2.3 Preparation and response

In the case of earthquakes, prediction (in a way that is helpful for emergency evacuation) remains a distant prospect – earthquakes are often instantaneous events (albeit followed by aftershocks) without currently discernible short-term warning signs. On the other hand, structural engineering has come a long way in developing the means by which buildings can be reinforced to withstand ground shaking: earthquakes don't kill people, buildings do. With volcanism it is the other way around: we do have means for forecasting eruptions and renewal of activity; and the crescendo towards climactic events often takes months. This allows time for evacuation, even if the damage to buildings (for instance from pyroclastic currents) can be very hard to engineer against. This partly explains the benefits of preparedness for volcanic eruptions. But it also exposes a major problem for decision-making during volcanic crises – their very prolongation, possibly for years, can inure the threatened community to an increasing volcanic risk while the scientists in charge are growing over-secure in their ability to forecast events. Much therefore needs to be done to develop the decision-support tools and frameworks that will enable civil protection authorities to define the extent of an exclusion zone and to call an evacuation despite the high levels of scientific uncertainty that surround any interpretation of volcano-monitoring data.

The eruption of El Chichón in México in 1982 tragically underlines the importance of preparedness and effective emergency response. The initial eruptions on 28 March 1982 prompted the spontaneous and confused flight of most villagers in the area. Over the following days, the volcano quietened and the authorities allowed many refugees to return home. Around 2000 people died during the climactic eruptions on 3 and 4 April. Another example is the 1985 eruption of Nevado del Ruíz in Colombia, which resulted in nearly 23,000 deaths due to inundation of settled areas by mudflows. This loss of life could readily have been averted had even a rudimentary (but effective) alarm system and evacuation plan been adopted at the time. It was at least an hour and a half after the onset of the eruption that the mudflows reached the town of Armero. A single telephone call from an observer high up in the valley could have allowed time for many, perhaps most, people to escape to higher ground.

One of the important issues in mitigating the effects of any new eruption of a dormant volcano is the timescale of unrest or precursory activity – such as earthquake swarms and increased gas emissions – ahead of a climactic event. In the case of Tambora (1815) it was about three years; at Pinatubo (1991) less than three months. Unfortunately, in the case of historic and ancient eruptions discussed in this book we have almost no information on whether there was a progression in such signals, for example, in respect of frequency and magnitude of explosive activity, felt earthquakes, or style or composition of gas emissions. Nor is it clear whether larger eruptions are presaged by longer build-up periods. It could even be the other way around: wide magma bodies in the crust, such as that which fed the Campanian Ignimbrite (Section 9.1), appear capable of very rapid eruption following an initial destabilisation [271].

In any case, it is not a good idea to wait for a crisis and then, within a matter of weeks, try to develop all the tools needed to evaluate risks, make decisions and communicate them to the authorities and public. On the other hand, progressive seismic shaking prior to a large explosive eruption might be just what is needed to prompt a spontaneous evacuation (as may have been the case at Akrotiri shortly before the Minoan eruption of Santorini; Section 9.3.2). But another problem that arises is that if volcanic unrest is prolonged then populations may become inured to the threat of a large event. In such circumstances, which have been observed in many volcanic crises in the past, the public can lose interest and concern.

There are many components to effective volcanic risk management and an adequate treatment would require a book in itself. Thus, my aim in this subsection is only to highlight the importance of preparedness in disaster response (in respect of mass evacuation and public health), and in particular of globalisation of preparedness for volcanic disasters. I will also refer intermittently to the volcanic catastrophe risk scenarios sketched in Section 14.1. The starting point for this discussion is that tools for sustained monitoring, risk analysis and decision-making are assumed to be in place, and that scientists have built bridges for effective and trusted communication with the political, military, civil protection, health and law enforcement authorities that will be responsible for preparing, testing and, if and when necessary, launching emergency plans. It also presumes good communications exist between stakeholders and the media.

Preparing for emergency evacuation

An effective long-term measure for reducing the human impacts of hazard events is to relocate people at risk. Of the many towns and cities that have been devastated by earthquakes or eruptions in the past, few were subsequently abandoned for good: León Vieja (Nicaragua) and Plymouth (Montserrat) come to mind. However, there is a plan to relocate the capital of Iran from Tehran in anticipation of the consequences of a future large earthquake. And, for a while at least, the idea of reducing exposure to volcanic threat was being taken seriously close to Vesuvius. In 2003, a fund was set up by the Campanian regional government to incentivise up to 15% of the more than half a million residents of the high-risk 'Red Zone' on the flanks of the volcano to move elsewhere. However, there was limited interest in the scheme and it was difficult to ensure that buildings vacated by the scheme would not simply be recycled into residential use. As things stand in Campania and elsewhere, timely mass evacuation is likely to remain the most effective means for drastically reducing loss of life during major volcanic emergencies.

Successful emergency evacuation plans must encompass communications, transport, lodging, medical care and protection of assets. But perhaps most essential is a compliant community – in fact multiple compliant communities since somewhere outside of the impact zone will be receiving guests. This can require decades of risk communication. In the municipalities threatened by Vesuvius, public awareness was being raised proactively via exhibitions and school education. The latter was seen as a particularly effective means of outreach since a relatively small number of teachers introducing information on hazards and risks into their curricula could reach tens of thousands of pupils. They, in turn, might then discuss what they had learned at school with their families, in theory spreading positive messages about risk reduction and individual and community responsibilities. Unfortunately, such imaginative programmes require constant support and promotion; it appears that efforts around Vesuvius have lapsed in recent years. Bearing in mind that the last eruption was in 1944, most of the people living on or near the volcano today have no direct experience of Vesuvius as an active volcano.

Mass evacuation takes time. But how much time is realistically going to be available between the call for evacuation and the anticipated eruption? Here the planner's dilemma is how late to leave it to call an evacuation: while confidence in the decision may increase as the evidence from monitoring stations for impending eruption mounts, the time window for carrying out the order narrows. This is again where probabilistic approaches suggest a way forward in deciding whether or not to evacuate a population [272]. Nevertheless, it is difficult to predict the time it would take to evacuate a large population – while small exercises can be carried out to develop the theory, there is little basis for scaling up the results to an evacuation of an entire municipality or city where traffic congestion can reveal very nonlinear behaviours. (A small-scale evacuation drill in 2006 in which a hundred residents of the 'Red Zone' on Vesuvius took part ran into the problem of a less than enthusiastic citizenry (it was raining on the day) and standstill traffic on the Naples–Pompeii highway.) A related issue is community tolerance of false alarms. Given the immense complexities in diagnosing a volcano's behaviour, even with access to state-of-the-art surveillance data, there is an inherent and large margin of error in eruption forecasts. In this respect, school education and media reporting could usefully aim to enhance public understanding of the nature and implications of scientific uncertainty.

The emergency plan for Vesuvius is reckoned on a warning time of two weeks but this timescale is very far from certain. It has also been based on a presumed orderly evacuation by zones according to proximity of the threat. Unfortunately, panic is a commonly observed behaviour during mass evacuations and needs to be accounted for in evacuation models and plans since it can be life threatening in itself. Another issue is noncompliance with evacuation orders. This becomes a very thorny issue at the interface between civic responsibility and human rights. Does someone have the right to ignore wilfully an officially sanctioned evacuation call? If so, does that individual have the right to make the same decision on behalf of his or her children?

Emergency plans should be flexible to accommodate very dynamic situations and unexpected complications, such as adverse weather conditions coinciding with the volcanic disaster (as witnessed in Luzon during the 1991 eruption of Pinatubo). They also crucially need to embrace the public health (including psychological health) issues associated with displacement of large numbers of people and the prospects of mass casualties in the event of an evacuation order that comes too late. Although they are relevant at all scales of disaster, emergency plans become exponentially complex for very high-impact scenarios when hundreds of thousands or millions of people are concerned.

In the case of an air-pollution disaster such as that outlined in Section 14.1.2, there is scope to put in place proactive public health measures. While large-scale volcanic clouds cannot be diverted, it is possible to reduce other sources of pollutants by the imposition, for instance, of traffic bans in cities. People can be advised to remain indoors to reduce exposure to acid aerosol if present in the atmosphere, and disposable dust masks can be used to limit inhalation of fine particles. Further research would be valuable to improve epidemiological analysis and modelling in this context: for example concerning the risk factors associated with variable exposure to mixtures of volcanic acid gases and aerosol. There is also scope to investigate the threat to agriculture of acid deposition from volcanic plumes during acute pollution episodes. Measures such as agricultural liming might well help, at least to combat the effects of soil acidification.

International frameworks

What constitutes an 'eruption that shook the world' depends on how one defines 'world'. For a Montserratian who remembers the island before Soufrière Hills volcano's rejuvenation in 1995 it must surely seem that the world has been ending in slow-motion while the eruption progressively displaced the population, annulled the island's capital and left two-thirds of the island off-limits. It is hard to imagine a more devastating turn of events for a small island community. And yet the eruption – though prolonged – has been relatively small (a total of M_e 5.4 for the period 1995–2010). The management of volcanic threats needs to be capable of operation at every scale from local to global. For the super-eruption and volcanogenic pollution scenarios sketched above, international cooperation is a prerequisite for effective risk management. (In the BBC's Supervolcano scenario in which Yellowstone erupts, one scene depicts Mexican border authorities overwhelmed by an exodus of North Americans!) Even the 'city on a volcano' narrative could well demand an international response at political, bureaucratic and technical levels, and hence benefit from an international level of preparedness.

The economic impacts of flight restrictions in effect during the 2010 Eyjafjallajökull eruption and the immense publicity that accompanied them further highlight the wider regional and global economic risks of volcanism and the limitations of applying the precautionary approach to risk reduction. It is clear that we need to innovate and integrate tools for volcano monitoring, modelling, risk analysis and decision support; to improve and harmonise procedures for emergency management; and reduce social vulnerability where it is most extreme. While trivial in terms of eruption magnitude, the Eyjafjallajökull crisis provides a powerful lesson in the benefits of international multi-agency, public and private sector cooperation. In a very severe crisis, many of the institutions relied on no longer work. The modern economy is extraordinarily intricate and, as the financial crisis since 2007 has exposed, not as robust as many had presumed. Monopolies and protectionism continue to distort trade and reduce options to evade crisis. These issues suggest the need to build more redundancy into global systems of trade and human mobility.

This brings us to the crux of the challenge of volcano catastrophe risk management: what international institutional models and organisations are relevant to addressing the threat?

In fact, there are numerous existing platforms to consider, and prior experience to draw on. They include UNESCO's Intergovernmental Oceanographic Commission (http://ioc-unesco.org) which promotes (among other things) regional tsunami warning networks, development of national disaster plans, community awareness programmes and evacuation drills. Another is the World Meteorological Organisation (WMO, another agency of the United Nations). Reading its 'vision and mission' statement it is easy to transpose 'meteorological' and 'volcanological', and instructive, therefore, to reproduce it here (http://www.wmo.int):

> The vision of WMO is to provide world leadership in expertise and international cooperation in weather, climate, hydrology and water resources and related environmental issues and thereby contribute to the safety and well-being of people throughout the world and to the economic benefit of all nations.
>
> The mission of WMO is to:
>
> Facilitate worldwide cooperation in the establishment of networks of stations for the making of meteorological observations as well as hydrological and other geophysical observations related to meteorology, and to promote the establishment and maintenance of centres charged with the provision of meteorological and related services;
>
> Promote the establishment and maintenance of systems for the rapid exchange of meteorological and related information;
>
> Promote standardization of meteorological and related observations and to ensure the uniform publication of observations and statistics;
>
> Further the application of meteorology to aviation, shipping, water problems, agriculture and other human activities;
>
> Promote activities in operational hydrology and to further close cooperation between Meteorological and Hydrological Services;

> Encourage research and training in meteorology and, as appropriate, in related fields, and to assist in coordinating the international aspects of such research and training.

The primary international body in volcanology is the International Association of Volcanology and Chemistry of the Earth's Interior (IAVCEI; http://www.iavcei.org), and while it does provide an umbrella for a number of commissions (including the World Organization of Volcano Observatories (WOVO; http://www.wovo.org)) it does not presently take a coordinating role in volcano monitoring and eruption warning. It is not going to do it anytime soon, either, but it took the national meteorological services decades to iron out the role of the WMO so there is no reason to dismiss the idea as impossible. The IAVCEI is constituted (via the intervening International Union of Geodesy and Geophysics (IUGG)) under the International Council for Science (ICSU), a non-governmental organisation founded in 1931. Then there are the Volcanic Ash Advisory Centres (VAACs), set up under the umbrella of the International Civil Aviation Organisation (ICAO), also a UN offshoot.

The purpose is not to proliferate an alphabet soup of acronyms but to demonstrate that there are existing pertinent agencies from which to build a global vision 'to provide world leadership in expertise and international cooperation in volcanology and related environmental issues and thereby contribute to the safety and well-being of people throughout the world and to the economic benefit of all nations'. If national governments do prove capable of reaching international consensus, resolve and cooperation to limit future damage of global warming then humankind will have demonstrated further its capacity to develop the institutional and public frameworks of discourse and action to cope with the threat and the occurrence even of a super-eruption.

Finally, if the more extreme eruption scenarios do come to pass, one agency in particular has considerable expertise in handling mass evacuation crises, namely the office of the United Nations High Commissioner for Refugees (UNHCR).

14.3 GLOBAL WARMING: FAKE VOLCANOES AND REAL ERUPTIONS

There are three main kinds of response to the impacts of global warming due to greenhouse-gas emissions. The first is to put up with climate change and adapt to shifting and shrinking biomes, rising sea level and more frequent meteorological extremes; the second is to curb use of fossil fuels; the third is to tamper further with the climate system – so-called geo-engineering. The last approach is particularly appropriate for us to consider since some of the techniques under consideration have a lot in common with volcanism. This section therefore reviews the proposal to seed aerosol in the stratosphere (to reduce the receipt of solar radiation at the Earth's surface). Lastly, since unchecked global warming melts glaciers and ice sheets and raises sea level, it is appropriate to consider whether it might thereby influence the frequency of eruptions.

14.3.1 Stratospheric geo-engineering

The main driver of Earth's climate is the uneven distribution of solar heating and radiative cooling from equator to poles. Sunlight that is absorbed in the atmosphere or at the Earth's surface is transformed into various forms of energy, redistributed between land, sea, ice and air, and ultimately returned to space via infrared radiation. One of the processes that would counter global warming at the Earth's surface due to greenhouse gases would be to reduce incoming sunlight. Proposals to achieve this include the installation of a network of giant solar reflectors in space, and the artificial generation of sulphate aerosols in the stratosphere. The latter approach, which could be achieved by burning sulphur or hydrogen sulphide in the stratosphere, would be a lot like a volcanic eruption such as that of Pinatubo in 1991, the difference being that the geo-engineering aim would require the equivalent of a 'Pinatubo' every four years or so! Indeed, the climatologist Tom Wigley (from the National Center for Atmospheric Research in Boulder, Colorado) suggested that '... the Mount Pinatubo eruption ... caused detectable short-term cooling ... but did not seriously disrupt the climate system. Deliberately adding aerosols or aerosol precursors to the stratosphere, so that the loading is similar to the maximum loading from the Mount Pinatubo eruption, should therefore present minimal climate risks' [273]. But how far does our knowledge of eruptions such as Pinatubo help us to reduce uncertainty in the potential impacts (and unintended consequences) of geo-engineering projects? Is this really a good idea?

In its favour is that of the various suggested geo-engineering schemes, the generation of stratospheric sulphate aerosol is generally considered the most feasible and cheapest. (In fact another geo-engineering approach that is somewhat analogous to volcanism is the proposed fertilisation of parts of the oceans with nutrients such as iron so as to stimulate plankton blooms, which extract carbon dioxide from the atmosphere.) Systems that have been envisioned for delivering artificial aerosol to the stratosphere include repetitive firing of artillery shells, a giant tower kept upright by the use of balloons with a hose to pump sulphur gases into the stratosphere and an 'exotic' space elevator linked to a geostationary satellite! But the use of military aircraft capable of stratospheric flight and either pumping sulphur gases from a tank or burning sulphur-rich fuel would be the cheapest option at a cost of a few billion dollars per year (less than 1% of the USA's annual budget for the military).

However, there are some significant unattractive side-effects of stratospheric geo-engineering, several of which can be surmised from the impacts of the Pinatubo eruption (Section 3.2) and numerous climate models of the effects of explosive volcanism. The first is reduced rainfall in Africa and Asia, which could significantly disrupt food production. The second is substantial ozone depletion. Rutgers University climatologist Alan Robock has raised several further arguments against stratospheric geo-engineering. These include: the reductions that could be expected in solar power production; an end to blue skies; the fact that it would do nothing to combat the acidification of the oceans that arises from high levels of carbon dioxide in the atmosphere (Section 6.5); and the moral hazard that the prospect of geo-engineering coming to the rescue of an overheating planet will thwart efforts to reach international consensus on reductions in carbon emissions [274]. Another effect would be particularly unwelcome to astronomers – all their efforts to build mountain-top observatories (to get above air pollution) would be compromised by a deliberate source of atmospheric pollution!

Some climate scientists have argued nevertheless that geo-engineering has to be considered an option to counteract dangerous climate change. They urge that research into the various feasible schemes should proceed rapidly, and should include real-world testing. Wigley, for example, maintains that 'If mitigation fails, either because we've underestimated the sensitivity of the climate system and/or because we've underestimated the technological and/or political challenges of reducing greenhouse-gas emissions, then we'll probably have to resort to some form of geoengineering' [275]. He is backed up by another atmospheric scientist, Ken Caldeira (from the Carnegie Institution for Science) who also urges planning for 'the worst case climate scenario': 'We may hope or even expect that we will collectively agree to delay some of this economic growth and development and invest instead in costlier energy systems that don't threaten Earth's climate. Nevertheless, prudence demands that we consider what we might do if cuts in carbon dioxide emissions prove too little or too late to avoid unacceptable climate damage'.

Robock and his colleagues counter that even the evaluation of stratospheric engineering would be highly risky. They argue that only a full-scale planet-wide and high dose test would yield meaningful data. This in itself would raise immense ethical issues given potential impacts on the Asian monsoon and thus on agricultural yields [276]. Furthermore, stopping such an experiment abruptly would then result in climate warming at a faster rate than if geo-engineering had not been carried out in the first place. Thus the density of sulphate aerosol would have to be decreased slowly to avoid ecological shocks. And if such a programme of investigation got underway, there would be strong commercial and political interests in maintaining it; whose responsibility would it be to operate the climate thermostat? Perhaps the aspect that should concern us most about any large-scale geo-engineering project would be the 'unknown unknowns' 'in which there lurks the ineradicable element of surprise' [260].

14.3.2 Could climate change trigger eruptions?

The evidence for volcanism resulting in climate change is unambiguous (Section 3.2). On the other hand, the possibility of a reverse connection – climate change triggering eruptions – has been seriously considered for more than 30 years [277]. The hypothesis was given further credence by evidence of enhanced volcanic turmoil evident in ice-core records coinciding with and following the end of the last glaciations [51]. Between 7000 and 5000 BCE, the Greenland GISP2 ice core reveals 18 eruptions with sulphate anomalies corresponding to concentrations in excess of 100 parts per billion (Section 4.2). In contrast, only five eruptions generated this level of sulphate deposition during the past two millennia. Furthermore, the increased eruption rate is apparent particularly for volcanoes that directly experienced the retreat of glaciers and thinning of ice sheets [278]. But how might climate change switch volcanoes on?

The answer is complex and involves feedbacks with other components of the climate system, but the key factor appears to be the removal of ice on volcanoes. Bearing in mind that, during the glaciations, ice thickness reached two kilometres over the Arctic and sub-Arctic landmasses gives a clue to the significance of deglaciation. The pressure beneath two kilometres of ice is equivalent to that beneath about 600 metres of rock. Thus, piling up ice at the surface adds a significant extra load on the underlying upper mantle, which is the source region for magma. On deglaciation, the thinning ice cover reduces pressure on the melting region of the mantle, thereby generating a strong pulse in melt production. This will be manifested by eruptions at the surface, with some time lag that depends on rates of deglaciation and rates of magma ascent in the crust. In Iceland, the delay between melting and eruptions appears to be very short, and the physical models for the process fit well with tephrochronological evidence that eruption rates increased 100-fold at the end of the last glaciation, around 12,500 years ago [279]. Unloading of ice cover on a volcano may also act to induce bubble formation in shallow magma chambers, thus triggering eruptions of systems that are already more or less primed for action.

As ice sheets wax and wane, sea level falls and rises, respectively. Given that nearly three fifths of active volcanoes form islands or occupy coastal sites, and most of the remainder are situated within 250 kilometres of the coastline, it is possible that many volcanoes worldwide are susceptible to the effects of the rapid global sea-level changes that accompany both glaciation and deglaciation. This could provide a mechanism by which even volcanoes far from the polar regions might respond to global climate change. Eruptions could be triggered by variations in the crustal stresses on magma reservoirs, changes in the water table, or edifice collapse induced by wave erosion. Modelling the physics of such processes is highly challenging, and much depends on the level of magma reservoirs in the crust, the exact location of the volcano relative to the sea, and whether its flanks are partially submerged and susceptible to wave erosion.

One study of the response of island-arc-type volcanoes, such as those in Japan, formulated the stress accumulation expected in the crust as a result of differential loading seaward and landward of a coastal volcano [280]. This suggested that during glacial periods (i.e. low sea level) the upper crust is under compression, inhibiting migration of magma to the surface, while the lower crust experiences stretching that favours accumulation of magma. Water loading during deglaciation reverses this stress distribution, with the effect of squeezing magma towards the surface where it can feed eruptions.

Much remains poorly constrained in these arguments, not least the great potential for observational bias in the tephra and ice-core records, and the uncertain link between frequency of eruptions and magma production rates. Nevertheless, the association between volcanism and glaciation / deglaciation appears robust. If deglaciation does induce enhanced volcanism, both at high-latitude volcanoes peering out from under the ice for the first time in millennia and at coastal volcanoes as far away as the tropics, could the enhanced eruptive emission of carbon dioxide accelerate deglaciation and explain why deglaciation is so rapid [278]? Much more empirical and theoretical work needs to be carried out to explore such possibilities, and the moderation of global volcanism through climate change remains a topic ripe for further research. Given that the timing of Quaternary glacial cycles is forced by subtle periodicities in the Earth's rotational dynamics (about its own axis and about the Sun), one intriguing possibility is that Earth's volcanoes are indirectly switched on and off (in a global statistical sense) by gravitational attractions of the Sun, Moon and other planets in the Solar System!

Finally, these arguments suggest the possibility that current and projected global warming and associated decreases in precipitation in some parts of the world will enhance rates and change styles of volcanism. Over the last century or two, the ice cover on many high-latitude and high-altitude volcanoes has reduced dramatically. For instance, ice cover has thinned at rates of more than half a metre per year atop Kilimanjaro, while the Vatnajökull ice cap in Iceland has lost 10% of its mass in the past century. Although these rates are perhaps only a tenth of those recorded for the end of the last glacial period, they could lead to a statistical increase in rates of volcanism in the future (on timescales of centuries, perhaps). Subglacial eruptions are best known today on Iceland. Typically, the weight of ice suppresses explosive activity. Instead, the huge quantities of thermal energy involved melt millions of cubic metres of ice. When the melt-water finally breaks out at the margin of the ice sheet, catastrophic floods ensue. However, if ice cover thins sufficiently, magmas will erupt increasingly explosively (thereby releasing gases directly into the atmosphere). It is thus possible that deglaciation over the next century or so will modify eruptive behaviour at some volcanoes, especially those with deep, ice-filled calderas such as those found in Chile [281].

14.4 SHAKEN BUT NOT STIRRED

Eruptions that Shook the World has aimed to show how evidence from very different fields (volcanology, geology, archaeology, anthropology, history, atmospheric science) can be applied to understand the impacts of major volcanic eruptions on the environment and global climate and, in particular, on human origins and society. It is easy enough to claim that volcanic eruptions led to the demise of civilisations but much harder to prove it. Reasoned judgements on the matter require exacting scrutiny and nuanced interpretation of all available evidence. Nevertheless, it is tempting to think that human society would have been different today had all the volcanoes been switched off 100,000 years ago.

Several chapters have dwelled on the 'push and pull' influences of volcanism. Volcanic regions have proved attractive to human settlement (Section 7.2), and there are compelling arguments that volcanoes and their eruptions have contributed in different ways to the evolution of human culture. Prehistoric and ancient communities flourished thanks to the lithic resources (for example, obsidian) and the 'rough' environments and diverse ecosystems associated with volcanic terrain, while occasional dustings of ash nourished agriculture through the supply of vital elements. On the other hand, major tephra eruptions have left vast areas sterile such that centuries elapsed before climax communities (and humans) returned.

On innumerable occasions, human survivors of volcanic disasters must have fled devastated homelands in search of new livelihoods. Such events might have spurred momentous migrations of our ancestors within and beyond the East African Rift Valley; in south and southeast Asia at the time of the Toba eruption; across Europe during the Middle to Upper Palaeolithic Transition; and across the western Pacific Ocean (the Lapita people). We have seen, too, how the stress of displacement might have inspired cognitive leaps and cultural innovations or, alternatively, led to famine, impoverishment and demise of material culture. In some cases, refugee crises were transformed into opportunities for monumental achievements (literally in the case of Teotihuacán) and expansion of territorial control.

Many past societies appear to have exceeded their capacity for adaptation in the face of natural disasters (reflecting the severity of impacts as well as cultural values, perceptions and social organisation). Furthermore, there are plenty of both ancient and modern examples in which disaster has shattered a society's world view thereby undermining or challenging the political leaders, elite or priesthood that have evidently failed the population by not foreseeing, forestalling or adequately responding to catastrophe. Eruptions can accelerate social, political and economic change – in either progressive or retrogressive ways depending on point of view (Section 5.1). Levels of social integration (tribe, authority, state or city), mobility, self-sufficiency, economic interdependence, external politics and social vulnerability may dictate a society's ability to withstand a volcanic disaster; or else collapse.

In any case, an obvious point not to lose sight of is that, notwithstanding the super-eruptions and lesser-magnitude catastrophes faced by our ancestors, we are still here! Humankind has yet to run into the buffers. This does not preclude the possibility of existential limits and should not make us feel complacent (especially in view of our extreme exploitation of non-renewable resources) but it surely gives some grounds for optimism.

While this book has thrown the spotlight on the coincidences of, and associations between, volcanic eruptions and abrupt environmental and social change, it has not been my intention to promote either a catastrophist or environmental determinist agenda. It is all too easy to sensationalise, focus on the 'disasters' and compile every last scrap of historical coincidence to exaggerate the evidence for catastrophe (meanwhile overlooking the innumerable instances of famine, pestilence and war that have occurred in the absence of major volcanic events). Sadly, volcanoes have been 'criminalised' in many epochs and cultures over the centuries, and continue to be portrayed in television documentaries as 'killers'. On the other hand, it is undeniable that volcanism has the potential to perturb global climate very suddenly and harshly (in common with asteroid or comet impacts); or (much more often) to devastate smaller regions with pyroclastics. While the effects of such disasters on societies have been profound and sometimes terminal, it is possible to argue that some outcomes have been beneficial when taking the very long view on human evolution and development.

Many communities today tolerate proximity to geophysical hazards: the perceived benefits outweigh often incalculable longer-term risks, and uprooting an individual family let alone a whole community or megacity is traumatic. Thus, the many colocations of people, flood plains, fault-lines and/or volcanoes (Tōkyō, Los Angeles, Naples, Istanbul and Port-au-Prince ...) pose considerable challenges. As the global human population heads towards ten billion, an increasing proportion of whom will likely live in poverty in cities and near coasts, it is undeniable that future large and very large volcanic eruptions pose vital management challenges for national governments and the global community. A major problem in risk management is that the most extreme possible scenarios become a kind of science fiction. On the other hand, considering only the 'worst' probable eruptions, earthquakes, tsunami and so on may fail to prepare us for more frequent events.

We have encountered abundant evidence of human resilience and flexibility when confronted by volcanic hazards. Societies that coped and adapted or found new means of production, interaction and protection brought benefits to their descendants (and to humankind). Altruism is another human condition that comes to the fore when disaster strikes - individuals, communities and governments will still stop what they are doing, forget old enmities and provide aid and compassion at times of the greatest crisis reflecting 'the inescapability of moral proximity over geographical distance' [259]. Maybe a super-eruption is just what humanity needs! As a shared global threat for which hazard prevention is beyond human control, a restless supervolcano might inspire such collective political action as to tackle more convincingly the major fractures in global society (conflict, the threat of nuclear exchange, poverty and economic and health inequalities).

Volcanic catastrophe risk should not be reduced to the product of the probability of the event and the magnitude of the associated losses. It should not be anticipated with eagerness, of course, but neither should its prospect be ignored nor viewed with fear. The human track record demonstrates that we have the capacity to manage volcanic threats with resolve, flexibility and creativity.

Appendix A: Large eruptions

The 25 largest documented Holocene eruptions (magnitude M_e of 6.5 and above). Based partly on reference 270 and the Smithsonian Institution's Global Volcanism Program database (http://www.volcano.si.edu).

	Volcano & location	Date[a]	Caldera diameter (km)	Altitude above sea level (m)	M_e	Sulphur yield (megatonnes)	Key reference	Image credit
	Fisher Caldera, Aleutian Islands	8700 BCE	8 × 16	1112	7.4[b]	–	282	J. Gardner, AVO / USGS
	Kuril Lake, Kamchatka	6450 BCE	8 × 14	7581	7.3	–	283	ASTER Volcano Archive, NASA JPL
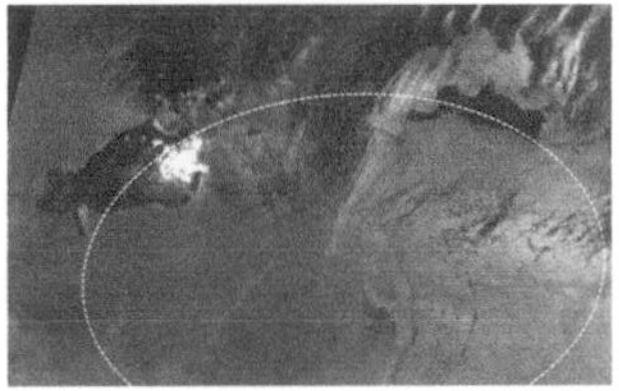	Kikai, Ryuku Islands	5480 BCE	17 × 20	704	7.2	–	284	ASTER Volcano Archive, NASA JPL

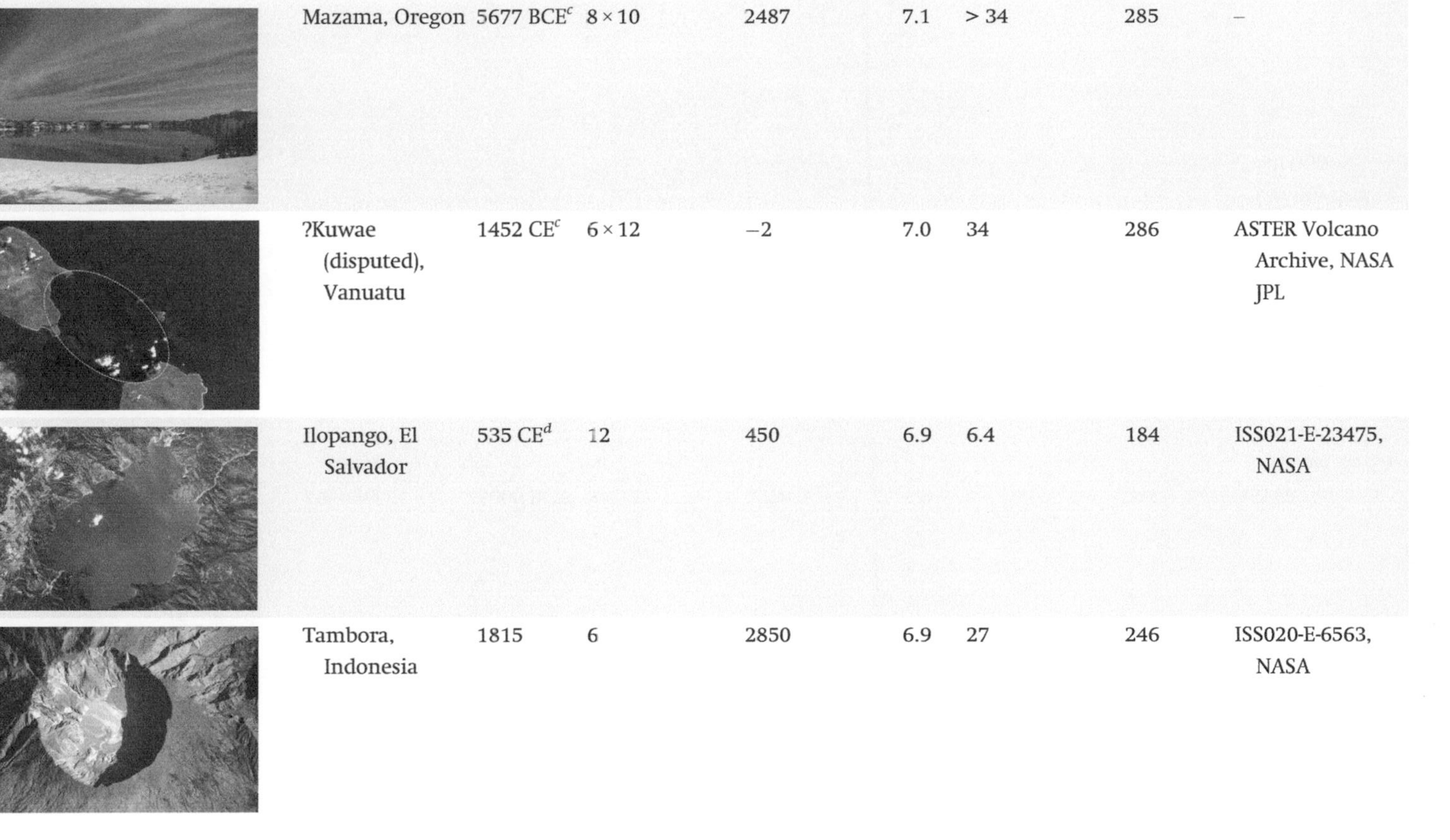	Mazama, Oregon	5677 BCE[c]	8 × 10	2487	7.1	> 34	285	–
	?Kuwae (disputed), Vanuatu	1452 CE[c]	6 × 12	−2	7.0	34	286	ASTER Volcano Archive, NASA JPL
	Ilopango, El Salvador	535 CE[d]	12	450	6.9	6.4	184	ISS021-E-23475, NASA
	Tambora, Indonesia	1815	6	2850	6.9	27	246	ISS020-E-6563, NASA

	Volcano & location	Date[a]	Caldera diameter (km)	Altitude above sea level (m)	M_e	Sulphur yield (megatonnes)	Key reference	Image credit
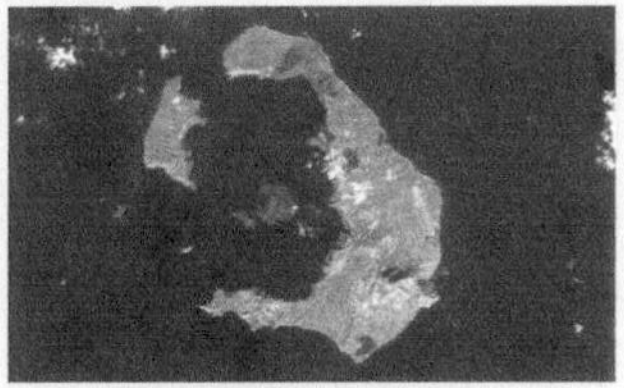	Santorini, Greece	1640 BCE	7 × 10[e]	367	6.9	64	287	ASTER Volcano Archive, NASA JPL
	Menengai, Kenya	6000 BCE[f]	8 × 12	2278	6.9	–	288	ASTER Volcano Archive, NASA JPL
	Okmok II, Aleutian Islands	50 BCE	9.3	1073	6.9	–	289	J. Reeder, Alaska Division of Geological and Geophysical Surveys

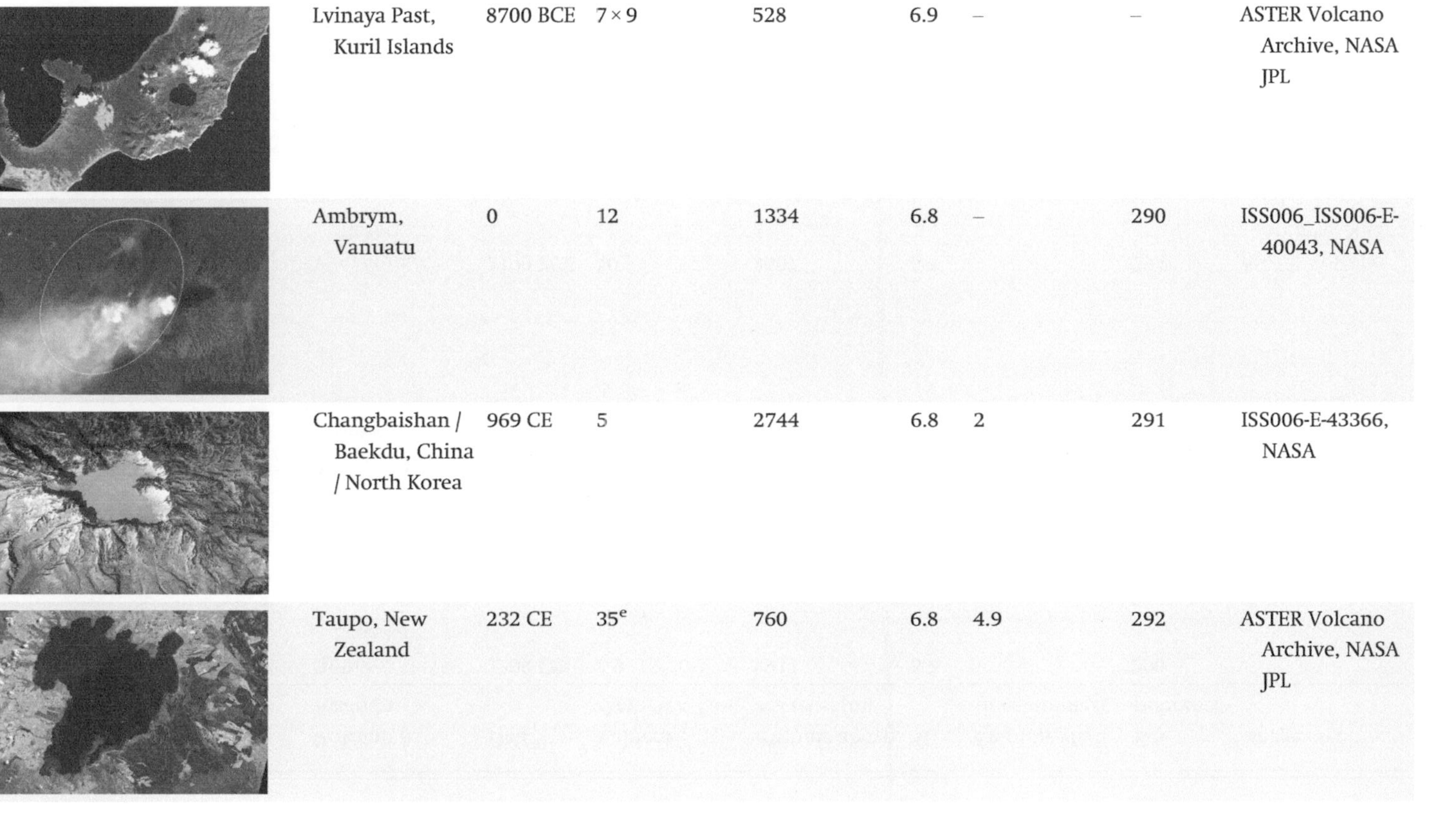

	Lvinaya Past, Kuril Islands	8700 BCE	7 × 9	528	6.9	–	–	ASTER Volcano Archive, NASA JPL
	Ambrym, Vanuatu	0	12	1334	6.8	–	290	ISS006_ISS006-E-40043, NASA
	Changbaishan / Baekdu, China / North Korea	969 CE	5	2744	6.8	2	291	ISS006-E-43366, NASA
	Taupo, New Zealand	232 CE	35[e]	760	6.8	4.9	292	ASTER Volcano Archive, NASA JPL

	Volcano & location	Date[a]	Caldera diameter (km)	Altitude above sea level (m)	M_e	Sulphur yield (megatonnes)	Key reference	Image credit
	Quilotoa, Ecuador	1200 CE	2.9	3914	6.7	–	200	–
	Aniakchak II, Alaska	1700 BCE	10	1341	6.7	–	293	M. Williams, National Park Service
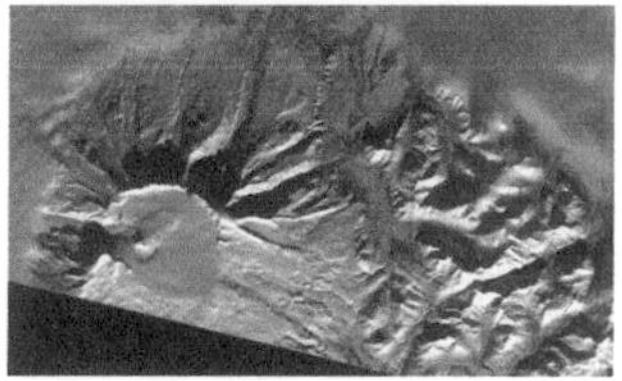	Veniaminof, Alaska	2100 BCE	10	2507	6.7	–	294	ASTER Volcano Archive, NASA JPL

	Okmok I, Aleutian Islands	7300 BCE	9.3[g]	1073	6.7	–	289	Jessica Larsen, AVO/UAF-GI
	Katmai, Alaska	1912	3 × 4	2047	6.5	2.7	295	Cyrus Read, AVO/USGS
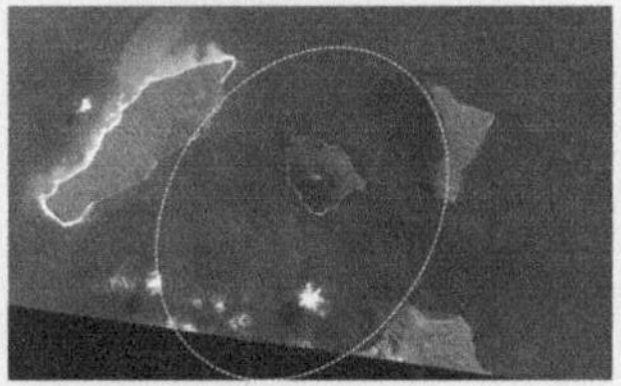	Krakatau, Indonesia	1883	7	813	6.5	5.4	296	ASTER Volcano Archive, NASA JPL
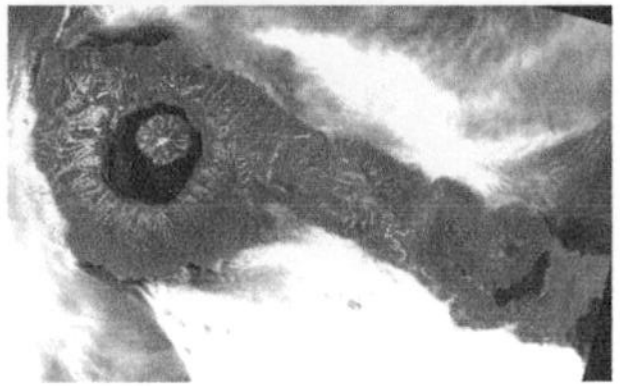	Tao-Rusyr Caldera, Kuril Islands	7300 BCE	7.5	1325	6.5	–	–	ASTER Volcano Archive, NASA JPL

	Volcano & location	Date[a]	Caldera diameter (km)	Altitude above sea level (m)	M_e	Sulphur yield (megatonnes)	Key reference	Image credit
	Aniakchak I, Alaska	5135 BCE	10[g]	1341	6.5	–	13	Game McGimsey, AVO/USGS
	Long Island, Papua New Guinea	~1550/ 1650	10 × 12.5	1280	6.5	–	297	ASTER Volcano Archive, NASA JPL
	Witori, Papua New Guinea	1400 BCE	5.5 × 7.5	742	6.5	–	70	ASTER Volcano Archive, NASA JPL

	Black Peak, Alaska	3000 BCE	3.5	1032	6.5	–	13	Jennifer Adleman, AVO/USGS

[a] Most dates are approximate (calibrated) radiocarbon dates

[b] The Fisher Caldera magnitude is very uncertain – the quoted value appears to be by far an upper limit

[c] Date based on an ice-core identification

[d] Date based on northern-hemisphere sightings of atmospheric optical phenomena

[e] Caldera formed by multiple ancient eruptions

[f] Very approximate younger end of age range based on stratigraphic correlations

[g] Present-day diameter following more recent caldera-forming event

Appendix B: Further reading and data sources

CHAPTER 1

Francis, P. & Oppenheimer, C. (2004) *Volcanoes*, Oxford: Oxford University Press.

Lockwood, J. & Hazlett, R. W. (2010) *Volcanoes: Global Perspectives*, Chichester: Wiley-Blackwell.

Parfitt, L. & Wilson, L. (2008) *Fundamentals of Physical Volcanology*, Chichester: Wiley-Blackwell.

Schmincke, H.-U. (2003) *Volcanism*, Berlin: Springer-Verlag.

Sigurdsson, H., Houghton, B., McNutt, S., Rymer, H. & Stix, J. (eds.) (1999) *Encyclopedia of Volcanoes*, San Diego: Academic Press.

CHAPTER 2

Baxter, P. J., Blong, R. & Neri, A. (eds.) (2008) Evaluating explosive eruption risk at European volcanoes - contributions from the EXPLORIS project, *Journal of Volcanology and Geothermal Research*, **178**, 331–592.

Blong, R. J. (1984) *Volcanic Hazards: A Sourcebook on the Effects of Eruptions*, Orlando, FL: Academic Press, Inc.

Branney, M. J. & Kokelaar, P. (2002) *Pyroclastic Density Currents and the Sedimentation of Ignimbrites*. Geological Society, London, Memoirs, **27**.

Cole, J. W., Milner, D. A. & Spinks, K. D. (2005) Calderas and caldera structures: a review, *Earth-Science Reviews*, **69**, 1–26.

Crisafulli, C. M., Swanson, F. J. & Dale, V. H. (2005) Overview of ecological responses to the eruption of Mount St. Helens: 1980–2005, in *Ecological Responses to the 1980 Eruption of Mount St. Helens*, New York, NY: Springer, pp. 287–299.

De Boer, J. Z. & Sanders, D. T. (2004) *Volcanoes in Human History: The Far-reaching Effects of Major Eruptions*, Princeton, NJ: Princeton University Press.

Martí, J. & Ernst, G. J. (2005) *Volcanoes and the Environment*, Cambridge: Cambridge University Press.

Thornton, I. (1996) *Krakatau: The Destruction and Reassembly of an Island Ecosystem*, Cambridge, MA: Harvard University Press.

CHAPTER 3

Deshler, T. (2008) A review of global stratospheric aerosol: measurements, importance, life cycle, and local stratospheric aerosol, *Atmospheric Research*, **90**, 223–232.

Lamb, H. H. (1995) *Climate, History and the Modern World*, London: Routledge.

Newhall, C. G. & Punongbayan, R. S. (eds.) (1996) *Fire and Mud – Eruptions and Lahars of Mount Pinatubo, Philippines*, Seattle and London: Philippine Institute of Volcanology and Seismology and the University of Washington Press. Also published online: http://www.pubs.usgs.gov/pinatubo/

Robock, A. & Oppenheimer, C. (eds.) (2003) Volcanism and the Earth's atmosphere, *Geophysical Monograph*, **139**.

Alan Robock's website (see Volcanic Eruptions and Climate PowerPoint): http://envsci.rutgers.edu/~robock/

CHAPTER 4

Alley, R. B. (2002) *The Two-mile Time Machine: Ice Cores, Abrupt Climate Change, and our Future*, Princeton, NJ: Princeton University Press.

Baillie, M. G. L. (1995) *A Slice Through Time: Dendrochronology and Precision Dating*, London: Routledge.

Cas, R. A. F. & Wright, J. V. (1988) *Volcanic Successions: Modern and Ancient*, London: Unwin Hyman.

Fisher, R. V. & Schmincke, H.-U. (1984) *Pyroclastic Rocks*, Berlin: Springer.

Rapp, G. R. & Hill, C. L. (2006) *Geoarchaeology: The Earth Science Approach to Archaeological Interpretation*, Newhaven, CT: Yale University Press.

Winchester, S. (2005) *Krakatoa: The Day the World Exploded*, New York, NY: Harper Perennial.

The Centre for Ice and Climate at the Niels Bohr Institute: http://icecores.dk

Data sources

Global Volcanism Program: http://www.volcano.si.edu

ASTER Volcano Archive (satellite images): http://ava.jpl.nasa.gov/default.htm

Holocene eruption database (see Table S1 in Auxiliary Material): http://www.agu.org/journals/jb/jb1006/2009JB006554/

Collapse Caldera Database: http://www.gvb-csic.es/CCDB.htm

A monthly and latitudinally varying volcanic forcing dataset in simulations of twentieth century climate: http://www.ncdc.noaa.gov/paleo/pubs/ammann2003/ammann2003.html

Climate Forcing Data (see Volcanic Aerosols): http://www.ncdc.noaa.gov/paleo/forcing.html

Volcanic Loading: The Dust Veil Index (1985): http://www.cdiac.esd.ornl.gov/ndps/ndp013.html

Ice-core Volcanic Index 2: http://www.climate.envsci.rutgers.edu/IVI2/

NOAA Ice Core Gateway: http://www.ngdc.noaa.gov/paleo/icgate.html

NOAA Tree Ring gateway: http://www.ngdc.noaa.gov/paleo/treering.html

Temperature maps reconstructed from tree rings (1400–1960): http://www.cru.uea.ac.uk/cru/people/briffa/temmaps

NASA Goddard Institute for Space Studies Surface Temperature Analysis (global temperature maps): http://data.giss.nasa.gov/gistemp/
NASA GISS Observed Land Surface Precipitation Data: http://data.giss.nasa.gov/precip_cru/

CHAPTER 5

Beard, M. (2008) *Pompeii: The Life of a Roman Town*, London: Profile Books.
Blong, R. J. (1982) *Time of Darkness: Local Legends and Volcanic Reality in Papua New Guinea*, Washington, DC: University of Washington Press.
Cashman, K. V. & Giordano, G. (eds.) (2008) Volcanoes and human history, *Journal of Volcanology and Geothermal Research*, **176**(3), 325–438.
Cruikshank, J. (2006) *Do Glaciers Listen? Local Knowledge, Colonial Encounters, and Social Imagination*, Vancouver: University of British Columbia Press.
Grattan, J. and Torrence, R. (eds.) (2007) *Living Under the Shadow: The Cultural Impacts of Volcanic Eruptions*, Walnut Creek, CA: Left Coast Press.
McCoy, F. W. & Heiken, G. (eds.) (2000) Volcanic hazards and disasters in human antiquity, *Geological Society of America, Special Paper*, **345**.
Sheets, P. D. (2005) *The Ceren Site: An Ancient Village Buried by Volcanic Ash in Central America*, Belmont, CA: Wadsworth Publishing Company.
Sheets, P. D. & Grayson, D. K. (eds.) (1979) *Volcanic Activity and Human Ecology*, New York, NY: Academic Press.
Torrence, R. & Grattan, J. (eds.) (2002) *Natural Disasters and Cultural Change*, London: Routledge.

CHAPTER 6

Alvarez, W. (2008) *T. rex and the Crater of Doom*, Princeton, NJ: Princeton University Press.
Bryan, S. E., Peate, D. W., Peate, I. U. *et al.* (2010) The largest volcanic eruptions on earth, *Earth-Science Reviews*, **102**, 207–229.
Hoffman, P. F., Kaufman, A. J., Halverson, G. P. & Schrag, D. P. (1998) A Neoproterozoic snowball Earth, *Science*, **281**, 1342–1346.
Lane, N. (2007) Mass extinctions: reading the book of death, *Nature*, **448**, 122–125.
Wignall, P. (2005) The link between large igneous province eruptions and mass extinctions, *Elements*, **1**, 293–297.
Large Igneous Provinces Commission: http://www.largeigneousprovinces.org
Discussion of the origin of 'hotspot' volcanism: http://www.mantleplumes.org

CHAPTER 7

Foley, R. & Gamble, C. (2009) The ecology of social transitions in human evolution, *Proceedings of the Royal Society B*, **364**, 3267–3279.
Hetherington, R. & Reid, R. G. B. (2010) *The Climate Connection: Climate Change and Modern Human Evolution*, Cambridge: Cambridge University Press.
Oppenheimer, S. (2004) *Out of Eden: The Peopling of the World*, London: Constable and Robinson.
Stringer, C. (2011) *Origin of our Species*, London: Allen Lane.

Tattersall, I. & Schwartz, J. H. (2009) Evolution of the genus *Homo*, *Annual Review of Earth and Planetary Sciences*, **37**, 67–92.
Wood, B. (2010) Reconstructing human evolution: Achievements, challenges, and opportunities, *Proceedings of the National Academy of Sciences*, **107**, 8902–8909.

CHAPTER 8

Petraglia, M. D. & Allchin, B. (eds.) (2007) *The Evolution and History of Human Populations in South Asia: Inter-disciplinary Studies in Archaeology, Biological Anthropology, Linguistics and Genetics*, Dordrecht: Springer.
Wark, D. A. & Miller, C. F. (eds.) (2008) Supervolcanoes, *Elements*, **4**(1).

CHAPTER 9

Adler, D. S. & Jőris, O. (eds.) (2008) Chronology of the Middle–Upper Paleolithic Transition in Eurasia, *Journal of Human Evolution*, **55**, 761–926.
Cline, E. H. (2010) *The Oxford Handbook of the Bronze Age Aegean*, Oxford: Oxford University Press.
Friedrich, W. L. (2009) *Santorini: Volcano, Natural History, Mythology*, Aarhus: Aarhus University Press.
Warburton, D. A. (2009) *Time's Up! Dating the Minoan Eruption of Santorini*, Aarhus: Aarhus University Press.

CHAPTER 10

Cowgill, G. L. (2000) The central Mexican highlands from the rise of Teotihuacan to the decline of Tula, in *Mesoamerica*, R. E. W. Adams & M. J. MacLeod (eds.), Cambridge: Cambridge University Press.
Evans, S. T. (2008) *Ancient Mexico and Central America: Archaeology and Culture History*, London and New York: Thames and Hudson.
A site dedicated to Teotihuacán: http://archaeology.la.asu.edu/teo/index.php

CHAPTER 11

Baillie, M. (1999) *Exodus to Arthur: Catastrophic Encounters with Comets*, London: Batsford Ltd.
Rosen, W. (2007) *Justinian's Flea: Plague, Empire, and the Birth of Europe*, New York, NY: Viking.
Sarris, P. (2009) *Economy and Society in the Age of Justinian*, Cambridge: Cambridge University Press.
Ward-Perkins, B. (2005) *The Fall of Rome and the End of Civilization*, Oxford: Oxford University Press.

CHAPTER 12

Laxness, H. (1946) *Independent People*, London: Vintage Classics.
Steingrímsson, J. (1788) Fullkomið skrif um Síðueld (A complete description on the Síða volcanic fire), translated by K. Kunz and published in 1998 as *Fires of*

the Earth: The Laki eruption 1783–1784 by the Nordic Volcanological Institute and the University of Iceland.

CHAPTER 13

Harington, C. R. (ed.) (1992) *The Year Without a Summer? World Climate in 1816*, Ottawa: Canadian Museum of Nature.
Post, J. D. (1977) *The Last Great Subsistence Crisis in the Western World*, Baltimore: The Johns Hopkins University Press.
Stommel, H. & Stommel, E. (1983) *Volcano Weather: The Story of 1816, The Year Without a Summer*, Newport: Seven Seas Press.

CHAPTER 14

Birkmann, J. (ed.) (2006) *Measuring Vulnerability to Natural Hazards: Towards Disaster Resilient Societies*, New York: United Nations University Press.
Bostrom, N. & Ćirković, M. (2008) *Global Catastrophic Risks*, Oxford: Oxford University Press.
Saarinen, T. F. & Sell, J. L. (1985) *Warning and Response to the Mount St. Helen's Eruption*, Albany: State University of New York Press.
Woo, G. (1999) *The Mathematics of Natural Catastrophes*, World Scientific Publishing Company.
International Tsunami Information Centre (UNESCO): http://ioc3.unesco.org/itic
The International Volcanic Health Hazard Network: http://www.ivhhn.org
Cities on Volcanoes Commission: http://cav.volcano.info

References

1. Scrope, G.P. (1862) *Volcanos*, London: Longman, Green, Longmans & Roberts.
2. Courtillot, V., Davaille, A., Besse, J. & Stock, J. (2003) Three distinct types of hotspots in the Earth's mantle, *Earth Planet. Sci. Lett.*, **205**, 295–308.
3. Lowenstern, J.B. & Hurwitz, S. (2008) Monitoring a supervolcano in repose: heat and volatile flux at the Yellowstone caldera, *Elements*, **4**, 35–40.
4. Le Bas, M.J., Le Maitre, R.W., Streckeisen, A. & Zanettin, B. (1986) A chemical classification of volcanic rocks based on the total alkali-silica diagram, *J. Petrol.*, **27**, 745–750.
5. Mason, B.G., Pyle, D.M. & Oppenheimer, C. (2004) The size and frequency of the largest explosive eruptions on Earth, *Bull.Volcanol.*, **66**, 735–748.
6. Witham, C.S. (2005) Volcanic disasters and incidents: a new database, *J. Volcanol. Geotherm. Res.*, **148**, 191–233.
7. Simkin, T., Siebert, L. & Blong, R. (2001) Volcano fatalities – lessons from the historical record, *Science*, **291**, 255.
8. Cronin, S.J., Hedley, M.J., Neall, V.E. & Smith, R.G. (1998) Agronomic impact of ash fallout from the 1995 and 1996 Ruapehu Volcano eruptions, New Zealand, *Environ. Geol.*, **34**, 21–30.
9. Horwell, C.J., Sparks, R.S.J., Brewer, T.S., Llewellin, E.W. & Williamson, B.J. (2003) Characterization of respirable volcanic ash from the Soufrière Hills volcano, Montserrat, with implications for human health hazards, *Bull. Volcanol.*, **65**, 346–362.
10. Cook, R.J., Barron, J.C., Papendick, R.I. & Williams, G.J. (1981) Impacts on agriculture of Mount St Helens eruption, *Science*, **211**, 16–22.
11. Rolett, B. & Diamond, J. (2004) Environmental predictors of pre-European deforestation on Pacific islands, *Nature*, **431**, 443–446.
12. Langmann, B., Zakšek, K., Hort, M. & Duggen, S. (2010) Volcanic ash as fertiliser for the surface ocean, *Atmos. Chem. Phys.*, **10**, 711–734.
13. Manville, V., Németh, K. & Kano, K. (2009) Source to sink: A review of three decades of progress in the understanding of volcaniclastic processes, deposits, and hazards, *Sed. Geol.*, **220**, 136–161.
14. Maeno, F. & Taniguchi, H. (2009) Sedimentation and welding processes of dilute pyroclastic density currents and fallout during a large-scale silicic eruption, Kikai caldera, Japan, *Sed. Geol.*, **220**, 227–242.
15. Delmelle, P., Delfosse, T. & Delvaux, B. (2003) Sulfate, chloride and fluoride retention in Andosols exposed to volcanic acid emissions, *Environ. Pollution*, **126**, 445–457.

16. Allibone, R. J., Cronin, S. J., Charley, D. T. *et al.* (2010) Dental fluorosis linked to degassing of Ambrym volcano, Vanuatu: a novel exposure pathway, *Environ. Geochem. Health*, doi: 10.1007/s10653-010-9338-2.
17. Longo, B. M., Rossignol, A. & Green, J. B. (2008) Cardiorespiratory health effects associated with sulphurous volcanic air pollution, *Public Health*, **122**, 809–820.
18. Tanguy, J.-C., Ribière, C., Scarth, A. & Tjetjep, W. S. (1998) Victims from volcanic eruptions: a revised database, *Bull.Volcanol.*, **60**, 137–144.
19. McCormick, M. P., Thomason, L. W. & Trepte, C. R. (1995) Atmospheric effects of the Mt Pinatubo eruption, *Nature*, **373**, 399–404.
20. Read, W. G., Froidevaux, L. & Waters, J. W. (1993) Microwave Limb Sounder measurements of stratospheric SO_2 from the Mt. Pinatubo eruption, *Geophys. Res. Lett.*, **20**, 1299–1302.
21. Minnis, P., Harrison, E. F., Stowe, L. L. *et al.* (1993) Radiative climate forcing by the Mount Pinatubo eruption, *Science*, **259**, 1411–1415.
22. Parker, D. E., Wilson, H., Jones, P. D., Christy, J. R. & Folland, C. K. (1996) The impact of Mount Pinatubo on world-wide temperatures, *Int. J. Climatol.*, **16**, 487–497.
23. Trenberth, K. E. & Dai, A. (2007) Effects of Mount Pinatubo volcanic eruption on the hydrological cycle as an analog of geoengineering, *Geophys. Res. Lett.*, **34**, L15702, doi:10.1029/2007GL030524.
24. Stenchikov, G., Robock, A., Ramaswamy, V. *et al.* (2002) Arctic Oscillation response to the 1991 Mount Pinatubo eruption: effects of volcanic aerosols and ozone depletion, *J. Geophys. Res.*, **107**(D24), 4803, 10.1029/2002JD0 02090.
25. Stenchikov, G., Delworth, T. L., Ramaswamy, V. *et al.* (2009) Volcanic signals in oceans, *J. Geophys. Res.*, **114**, D16104, doi:10.1029/2008JD011673.
26. Gleckler, P. J., Wigley, T. M., Santer, B. D. *et al.* (2006) Volcanoes and climate: Krakatoa's signature persists in the ocean, *Nature*, **439**, 675.
27. Mercado, L. M., Bellouin, N., Sitch, S. *et al.* (2009) Impact of changes in diffuse radiation on the global land carbon sink, *Nature*, **458**, 1014–1017.
28. Gu, L., Baldocchi, D., Verma, S. B. *et al.* (2002) Advantages of diffuse radiation for terrestrial ecosystem productivity, *J. Geophys. Res.*, **107**(D6), 4050, doi:10.1029/2001JD001242.
29. Timmreck, C., Lorenz, S. J., Crowley, T. J. *et al.* (2009) Limited temperature response to the very large AD 1258 volcanic eruption, *Geophys. Res. Lett.*, **36**, L21708, doi:10.1029/2009GL040083.
30. Kravitz, B., Robock, A. & Bourassa, A. (2010) Negligible climatic effects from the 2008 Okmok and Kasatochi volcanic eruptions, *J. Geophys. Res.*, **115**, D00L05, doi:10.1029/2009JD013525.
31. Oman, L., Robock, A., Stenchikov, G., Schmidt, G. A. & Ruedy, R. (2005) Climatic response to high-latitude volcanic eruptions, *J. Geophys. Res.*, **110**, D13103, doi:10.1029/2004JD005487.
32. Graf, H.-F. & Timmreck, C. (2001) A general climate model simulation of the aerosol radiative effects of the Laacher See eruption, *J. Geophys. Res.*, **106**, 14,747–14,756.
33. Kravitz, B. & Robock, A. (2011) The climate effects of high latitude volcanic eruptions: role of time of year, *J. Geophys. Res.*, **116**, doi:10.1029/2010JD014448.
34. Schmidt, A., Carslaw, K. S., Mann, G. W. *et al.* (2010) The impact of the 1783–1784 AD Laki eruption on global aerosol formation processes and cloud condensation nuclei, *Atmos. Chem. Phys.*, **10**, 6025–6041.

35. Timmreck, C. & Graf, H.-F. (2006) The initial dispersal and radiative forcing of a northern hemisphere mid-latitude super volcano: a model study, *Atmos. Chem. Phys.*, **6**, 35–49, doi:10.5194/acp-6-35-2006.
36. Crowley, T. J. (2000) Causes of climate change over the past 1000 years, *Science*, **289**, 270–277.
37. Carey, S. & Sigurdsson, H. (1989) The intensity of plinian eruptions, *Bull. Volcanol.*, **51**, 28–40.
38. Pyle, D. M. (1989) The thickness, volume and grainsize of tephra fall deposits, *Bull.Volcanol.*, **51**, 1–15.
39. Walker, G. P. L. (1980) The Taupo pumice: product of the most powerful known (ultraplinian) eruption? *J.Volcanol. Geotherm. Res.*, **8**, 69–94.
40. Carey, S. N. & Sparks, R. S. J. (1986) Quantitative models of the fallout and dispersal of tephra from volcanic eruption columns, *Bull.Volcanol.*, **48**, 109–125.
41. Turney, C. S. M., Harkness, D. D. & Lowe, J. J. (1997) The use of microtephra to correlate Late-glacial lake sediment successions in Scotland, *J. Quat. Sci.*, **12**, 525–531.
42. Blockley, S. P. E., Lane, C. S., Lotter, A. F. & Pollard, A. M. (2007) Evidence for the presence of the Vedde Ash in central Europe, *Quat. Sci. Rev.*, **26**, 3030–3036.
43. Pearce, N. J. G., Bendall, C. A. & Westgate, J. A. (2008) Comment on "Some numerical considerations in the geochemical analysis of distal microtephra" by A. M. Pollard, S. P. E. Blockley & C. S. Lane, *Appl. Geochem.*, **23**, 1353–1364.
44. Wulf, S., Kraml, M., Brauer, A., Keller, J. & Negendank, J. F. W. (2004) Tephrochronology of the 100 ka lacustrine sediment record of Lago Grande di Monticchio (southern Italy), *Quat. Int.*, **122**, 7–30.
45. Devine, J. D., Sigurdsson, H., Davis, A. N. & Self, S. (1984) Estimates of sulfur and chlorine yield to the atmosphere from volcanic eruptions and potential climatic effects, *J. Geophys. Res.*, **89**(B7), 6309–6325, doi:10.1029/JB089iB07p06309.
46. Scaillet, B. & Pichavant, M. (2003) Experimental constraints on volatile abundance in arc magmas and their implications for degassing processes, *Geol. Soc., London, Spec. Publ.*, **213**, 23–52.
47. Wolff, E. W., Barbante, C., Becagli, S. *et al.* (2010) Changes in environment over the last 800,000 years from chemical analysis of the EPICA Dome C ice core, *Quat. Sci. Rev.*, **29**, 285–295.
48. Hammer, C. U., Clausen, H. B. & Dansgaard, W. (1980) Greenland ice sheet evidence of postglacial volcanism and its climatic impact, *Nature*, **288**, 230–235.
49. Steffensen, J. P., Andersen, K. K., Bigler, M. *et al.* (2008) High-resolution Greenland ice core data show abrupt climate change happens in a few years, *Science*, **321**, 680–684.
50. Zielinski, G. A., Mayewski, P. A., Meeker, L. D. *et al.* (1994) Record of volcanism since 7000 B.C. from the GISP2 Greenland ice core and implications for the volcano–climate system, *Science*, **264**, 948–952.
51. Zielinski, G. A., Mayewski, P. A., Meeker, L. D., Whitlow, S. & Twickler, M. S. (1996) A 110,000-yr record of explosive volcanism from the GISP2 (Greenland) ice core, *Quat. Res.*, **45**, 109–118.
52. Traversi, R., Becagli, S., Castellano, E. *et al.* (2009) Sulfate spikes in the deep layers of EPICA-Dome C ice core: evidence of glaciological artifacts. *Env. Sci. Technol.*, **43**, 8737–8743.

53. Dai, J., Mosley-Thompson, E. & Thompson, L. G. (1991) Ice core evidence for an explosive tropical volcanic eruption 6 years preceding Tambora, *J. Geophys. Res.*, **96**, 17,361–17,366.
54. de Silva, S. L. & Zielinski, G. A. (1998) Global influence of the AD 1600 eruption of Huaynaputina, Peru, *Nature*, **393**, 455–458.
55. Abbott, P. M., Davies, S. M., Steffensen, J.-P. *et al.* A detailed framework of Marine Isotope Stage 4 and 5 volcanic events recorded in two Greenland ice-cores, *Quat. Sci. Rev.*, in review.
56. Gao, C., Robock, A. & Ammann, C. (2008) Volcanic forcing of climate over the past 1500 years: an improved ice core-based index for climate models, *J. Geophys. Res.*, **113**, D23111, doi:10.1029/2008JD010239.
57. LaMarche, V. C., Jr. & Hirschboeck, K. K. (1984) Frost rings in trees as records of major volcanic eruptions, *Nature*, **307**, 121–126.
58. Baillie, M. G. L. & Munro, M. A. R. (1988) Irish tree rings, Santorini and volcanic dust veils, *Nature*, **332**, 344–346.
59. Salzer, M. W. & Hughes, M. K. (2007) Bristlecone pine tree rings and volcanic eruptions over the last 5000 years, *Quat. Res.*, **67**, 57–68.
60. Briffa, K. R., Jones, P. D., Schweingruber, F. H. & Osborn, T. J. (1998) Influence of volcanic eruptions on northern hemisphere summer temperatures over 600 years, *Nature*, **393**, 450–455.
61. Briffa, K. R., Osborn, T. J. & Schweingruber, F. H. (2004) Large-scale temperature inferences from tree rings: a review, *Global Planet. Change*, **40**, 11–26.
62. Allison, P. M. (2002) Recurring tremors: the continuing impact of the AD 79 eruption of Mt Vesuvius, in R. Torrence and J. Grattan (eds.), *Natural Disasters and Cultural Change*, London: Routledge, pp. 107–125.
63. Khalidi, L., Oppenheimer, C., Gratuze, B. *et al.* (2010) Obsidian sources in highland Yemen and their relevance to archaeological research in the Red Sea region, *J. Archaeol. Sci.*, **37**, 2332–2345.
64. Sheets, P. (2008) Armageddon to the Garden of Eden: explosive volcanic eruptions and societal resilience in ancient Middle America, in D. Sandweiss & J. Quilter (eds.), *El Niño: Catastrophism, and Culture Change in Ancient America*, Washington, DC: Harvard University Press, pp. 167–186.
65. Specht, J. & Torrence, R. (2007) Lapita all over: land-use on the Willaumez Peninsula, Papua New Guinea, *Terra Australis*, **26**, 71–96.
66. Torrence, R., Neall, V. & Boyd, W. E. (2009) Volcanism and historical ecology on the Willaumez Peninsula, Papua New Guinea, *Pacific Sci.*, **63**, 507–535.
67. Parr, J. F., Boyd, W. E., Harriott, V. & Torrence, R. (2009) Human adaptive responses to catastrophic landscape disruptions during the Holocene, Numundo, PNG, *Geogr. Res.*, **47**, 155–174.
68. Neall, V. E., Wallace, R. C. & Torrence, R. (2008) The volcanic environment for 40,000 years of human occupation on the Willaumez Isthmus, West New Britain, Papua New Guinea, *J. Volcanol. Geotherm. Res.*, **176**, 330–343.
69. Lentfer, C. & Torrence, R. (2007) Holocene volcanic activity, vegetation succession, and ancient human land use: unraveling the interactions on Garua Island, Papua New Guinea, *Rev. Palaeobotany Palynol.*, **143**, 83–105.
70. McKee, C. O., Neall, V. E. & Torrence, R. (2011) A remarkable pulse of large-scale volcanism on New Britain Island, Papua New Guinea, *Bull. Volcanol.*, **73**, 27–37.
71. Rodolfo, K. S. & Umbal, J. V. (2008) A prehistoric lahar-dammed lake and eruption of Mount Pinatubo described in a Philippine aborigine legend, *J. Volcanol. Geotherm. Res.*, **176**, 432–437.
72. Frierson, P. (1991) *The Burning Island: A Journey Through Myth and History in Volcano Country, Hawai'i*, San Francisco: Sierra Club Books.

73. Swanson, D. A. (2008) Hawaiian oral tradition describes 400 years of volcanic activity at Kīlauea, *J. Volcanol. Geotherm. Res.*, **176**, 427–431.
74. Mandeville, C. W., Webster, J. D., Tappen, C. *et al.* (2009) Stable isotope and petrologic evidence for open-system degassing, *Geochim. Cosmochim. Acta*, **73**, 2978–3012.
75. Clark, E. E. (1953) *Indian Legends of the Pacific Northwest*, Berkeley, CA: University of California Press.
76. Symons, G. J. (ed.) (1888) *The Eruption of Krakatoa and Subsequent Phenomena*, London: Harrison & Sons.
77. von Helmholtz, R. (1883) The remarkable sunsets, *Nature*, **29**, 130.
78. Lamb, H. H. (1970) Volcanic dust in the atmosphere with a chronology and assessment of its meteorological significance, *Philos. Trans. R. Soc. London A*, **266**, 425–533.
79. Stothers, R. B. & Rampino, M. R. (1983) Volcanic eruptions in the Mediterranean before AD 630 from written and archaeological sources, *J. Geophys. Res.*, **88**, 6357–6371.
80. Stothers, R. B. (2002) Cloudy and clear stratospheres before A.D. 1000 inferred from written sources, *J. Geophys. Res.*, **107**, 4718, 10.1029/2002JD002105.
81. Mellaart, J. (1967) *Catal Huyuk, a Neolithic Town in Anatolia*, New York, NY: McGraw Hill.
82. Meece, S. (2006) A bird's eye view - of a leopard's spots: the Çatalhöyük 'map' and the development of cartographic representation in prehistory, *Anatolian Stud.*, **56**, 1–16.
83. Zerefos, C. S., Gerogiannis, V. T., Balis, D., Zerefos, S. C. & Kazantzidis, A. (2007) Atmospheric effects of volcanic eruptions as seen by famous artists and depicted in their paintings, *Atmosph. Chem. Phys.*, **7**, 4027–4042.
84. Wiart, P. A. M. & Oppenheimer, C. (2000) Largest known historic eruption in Africa: Dubbi volcano, Eritrea, 1861, *Geology*, **28**, 291–294.
85. Schmincke, H.-U., Kutterolf, S., Perez, W. *et al.* (2009) Walking through volcanic mud: the 2,100-year-old Acahualinca footprints (Nicaragua), *Bull. Volcanol.*, **71**, 479–493.
86. Rampino, M. R. (2010) Mass extinctions of life and catastrophic flood basalt volcanism, *Proc. Natl. Acad. Sci. USA*, **107**, 6555–6556.
87. Campbell, I. H. (2005) Large igneous provinces and the mantle plume hypothesis, *Elements*, **1**, 265–269.
88. Bryan, S. E. & Ernst, R. E. (2008) Revised definition of large igneous provinces (LIPs), *Earth Sci. Rev.*, **86**, 175–202.
89. Self, S., Blake, S., Sharma, K., Widdowson, M. & Sephton, S. (2008) Sulfur and chlorine in Late Cretaceous Deccan magmas and eruptive gas release, *Science*, **319**, 1654–1657.
90. Stothers, R. B. (1993) Flood basalts and extinction events, *Geophys. Res. Lett.*, **20**, 1399–1402.
91. Christenson, G. L., Collins, G. S., Morgan, J. V. *et al.* (2009) Mantle deformation beneath the Chicxulub impact crater, *Earth Planet. Sci. Lett.*, **284**, 249–257.
92. Schulte, P., Alegret, L., Arenillas, I. *et al.* (2010) The Chicxulub asteroid impact and mass extinction at the Cretaceous–Paleogene boundary, *Science*, **327**, 1214–1218.
93. Kring, D. A. (2007) The Chicxulub impact event and its environmental consequences at the Cretaceous–Tertiary boundary, *Palaeogeogr. Palaeoclimatol. Palaeoecol.*, **255**, 4–21.

94. Keller, G., Adatte, T., Berner, Z. *et al.* (2007) Chicxulub impact predates K–T boundary: new evidence from Brazos, Texas, *Earth Planet. Sci. Lett.*, **255**, 339–356.
95. Chenet, A.-L., Quidelleur, X., Fluteau, F., Courtillot, V. & Bajpai, S. (2007) ^{40}K–^{40}Ar dating of the Main Deccan large igneous province: further evidence of KTB age and short duration, *Earth Planet. Sci. Lett.*, **263**, 1–15.
96. Schulte, P., Speijer, R. P., Brinkuis, H. *et al.* (2008) Comment on the paper ‘Chicxulub impact predates K–T boundary: new evidence from Brazos, Texas’ by Keller *et al.* (2007), *Earth Planet. Sci. Lett.*, **269**, 614–620.
97. Keller, G., Adatte, T., Baum, G. & Berner, Z. (2008) Reply to ‘Chicxulub impact predates K–T boundary: new evidence from Brazos, Texas’ Comment by Schulte *et al.*, *Earth Planet. Sci. Lett.*, **269**, 621–629.
98. Sills, J. (ed.) (2010) Letters, *Science*, **328**, 973–976.
99. Jones, A. P., Price, G. D., Price, N. J., DiCarli, P. S. & Clegg, R. A. (2002) Impact induced melting and the development of large igneous provinces, *Earth Planet. Sci. Lett.*, **202**, 551–561.
100. Courtillot, V. and Olsen, P. (2007) Mantle plumes link magnetic superchrons to Phanerozoic mass depletion events, *Earth Planet. Sci. Lett.*, **260**, 495–504.
101. Knoll, A. H., Bambach, R. K., Payne, J. L., Pruss, S. & Fischer, W. W. (2007) Paleophysiology and end-Permian mass extinction, *Earth Planet. Sci. Lett.*, **256**, 295–313.
102. Ries, J. B., Cohen, A. L. & McCorkle, D. C. (2009) Marine calcifiers exhibit mixed responses to CO_2-induced ocean acidification, *Geology*, **37**, 1131–1134.
103. Wille, M., Nägler, T. F., Lehmann, B., Schröder, S. & Kramers, J. D. (2008) Hydrogen sulphide release to surface waters at the Precambrian/Cambrian boundary, *Nature*, **453**, 767–769.
104. Whiteside, J. H., Olsen, P. E., Eglington, T., Brookfield, M. E. & Sambrotto, R. N. (2010) Compound-specific carbon isotopes from Earth’s largest flood basalt eruptions directly linked to the end-Triassic mass extinction, *Proc. Nat. Acad. Sci.*, **107**, 6721–6725.
105. Cockell, C. S. (1999) Crises and extinction in the fossil record; a role for ultraviolet radiation, *Paleobiology*, **25**, 212–225.
106. Vogelmann, A. M., Ackerman, T. P. & Turco, R. P. (1992) Enhancements in biologically effective ultraviolet radiation following volcanic eruptions, *Nature*, **359**, 47–49.
107. Cather, S. M., Dunbar, N., McDowell, F. W., McIntosh, W. C. & Schole, P. A. (2009) Climate forcing by iron fertilization from repeated ignimbrite eruptions: the icehouse–silicic large igneous province (SLIP) hypothesis, *Geosphere*, **5**, 315–324.
108. Stern, R. J., Avigad, D., Miller, N. & Beyth, M. (2008) From volcanic winter to snowball Earth: an alternative explanation for Neoproterozoic biosphere stress, in Y. Dilek, H. Furnes & K. Muehlenbachs (eds.), *Links Between Geological Processes, Microbial Activities & Evolution of Life*, Berlin: Springer, pp. 313–337.
109. King, G. & Bailey, G. (2006) Tectonics and human evolution, *Antiquity*, **80**, 265–286.
110. Ukstins Peate, I., Baker, J. A., Kent, A. J. R. *et al.* (2003) Correlation of Indian Ocean tephra to individual Oligocene silicic eruptions from Afro-Arabian flood volcanism, *Earth Planet. Sci. Lett.*, **211**, 311–327.
111. Pik, R., Marty, B., Carignan, J., Yirgu, G. & Ayalew, T. (2009) Timing of East African Rift development in southern Ethiopia: implication for mantle plume activity and evolution of topography, *Geology*, **36**, 167–170.

112. Biggs, J., Anthony, E.Y. & Ebinger, C.J. (2009) Multiple inflation and deflation events at Kenyan volcanoes, East African Rift, *Geology*, **37**, 979–982.
113. Raichlen, D.A., Gordon, A.D., Harcourt-Smith, W.E.H., Foster, A.D. & Haas, W.R. Jr. (2010) Laetoli footprints preserve earliest direct evidence of human-like bipedal biomechanics, *PLoS ONE*, **5**(3): e9769. doi:10.1371/journal.pone.0009769.
114. Sauer, C.O. (1962) Seashore – primitive home of man? *Proc. Am. Philos. Soc.*, **106**, 41–47.
115. King, G., Bailey, G. & Sturdy, D. (1994) Active tectonics and human survival strategies, *J. Geophys. Res.*, **99**(B10), 20,063–20,078, doi:10.1029/94JB00280.
116. McDougall, I., Brown, F.H. & Fleagle, J.G. (2005) Stratigraphic placement and age of modern humans from Kibish, Ethiopia, *Nature*, **433**, 733–736.
117. Basell, L.S. (2008) Middle Stone Age (MSA) site distributions in eastern Africa and their relationship to Quaternary environmental change, refugia and the evolution of *Homo sapiens*, *Quat. Sci. Rev.*, **27**, 2484–2498.
118. Mohr, P., Mitchell, J.G. & Raynolds, R.G.H. (1980) Quaternary volcanism and faulting at O'a caldera, Central Ethiopian Rift, *Bull. Volcanol.*, **43**, 173–189.
119. Grün, R., Stringer, C., McDermott, F. *et al.* (2005) U-series and ESR analyses of bones and teeth relating to the human burials from Skhul, *J. Human Evolution*, **49**, 316–334.
120. Oppenheimer, S. (2009) The great arc of dispersal of modern humans: Africa to Australia, *Quat. Int.*, **202**, 2–13.
121. Endicott, P., Ho, S.Y.W., Metspalu, M. & Stringer, C. (2009) Evaluating the mitochondrial timescale of human evolution, *Trends Ecol. Evol.*, **24**, 515–521.
122. Soares, P., Ermini, L., Thomson, N. *et al.* (2009) Correcting for purifying selection: an improved human mitochondrial molecular clock, *Am. J. Hum. Gen.*, **84**, 740–759.
123. Green, R.E., Krause, J., Briggs, A.W. *et al.* (2010) A draft sequence of the Neandertal genome, *Science*, **328**, 710–722.
124. Ambrose, S.H. (1998) Late Pleistocene human population bottlenecks, volcanic winter, and differentiation of modern humans, *J. Hum. Evol.*, **34**, 623–651.
125. Rose, W.I. & Chesner, C.A. (1987) Dispersal of ash in the great Toba eruption, 75 kyr, *Geology*, **15**, 913–917.
126. Vazquez, J.A. & Reid, M.R. (2004) Probing the accumulation history of the voluminous Toba magma, *Science*, **305**, 991–994.
127. Chesner, C.A. & Rose, W.I. (1991) Stratigraphy of the Toba Tuffs and the evolution of the Toba Caldera Complex, Sumatra, Indonesia, *Bull. Volcanol.*, **53**, 343–356.
128. Rose, W.I. & Chesner, C.A. (1990) Worldwide dispersal of ash and gases from Earth's largest known eruption: Toba, Sumatra, 75 kyr, *Palaeogeogr. Palaeoclimatol. Palaeoecol.*, **89**, 269–275.
129. Baines, P.G. & Sparks, R.S.J. (2005) Dynamics of giant volcanic ash clouds from supervolcanic eruptions, *Geophys. Res. Lett.*, **32**, L24808, doi:10.1029/2005GL024597.
130. Herzog, M. & Graf, H.-F. (2010) Applying the three-dimensional model ATHAM to volcanic plumes: dynamic of large co-ignimbrite eruptions and associated injection heights for volcanic gases, *Geophys. Res. Lett.*, **37**, L19807, doi:10.1029/2010GL044986.
131. Ledbetter, M. & Sparks, R.S.J. (1979) Duration of large-magnitude explosive eruptions deduced from graded bedding in deep-sea ash layers, *Geology*, **7**, 240–244.

132. Carey, S. (1997) Influence of convective sedimentation on the formation of widespread tephra fall layers in the deep sea, *Geology*, **25**, 839–842.
133. Weisner, M., Wang, Y. & Zheng, L. (1995) Fallout of volcanic ash to the deep South China Sea induced by the 1991 eruption of Mount Pinatubo, Philippines, *Geology*, **23**, 885–888.
134. Zielinski, G. A., Mayewski, P. A., Meeker, L. D. *et al.* (1996) Potential atmospheric impact of the Toba mega-eruption ~71,000 years ago, *Geophys. Res. Lett.*, **23**(8), 837–840.
135. Scaillet, B., Clemente, B., Evans, B. W. & Pichavant, M. (1998) Redox control of sulphur degassing in silicic magmas, *J. Geophys. Res.*, **103**, 23,937–23,949.
136. Chesner, C. A. & Luhr, J. F. (2010) A melt inclusion study of the Toba Tuffs, Sumatra, Indonesia, *J. Volcanol. Geotherm. Res.*, **197**, 259–278.
137. Niemeier, U., Timmreck, C., Graf, H.-F. *et al.* (2009) Initial fate of fine ash and sulfur from large volcanic eruptions, *Atmos. Chem. Phys.*, **9**, 9043–9057, doi:10.5194/acp-9-9043-2009.
138. Rampino, M. R. & Self, S. (1992) Volcanic winter and accelerated glaciation following the Toba super-eruption, *Nature*, **359**, 50–52.
139. Rampino, M. R. & Ambrose, S. H. (2000) Volcanic winter in the Garden of Eden: the Toba super-eruption and the Late Pleistocene human population crash, *Geol. Soc. Am. Spec. Paper*, **345**, 71–82.
140. Rampino, M. R. & Self, S. (1993) Climate–volcanism feedback and the Toba eruption of ~74,000 years ago, *Quat. Res.*, **40**, 269–280.
141. Jones, G. S., Gregory, J. M., Stott, P. A., Tett, S. F. & Thorpe, R. B. (2005) An AOGCM simulation of the climate response to a volcanic super-eruption, *Clim. Dyn.*, **25**, 725–738.
142. Robock, A., Ammann, C. M., Oman, L. *et al.* (2009) Did the Toba volcanic eruption of ~74 ka B.P. produce widespread glaciation? *J. Geophys. Res.*, **114**, D10107, doi:10.1029/2008JD011652.
143. Timmreck, C., Graf, H.-F., Lorenz, S. J. *et al.* (2010) Aerosol size confines climate response to volcanic super-eruptions, *Geophys. Res. Lett.*, **37**, L24705, doi:10.1029/2010GL045464.
144. Williams, M. A. J., Ambrose, S. H., van der Kaars, S. *et al.* (2009) Environmental impact of the 73 ka Toba super-eruption in South Asia, *Palaeogeogr. Palaeoclimatol. Palaeoecol.*, **284**, 295–314.
145. Ambrose, S. H. (2003) Did the super-eruption of Toba cause a human population bottleneck? Reply to Gathorne-Hardy and Harcourt-Smith, *J. Hum. Evol.*, **45**, 231–237.
146. Rossano, M. J. (2009) Ritual behaviour and the origins of modern cognition, *Cambridge Archaeol. J.*, **19**, 243–256.
147. Gagneux, P., Wills, C., Gerloff, U. *et al.* (1999) Mitochondrial sequences show diverse evolutionary histories of African hominoids, *Proc. Natl. Acad. Sci.*, **96**, 5077–5082.
148. Louys, J. (2007) Limited effect of the Quaternary's largest super-eruption (Toba) on land mammals from Southeast Asia, *Quat. Sci. Rev.*, **26**, 3108–3117.
149. Brumm, A., Jensen, G. M., van den Bergh, G. D. *et al.* (2010) Hominins on Flores, Indonesia, by one million years ago, *Nature*, **464**, 748–752.
150. Jones, S. C. (2010) Palaeoenvironmental response to the ~74 ka Toba ash-fall in the Jurreru and Middle Son valleys in southern and north-central India, *Quat. Res.*, **73**, 336–350.
151. Haslam, M., Clarkson, C., Petraglia, M. *et al.* (2010) The 74 ka Toba super-eruption and southern Indian hominins: archaeology, lithic technology and environments at Jwalapuram Locality 3, *J. Archaeol. Sci.* **37**, 3370–3384.

152. Riede, F. (2008) The Laacher See eruption (12,920 BP) and material culture change at the end of the Allerød in Northern Europe, *J. Archaeol. Sci.*, **35**, 591–599.
153. Banks, W. E., d'Errico, F., Peterson, A. T. *et al.* (2008) Neanderthal extinction by competitive exclusion, *PLoS ONE*, **3**(12), e3972, doi:10.1371/journal.pone.0003972.
154. Chazan, M. (2010) Technological perspectives on the Upper Paleolithic, *Evol. Anthropol.*, **19**, 57–65.
155. Sinitsyn, A. A. (2003) A Palaeolithic 'Pompeii' at Kostenki, Russia, *Antiquity*, **77**, 9–14.
156. Hoffecker J. F., Holliday, V. T., Anikovich, M. V. *et al.* (2008) From the Bay of Naples to the River Don: the Campanian Ignimbrite eruption and the Middle to Upper Paleolithic transition in Eastern Europe, *J. Hum. Evol.*, **55**, 858–870.
157. Fedele, F. G., Giaccio, B. & Hajdas, I. (2008) Timescales and cultural process at 40,000 BP in the light of the Campanian Ignimbrite eruption, western Eurasia, *J. Hum. Evol.*, **55**, 834–857.
158. Fedele, F. G., Giaccio, B., Isaia, R. & Orsi, G. (2002) Ecosystem impact of the Campanian Ignimbrite eruption in Late Pleistocene Europe, *Quat. Res.*, **57**, 420–424.
159. Golovanova, L. V., Doronichev, V. B., Cleghorn, N. E. *et al.* (2010) Significance of ecological factors in the Middle to Upper Paleolithic Transition, *Curr. Anthropol.*, **51**, 655–691.
160. Schmincke, H.-U., Park, C. & Harms, E. (2009) Evolution and environmental impacts of the eruption of Laacher See volcano (Germany) 12 900 a BP, *Quat. Int.*, **61**, 61–72.
161. Baales, M. (2006) Final Palaeolithic environment and archaeology in the central Rhineland (Rhineland-Palatinat, western Germany): conclusions of the last 15 years of research, *L'Anthropologie*, **110**, 418–444.
162. Graf, H.-F. & Timmreck, C. (2001) A general climate model simulation of the aerosol radiative effects of the Laacher See eruption (10,900 B.C.), *J. Geophys. Res.*, **106**, 14,747–14,756, doi:10.1029/2001JD900152.
163. de Klerk, P., Janke, W., Kühn, P. & Theuerkauf, M. (2008) Environmental impact of the Laacher See eruption at a large distance from the volcano: integrated palaeoecological studies from Vorpommern (NE Germany), *Palaeogeogr. Palaeoclimatol. Palaeoecol.*, **270**, 196–214.
164. Henrich, J. (2004) Demography and cultural evolution: how adaptive cultural processes can produce maladaptive losses: the Tasmanian case, *Am. Antiquity*, **69**, 197–214.
165. Powell, A., Shennan, S. & Thomas, M. G. (2009) Late Pleistocene demography and the appearance of modern human behaviour, *Science*, **324**, 1298–1301.
166. Sigurdsson, H., Carey, S., Alexandri, M. *et al.* (2006) Marine investigations of Greece's Santorini volcanic field, *EOS Trans. Am. Geophys. Union*, **87**, 337–342.
167. Bietak, M. (2004) Review of Manning's 'A test of time', *Bibliotheca Orientalis* **61**, 200–222.
168. Ramsey, C. B., Manning, S. W. & Galimberti, M. (2004) Dating the volcanic eruption at Thera, *Radiocarbon*, **46**, 325–344.
169. Friedrich, W. L., Kromer, B., Friedrich, M. *et al.* (2006) Santorini eruption radiocarbon dated to 1627–1600 B.C., *Science*, **312**, 548.
170. Bronk Ramsey, C., Dee, M. W., Rowland, J. M. *et al.* (2010) Radiocarbon-based chronology for Dynastic Egypt, *Science*, **328**, 1554–1557.

171. Pearson, C. L., Dale, D. S., Brewer, P. W. *et al.* (2009) Dendrochemical analysis of a tree-ring growth anomaly associated with the Late Bronze Age eruption of Thera, *J. Archaeol. Sci.*, **36**, 1206–1214.
172. Wiener, M. H. & Allen, J. P. (1998) Separate lives: the Ahmose Tempest Stela and the Theran eruption, *J. Near Eastern Stud.*, **57**, 1–28.
173. McCoy, F. W. & Heiken, G. (2000) Tsunami generated by the Late Bronze Age eruption of Thera (Santorini), Greece, *Pure Appl. Geophys.*, **157**, 1227–1256.
174. Bruins, H. J., MacGillivray, J. A., Synolakis, C. E. *et al.* (2008) Geoarchaeological tsunami deposits at Palaikastro (Crete) and the Late Minoan IA eruption of Santorini, *J. Archaeol. Sci.*, **35**, 191–212.
175. Driessen, J. (2002) Towards an archaeology of crisis: defining the long-term impact of the Bronze Age Santorini eruption, in R. Torrence and J. Grattan (eds.) *Natural Disasters and Cultural Change*, London: Routledge, pp. 250–263.
176. Bicknell, P. (2000) Late Minoan IB marine ware, the marine environment of the Aegean and the Bronze Age eruption of Thera volcano, *Geol. Soc. London Spec. Publ.*, **171**, 95–103.
177. Plunket, P. & Uruñuela, G. (1998) Preclassic household patterns preserved under volcanic ash at Tetimpa, Puebla, *Latin Am. Antiquity*, **9**, 287–309.
178. Plunket, P. & Uruñuela, G. (2000) The quick and the dead: decision making in the abandonment of Tetimpa, *Mayab*, **13**, 78–87.
179. Plunket, P. & Uruñuela, G. (2008) Mountain of sustenance, mountain of destruction: the prehispanic experience with Popocatépetl volcano, *J. Volcanol. Geotherm. Res.*, **170**, 111–120.
180. Plunket, P. & Uruñuela, G. (2006) Social and cultural consequences of a late Holocene eruption in central Mexico, *Quat. Int.*, **151**, 19–28.
181. Plunket, P. and Uruñuela, G. (1998) Appeasing the volcano gods, *Archaeology*, **54**, 36–42.
182. Panfil, M. S., Gardner, T. W. & Hirth, K. G. (1999) Late Holocene stratigraphy of the Tetimpa archaeological sites, northeast flank of Popocatépetl Volcano, central Mexico, *Geol. Soc. Am. Bull.*, **111**, 204–218.
183. Kutterolf, S., Freundt, A. & Peréz, W. (2008) Pacific offshore record of Plinian arc volcanism in central America: 2. Tephra volumes and erupted masses, *Geochem. Geophys. Geosystems*, **9**, Q02S02, doi:10.1029/2007GC001791.
184. Mehringer, P. J., Sarna-Wojcicki, A. M., Wollwage, L. K. & Sheets, P. (2005) Age and extent of the Ilopango TBJ tephra inferred from a Holocene chronostratigraphic reference section, Lago de Yojoa, Honduras, *Quat. Res.* **63**, 199–205.
185. Dull, R. A. (2004) An 8000-year record of vegetation, climate, and human disturbance from the Sierra de Apaneca, El Salvador, *Quat. Res.*, **61**, 159–167.
186. Price, T. D., Burton, J. H., Sharer, R. J. *et al.* (2010) Kings and commoners at Copán: isotopic evidence for origins and movement in the Classic Maya period, *J. Anthropol. Archaeol.*, **29**, 15–32.
187. Dull, R. A., Southon, J. R. & Sheets, P. (2001) Volcanism, ecology and culture: a reassessment of the Volcán Ilopango TBJ eruption in the southern Maya realm, *Latin Am. Antiquity*, **12**, 25–44.
188. Pfister, C. (2010) The vulnerability of past societies to climatic variation: a new focus for historical climatology in the twenty-first century, *Climatic Change*, **200**, 25–31.
189. Stothers, R. B. (1984) Mystery cloud of AD 536, *Nature*, **307**, 344–345.
190. Larsen, L. B., Vinther, B. M., Briffa, K. R. *et al.* (2008) New ice core evidence for a volcanic cause of the A.D. 536 dust veil, *Geophys. Res. Lett.*, **35**, L04708, doi:10.1029/2007GL032450.

191. Dull, R., Southon, J. R., Kutterolf, S. *et al.* (2010) Did the TBJ Ilopango eruption cause the AD 536 event? American Geophysical Union Fall Meeting, Abstract #V13C-2370.
192. Drancourt, M., Roux, V., Dang, L. V. *et al.* (2004) Genotyping, Orientalis-like *Yersinia pestis*, and plague pandemics, *Emerging Infectious Diseases*, **10**, 1585–1592.
193. Heather, P. (1995) The Huns and the end of the Roman Empire in Western Europe, *English Historical Rev.*, **110**, 4–41.
194. Dijkstra, J. H. F. (2004) A cult of Isis at Philae after Justinian? Reconsidering *P. Cair. Masp.* I 67004, *Zeit. Papyrologie Epigraphik*, **146**, 137–154.
195. Sarris, P. (2002) The Justinianic plague: origins and effects, *Continuity Change*, **17**, 169–182.
196. Baillie, M. G. L. (1994) Dendrochronology raises questions about the nature of the AD 536 dust-veil event, *The Holocene*, **4**, 212–217.
197. Fei, J., Zhou, J. & Hou, Y. (2007) Circa A.D. 626 volcanic eruption, climatic cooling, and the collapse of the Eastern Turkic Empire, *Climatic Change*, **81**, 469–475.
198. Palais, J. M., Germani, M. S. & Zielinski, G. A. (1992) Interhemispheric transport of volcanic ash from a 1259 A.D. volcanic eruption to the Greenland and Antarctic ice sheets, *Geophys. Res. Lett.*, **19**, 801–804.
199. Kellerhals, T., Tobler, L., Brütsch, S. *et al.* (2010) Thallium as a tracer for preindustrial volcanic eruptions in an ice core record from Illimani, Bolivia, *Environ. Sci. Technol.*, **44**, 888–893.
200. Mothes, P. A. & Hall, M. L. (2008) The Plinian fallout associated with Quilotoa's 800 yr BP eruption, Ecuadorian Andes, *J. Volcanol. Geotherm. Res.*, **176**, 56–69.
201. Stothers, R. B. (2000) Climatic and demographic consequences of the massive volcanic eruption of 1258, *Climatic Change*, **45**, 361–374.
202. Jones, P. D., Briffa K. R., Barnett T. P. & Tett, S. F. B. (1998) High-resolution palaeoclimatic records for the last millennium: interpretation, integration and comparison with General Circulation Model control-run temperatures, *The Holocene*, **8**, 455–471.
203. Emile-Geay, J., Seager, R., Cane, M. A., Cook, E. R. & Haug, G. H. (2008) Volcanoes and ENSO over the past millennium, *J. Climate*, **21**, 3134–3148.
204. Crowley, T. J., Zielinski, G., Vinther, B. *et al.* (2008) Volcanism and the Little Ice Age, *PAGES News*, **16**, 22–23.
205. Schneider, D. P., Ammann, C. M., Otto-Bliesner, B. L. & Kaufman, S. S. (2009) Climate response to large, high-latitude and low-latitude volcanic eruptions in the Community Climate System Model, *J. Geophys. Res.*, **114**, D15101, doi:10.1029/2008JD011222.
206. Jackson, P. (1978) The dissolution of the Mongol empire, *Central Asiatic J.*, **22**, 186–244. Also published in Jackson, P. (2009) *Studies on the Mongol Empire and Early Muslim India*, Burlington, VT: Ashgate.
207. Morgan, D. (2009) The decline and fall of the Mongol Empire, *J. R. Asiatic Soc.*, **19**, 427–437.
208. D'Arrigo, R., Jacoby, G., Frank, D. & Pederson, N. (2001) Spatial response to major volcanic events in or about AD 536, 934 and 1258: frost rings and other dendrochronological evidence from Mongolia, *Climatic Change*, **49**, 239–246.
209. Fletcher, J. (1986) The Mongols: ecological and social perspectives, *Harvard J. Asiatic Stud.*, **46**, 11–50.
210. Thordarson, Th. & Larsen, G. (2007) Volcanism in Iceland in historical time: volcano types, eruption styles and eruptive history, *J. Geodynamics*, **43**, 118–152.

211. Thordarson, Th., Larsen, G., Steinþórsson, S. & Self, S. (2003) The 1783–1785 A. D. Laki-Grímsvötn eruptions II: Appraisal based on contemporary accounts, *Jökull*, **53**, 11–48.
212. Thordarson, Th. & Self, S. (1993) The Laki (Skaftár Fires) and Grímsvötn eruptions in 1783–1785, *Bull.Volcanol.*, **55**, 233–263.
213. Guilbaurd, M.-N., Self, S., Thordarson, Th. & Blake, S. (2005) Morphology, surface structures, and emplacement of lavas produced by Laki, A.D. 1783–1784, *Geol. Soc. Am. Spec. Paper*, **396**, 81–102.
214. Hamilton, C.W., Thordarson, Th. & Fagents, S.A. (2010) Explosive lava-water interactions I: architecture and emplacement chronology of volcanic rootless cone groups in the 1783–1784 Laki lava flow, Iceland, *Bull. Volcanol.*, **72**, 449–467.
215. Thordarson, Th. & Self, S. (2003) Atmospheric and environmental effects of the 1783–1784 Laki eruption: a review and reassessment, *J. Geophys. Res.*, **108**(D1), 4011, doi:10.1029/2001JD002042.
216. Franklin, B. (1785) Meteorological imaginations and conjectures, *Memoirs of the Literary and Philosophical Society of Manchester*, **2**, 373–377.
217. van Swinden, J. H. (1785) Observationes nebulam, quae mense Junio 1783 Apparuit, specantes in *Ephemerides Societatis Meteorologicae Palatinae*, translated by Linteman, S. & Thordarson, T., *Jökull*, **50**, 73–80.
218. Met Office Hadley Centre Temperature (HadCET) datasets. See http://hadobs.metoffice.com/hadcet/
219. D'Arrigo, R., Mashig, E., Frank, D., Jacoby, G. & Wilson, R. (2004) Reconstructed warm season temperatures for Nome, Seward Peninsula, Alaska, *Geophys. Res. Lett.*, **31**, L09202, doi:10.1029/2004GL019756.
220. Grattan, J., Brayshay, M. & Sadler, J. (1998) Modelling the distal impacts of past volcanic gas emissions, *Quaternaire*, **9**, 25–35.
221. Brázdil, R., Demarée, G. R., Deutsch, M. *et al.* (2010) European floods during the winter 1783/1784: scenarios of an extreme event during the 'Little Ice Age', *Theor. Appl. Climatol.*, **100**, 163–189.
222. Elleder, L. (2010) Reconstruction of the 1784 flood hydrograph for the Vltava River in Prague, Czech Republic, *Global Planet. Change*, **70**, 117–124.
223. Oman, L., Robock, A., Stenchikov, G. L. *et al.* (2006) Modeling the distribution of the volcanic aerosol cloud from the 1783–1784 Laki eruption, *J. Geophys. Res.*, **111**, D12209, doi:10.1029/2005JD006899.
224. Oman, L., Robock, A., Stenchikov, G. L. & Thordarson, Th. (2006) High-latitude eruptions cast shadow over the African monsoon and the flow of the Nile, *Geophys. Res. Lett.*, **33**, L18711, doi:10.1029/2006GL027665.
225. Eggertsson, T. (1998) Sources of risk, institutions for survival, and a game against nature in premodern Iceland, *Explor. Econ. Hist.*, **35**, 1–30.
226. Eggertsson, T. (1996) No experiments, monumental disasters: Why it took a thousand years to develop a specialized fishing industry in Iceland, *J. Econ. Behavior Organisation*, **30**, 1–23.
227. Thorarinsson, S. (1979) On the damage caused by volcanic eruptions with special reference to tephra and gases, in P. D. Sheets & D. K. Grayson (eds.), *Volcanic Activity and Human Ecology*, New York, NY: Academic Press, pp. 125–160.
228. Vasey, D. E. (1991) Population, agriculture, and famine - Iceland, 1784–1785, *Hum. Ecol.*, **19**, 323–350.
229. Wrigley, E. A. & Schofield, R. S. (1989) *The Population History of England 1541–1871: A Reconstruction*, Cambridge: Cambridge University Press.
230. Whitam, C. S. & Oppenheimer, C. (2005) Mortality in England during the 1783–4 Laki Craters eruption, *Bull. Volcanol.*, **67**, 15–26.

231. Grattan, J., Rabartin, R., Self, S. & Thordarson, Th. (2005) Volcanic air pollution and mortality in France 1783–1784, *C. R. Geosci.*, **337**, 641–651.
232. Carus, W. (1847) *Memoirs of the Life of the Rev. Charles Simeon*, London: Hatchard and Son.
233. Volney, M. C.-F. (1787) *Travels Through Syria and Egypt in the Years 1783, 1784, and 1785*, London: G. G. J. & J. Robinson.
234. Grove, R. H. (2007) The great El Niño of 1789–93 and its global consequences: reconstructing an extreme climate event in world environmental history, *Mediev. Hist. J.*, **10**, 75–98.
235. Yasui, M. & Koyaguchi, T. (2004) Sequence and eruptive style of the 1783 eruption of Asama Volcano, central Japan: a case study of an andesitic explosive eruption generating fountain-fed lava flow, pumice fall, scoria flow and forming a cone, *Bull. Volcanol.*, **66**, 243–262.
236. Le Roy Ladurie, E. & Daux, V. (2008) The climate in Burgundy and elsewhere, from the fourteenth to the twentieth century, *Interdiscipl. Sci. Rev.*, **33**, 10–24.
237. Kington, J. A. (1980) Daily weather mapping from 1781: a detailed synoptic examination of weather and climate during the decade leading up to the French Revolution, *Climatic Change*, **3**, 7–36.
238. Thordarson, Th., Miller, D. J., Larsen, G., Self, S. & Sigurdsson, H. (2001) New estimates of sulfur degassing and atmospheric mass-loading by the 934 AD Eldgjá eruption, Iceland, *J. Volcanol. Geotherm. Res.*, **108**, 33–54.
239. Stanza from an epic poem (*syair*) from Sumbawa compiled in Malay around 1830. Chambert-Loir, H. (ed.) (1982) *Syair kerajaan Bima*, Jakarta and Bandung: Ecole Francaise d'Extrême-Orient.
240. de Jong Boers, B. (1995) Mount Tambora in 1815: a volcanic eruption in Indonesia and its aftermath, *Indonesia*, **60**, 37–60.
241. Radermacher, Korte beschrijving van het eiland Celebes ende eilanden Floris, Sumbauwa, Lombok en Bali, 1786, p. 186. Translated in de Jong Boers, B. (1995) Mount Tambora in 1815: a volcanic eruption in Indonesia and its aftermath, *Indonesia*, **60**, 37–60.
242. Raffles, T. S. (1817) *The History of Java*, London: Black, Parbury & Allen.
243. Raffles, T. S. (1830) Memoir of the life and public services of Sir Thomas Stamford Raffles, F. R. S. &c., particularly in the government of Java, 1811–1816, and of Bencoolen and its dependencies, 1817–1824: with details of the commerce and resources of the eastern archipelago, and selections from his correspondence, London: John Murray.
244. Crawfurd, J. (1856) *A Descriptive Dictionary of the Indian Islands and Adjacent Countries*, London, Bradbury and Evans.
245. Sigurdsson, H. & Carey, S. (1989) Plinian and co-ignimbrite tephra fall from the 1815 eruption of Tambora volcano, *Bull. Volcanol.*, **51**, 243–270.
246. Self, S., Rampino, M. R., Newton, M. S. & Wolff, J. A. (1984) Volcanological study of the great Tambora eruption of 1815, *Geology*, **12**, 659–663.
247. Self, S., Gertisser, R., Thordarson, Th., Rampino, M. R. & Wolff, J. A. (2004) Magma volume, volatile emissions, and stratospheric aerosols from the 1815 eruption of Tambora, *Geophys. Res. Lett.*, **31**, L20608, doi:10.1029/2004GL020925.
248. Baron, W. R. (1992) 1816 in perspective: the view from the northeastern United States, in C. R. Harington (ed.), *The Year Without a Summer? World Climate in 1816*, Ottawa: Canadian Museum of Nature, pp. 124–144.
249. Stommel, H. M. & Stommel, E. (1983) *Volcano Weather: The Story of 1816, the Year Without a Summer*, Newport, RI: Seven Seas Press.

250. Dewey (1821) Results of meteorological observations made at Williamstown, Massachusetts, *Mem. Am. Acad. Arts Sci.*, **4**, 387–392.
251. Surmieda, M. R., Lopez, J. M., Abad-Viola, G. *et al.* (1992) Surveillance in evacuation camps after the eruption of Mt. Pinatubo, Philippines, *CDC Surveillance Summaries, CDC Morbidity Mortality Weekly Rep.*, **41**, 9–12.
252. Petroeschevsky, W. A. (1949) A contribution to the knowledge of the unung Tambora (Sumbawa), *Tijdschrift van het Koninklijk Nederlands Aardrijkskundig Genootschap*, **66**, 688–703.
253. Goethals, P. R. (1961) *Aspects of Local Government in a Sumbawan Village (Eastern Indonesia)*, Ithaca, NY: Southeast Asia Programme, Department of Southeastern Studies, Cornell University.
254. Fries, A. L. (1947) *Records of the Moravians in North Carolina 1752–1879*, Raleigh, NC: Edwards & Broughton, vol. 7, pp. 3294–3313.
255. Clausewitz, C. von (1922) *Politische Schriften und Briefe*, H. Rothfels (ed.), Munich, Drei Masken Verlag, pp. 189–191.
256. Pant, G. B., Parthasarathy, B. & Sontakke, N. A. (1992) Climate over India during the first quarter of the nineteenth century, in C. R. Harington (ed.), *The Year Without a Summer? World Climate in 1816*, Ottawa: Canadian Museum of Nature, pp. 429–435.
257. Harty, W. (1820) *An Historic Sketch of the Causes, Progress, Extent, and Mortality of the Contagious Fever Epidemic in Ireland During the Years 1817, 1818 and 1819*, London: Royal Geographical Society, Manuscripts Collection, pp. 113–115.
258. Webb, P. (2002) Emergency relief during Europe's famine of 1817 anticipated crisis-response mechanisms of today, *J. Nutr.*, **132**, 2092S–2095S.
259. Beck, U. (2009) *World Risk Society*, Cambridge: Polity Press.
260. Adger, W. N., Hughes, T. P., Folke, C., Carpenter, S. R. & Rockström, J. (2005) Social-ecological resilience to coastal disasters, *Science*, **309**, 1036–1039.
261. Self, S. & Blake, S. (2008) Consequences of explosive supereruptions, *Elements*, **4**, 41–46.
262. Chu, R., Helmberger, D. V., Sun, D., Jackson, J. M. & Zhu, L. (2010) Mushy magma beneath Yellowstone, *Geophys. Res. Lett.*, **37**, L01306, doi:10.1029/2009GL041656.
263. Wilson, C. J. N. & Hildreth, W. (1997) The Bishop Tuff: new insights from eruptive stratigraphy, *J. Geol.*, **105**, 407–440.
264. Jones, M. T., Sparks, R. S. J. & Valdes, P. J. (2007) The climatic impact of supervolcanic ash blankets, *Clim. Dynam.*, **29**, 553–564.
265. Rampino, M. R. (2002) Supereruptions as a threat to civilizations on Earth-like planets, *Icarus*, **156**, 562–569.
266. White, G. F. & Haas, J. E. (1975) *Assessment of Research on Natural Hazards*, Cambridge, MA: MIT Press.
267. Button, G. (2010) *Disaster Culture: Knowledge and Uncertainty in the Wake of Human and Environmental Catastrophe*, Walnut Creek, CA: Left Coast Press, Inc.
268. Aspinall, W. P., Woo, G., Voight, B. V. & Baxter, P. J. (2003) Evidence-based volcanology: application to eruption crises, *J. Volcanol. Geotherm. Res.*, **128**, 273–285.
269. Sornette, D. (2009) Dragon kings, black swans and the prediction of crises, *Int. J. Terraspace Sci. Eng.*, **2**, 1–18.
270. Deligne, N. I., Coles, S. G. & Sparks, R. S. J. (2010) Recurrence rates of large explosive volcanic eruptions, *J. Geophys. Res.*, **115**, B06203, doi:10.1029/2009JB006554.
271. Pappalardo, L., Ottolini, L. & Mastrolorenzo, G. (2008) The Campanian Ignimbrite (southern Italy) geochemical zoning: insight on the generation

of a super-eruption from catastrophic differentiation and fast withdrawal, *Contrib. Mineral. Petrol.*, **156**, 1–26.

272. Woo, G. (2008) Probabilistic criteria for volcano evacuation decisions, *Nat. Hazards*, **45**, 87–97.
273. Wigley, T. M. L. (2006) A combined mitigation/geoengineering approach to climate stabilization, *Science*, **314**, 452–454.
274. Robock, A. (2008) 20 reasons why geoengineering may be a bad idea, *Bull. Atomic Scientists*, **64**, 14–18.
275. Has the time come for geoengineering? See http://www.thebulletin.org/web-edition/roundtables/has-the-time-come-geoengineering
276. Robock, A., Bunzl, M., Kravitz, B. & Stenchiko, G. L. (2010) A test for geoengineering? *Science*, **327**, 530–531.
277. Rampino, M. R., Self, S. & Fairbridge, R. W. (1979) Can rapid climate change cause volcanic eruptions? *Science*, **206**, 826–829.
278. Huybers, P. & Langmuir, C. (2009) Feedback between deglaciation, volcanism, and atmospheric CO_2, *Earth Planet. Sci. Lett.*, **286**, 479–491.
279. Maclennan, J., Jull, M., McKenzie, D., Slater, L. & Grönvold, K. (2002) The link between volcanism and deglaciation in Iceland, *Geochem. Geophys. Geosystems*, **3**(11), 1062, doi:10.1029/2001GC000282.
280. Nakada, M. & Yokose, H. (1992) Ice age as a trigger of active Quaternary volcanism and tectonism, *Tectonophysics*, **212**, 321–329.
281. Tuffen, H. (2010) How will melting of ice affect volcanic hazards in the twenty-first century? *Philos. Trans. R. Soc. A*, **368**, 2535–2558.
282. Stelling, P., Gardner, J. E. & Begét, J. (2005) Eruptive history of Fisher Caldera, Alaska, USA, *J. Volcanol. Geotherm. Res.*, **139**, 163–183.
283. Ponomareva, V. V., Kyle, P. R., Melekestsev, I. *et al.* (2004) The 7600 (^{14}C) year BP Kurile Lake caldera-forming eruption, Kamchatka, Russia: stratigraphy and field relationships, *J. Volcanol. Geotherm. Res.*, **136**, 199–222.
284. Maeno, F. & Taniguchi, H. (2007) Spatiotemporal evolution of a marine caldera-forming eruption, generating a low-aspect ratio pyroclastic flow, 7.3 ka, Kikai caldera, Japan: implication from near-vent eruptive deposits, *J. Volcanol. Geotherm. Res.*, **167**, 212–238.
285. Bacon, C. R. & Lanphere, M. A. (2006) Eruptive history and geochronology of Mount Mazama and the Crater Lake region, Oregon, *Geol. Soc. Am. Bull.*, **118**, 1131–1159.
286. Witter, J. B. & Self, S. (2006) The Kuwae (Vanuatu) eruption of AD 1452: potential magnitude and volatile release, *Bull. Volcanol.*, **69**, 301–318.
287. Druitt, T. H., Edwards, L., Mellors, R. M. *et al.* (1999) Santorini Volcano, *Geol. Soc. London, Memoir*, **19**.
288. Macdonald, R. & Scaillet, B. (2006) The central Kenya peralkaline province: insights into the evolution of peralkaline salic magmas, *Lithos*, **91**, 59–73.
289. Burgisser, A. (2005) Physical volcanology of the 2,050 BP caldera-forming eruption of Okmok caldera, Alaska, *Bull. Volcanol.*, **67**, 497–525.
290. Robin, C., Eissen, J.-P. & Monzier M. (1993) Giant tuff cone and 12-km-wide associated caldera at Ambrym volcano (Vanuatu, New Hebrides arc), *J. Volcanol. Geotherm. Res.*, **55**, 225–238
291. Horn, S. & Schmincke, H.-U. (2000) Volatile emissions during the eruption of Baitoushan volcano (China/North Korea) *ca.* 969 AD, *Bull. Volcanol.*, **61**, 537–555.
292. Walker, G. P. L. (1980) The Taupo pumice: product of the most powerful known (ultraplinian) eruption, *J. Volcanol. Geotherm. Res.*, **8**, 69–94.
293. Begét, J. E., Mason, O. K. & Andersen, P. M. (1992) Age, extent and climatic significance of the *c.* 3400 BP Aniakchak tephra, western Alaska, USA, *Holocene*, **2**, 51–56.

294. Miller, T. P. & Smith, R. L. (1997) Late Quaternary caldera-forming eruptions in the eastern Aleutian arc, Alaska, *Geology*, **15**, 434–438.
295. Hildreth, W. (1983) The compositionally zoned eruption of 1912 in the Valley of Ten Thousand Smokes, Katmai National Park, Alaska, *J. Volcanol. Geotherm. Res.*, **18**, 1–56.
296. Self, S. & Rampino, M. R. (1981) The 1883 eruption of Krakatau, *Nature*, **294**, 699–704.
297. Pain, C. F., Blong, R. J. & McKee, C. O. (1981) Pyroclastic deposits and eruptive sequences on Long Island, Papua New Guinea. 1. Lithology, stratigraphy, and volcanology, *Geol. Survey Papua New Guinea, Memoirs*, **10**, 101–107.

Index

Page numbers in **bold type** refer to figures.